ADVANCES IN MEDICINAL CHEMISTRY

Volume 4 • 1999

ADVANCES IN
MEDICINAL CHEMISTRY

Editors: BRUCE E. MARYANOFF
ALLEN B. REITZ
Drug Discovery
R. W. Johnson Pharmaceutical
Research Institute

VOLUME 4 • 1999

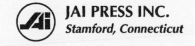

JAI PRESS INC.
Stamford, Connecticut

ISBN: 0-7623-0064-7
ISBN: 1067-5698

Transferred to digital printing 2005
Printed and bound by Antony Rowe Ltd, Eastbourne

CONTENTS

FARNESYL TRANSFERASE INHIBITORS: DESIGN OF A
NEW CLASS OF CANCER CHEMOTHERAPEUTIC
AGENTS

LIST OF CONTRIBUTORS

M. Altorfer
Institute of Organic Chemistry
University of Zürich
Zürich, Switzerland

Maria Alvim-Gaston
Department of Medicinal Chemistry
University of Mississippi
University, Mississippi

Mitchell A. Avery
Department of Medicinal Chemistry
University of Mississippi
University, Mississippi

Mary Pat Beavers
Drug Discovery
R. W. Johnson Pharmaceutical Research
Institute
Raritan, New Jersey

Brian D. Blackhart
COR Therapeutics, Inc.
South San Francisco, California

Richard D. Cramer
Tripos, Inc.
St. Louis, Missouri

Subrata Chakravarty
Department of Chemistry
State University of New York at Stony Brook
Stony Brook, New York

Christopher J. Dinsmore
Department of Medicinal Chemistry
Merck Research Laboratories
West Point, Pennsylvania

Allan J. Ferguson
Tripos, Inc.
St. Louis, Missouri

Scott D. Kuduk   Laboratory of Bioorganic Chemistry
                                        Sloan Kettering Institute for Cancer Research
                                        New York, New York

Bruce E. Maryanoff   Drug Discovery
                                        R. W. Johnson Pharmaceutical Research
                                        Institute
                                        Spring House, Pennsylvania

Dat Nguyen   COR Therapeutics, Inc.
                                        South San Francisco, California

D. Obrecht   Polyphor AG
                                        Zürich, Switzerland

Iwao Ojima   Department of Chemistry
                                        State University of New York at Stony Brook
                                        Stony Brook, New York

J. A. Robinson   Department of Medicinal Chemistry
                                        University of Zürich
                                        Zürich, Switzerland

Theresa M. Williams   Department of Medicinal Chemistry
                                        Merck Research Laboratories
                                        West Point, Pennsylvania

John R. Woolfrey   Department of Medicinal Chemistry
                                        University of Mississippi
                                        University, Mississippi

PREFACE

Advances in Medicinal Chemistry brings interesting, personalized accounts of drug discovery and drug development into the limelight. This fourth volume in the series is comprised of six chapters on a wide range of topics in medicinal chemistry, including molecular modeling, structure-based drug design, organic synthesis, peptide conformational analysis, biological assessment, structure–activity correlation, and lead optimization. The first account presents an exciting story about amino acid-based peptide mimetics corresponding to β-turn, loop, and helical motifs in proteins as a probe of ligand–receptor and ligand–enzyme molecular interactions. An integrated approach that involves structural knowledge from libraries of conformationally constrained peptides and combinatorial chemistry is discussed.

The second chapter addresses new facets of the medicinal chemistry of the important anticancer drug Taxol® (paclitaxel). Ojima and coworkers explore, in particular, the structure–activity relationship associated with the 3-phenylisoserine side chain, synthetically exploiting their "β-Lactam Synthon Method". Their research has led to, among other things, a series of noteworthy "second-generation" taxoid anticancer agents.

The third chapter is a tour-de-force account of the search for new drugs for the treatment of malaria based on the natural product artemisinin. Malaria continues to inspire new drug discovery research because of its increased incidence in developing countries, as well as the development of resistance to conventionally administered drugs. Projections show that this disease will gain a greater biological range as the Earth gradually warms in the future. The University of Mississippi group has combined traditional synthetic methods, the use of mechanistic probes, and molecular modeling to improve our understanding in this area.

The fourth chapter applies computational chemistry to the evaluation of compound libraries for biological testing. To the trained medicinal chemist, the question of molecular diversity can often be answered by using subjective criteria based on intuition and experience. To put this analysis on a more objective and reproducible footing, workers at Tripos have developed computer algorithms to design the proprietary virtual ChemSpace and synthesized Optiverse libraries. Comparisons with current drugs and those in development offer evidence that this approach is moving in the right direction.

The fifth chapter describes the construction of a three-dimensional molecular model of the human thrombin receptor, the first protease-activated G-protein coupled receptor (PAR-1), as a means to explore the intermolecular contacts involved in agonist peptide recognition. Site-directed mutagenesis data coupled with modeling results led to a suggestion of a potential ligand–receptor mode of interaction, which might be useful for the design of PAR-1 antagonist ligands.

Finally, workers at Merck describe their research on inhibitors of farnesyl transferase as a potential treatment for human cancers. This chapter reveals today's medicinal chemistry at its best, with many different research facets being brought into focus, including: screening of compound libraries, peptidomimetic research to simplify an active peptide agent, computer-assisted modeling, NMR and X-ray structure analysis of enzyme-bound inhibitors, and directed synthesis of single compounds or libraries for biological testing.

We thank the contributing authors for their superb chapters and the R. W. Johnson Pharmaceutical Research Institute for support of our editorial efforts.

The R. W. Johnson Pharmaceutical Research Institute   Bruce E. Maryanoff
Spring House, Pennsylvania 19477   Allen B. Reitz
                                                                                         Series Editors

NOVEL PEPTIDE MIMETIC BUILDING BLOCKS AND STRATEGIES FOR EFFICIENT LEAD FINDING

D. Obrecht, M. Altorfer, and J. A. Robinson

Advances in Medicinal Chemistry
Volume 4, pages 1–68
Copyright © 1999 by JAI Press Inc.
All rights of reproduction in any form reserved.
ISBN: 0-7623-0064-7

1

I. INTRODUCTION

A. Sources for Finding Lead Compounds

Due to the enormous progress over recent years in genome sciences, an increasing number of biologically relevant target proteins (e.g. receptors, enzymes, transcription factors, modulators, chaperones) have become available in pure form for detailed studies of their structure, mechanism of action, and function in living systems. Numerous crystal structures of pharmacologically relevant proteins [e.g. platelet derived growth factor (PDGF),[1,2] nerve growth factor (NGF),[3] HIV- proteinase,[4] collagenase,[5] insulin receptor kinase domain,[6,7] calcineurin/FKBP12-FK506 complex[8]] have initiated drug development programs that turned out to be very successful.

This literal burst of novel biological targets has also created a need for novel sources of new organic molecules for screening, and also for more efficient screening technologies. Combinatorial and parallel synthetic chemistry have recently emerged as promising approaches to satisfy the increasing demand for new families of novel compounds.

The interactions of the four key elements that are needed to carry out successful screening and lead finding are shown schematically in Figure 1. In addition to molecular biology, genomic sciences, and chemistry, which provide novel targets and the necessary biomolecules for high-throughput screening (HTS), efficient data management has also become a key factor.

The last 15 years of drug discovery have been heavily influenced by structural knowledge of target proteins and molecular modeling techniques, which allowed in some cases the design of tailor-made ligands. This so-called "rational drug design" approach, albeit quite successful, has recently been critically analyzed and reviewed in many drug discovery programs, and elements of random design have again become quite fashionable. In this account, we would like to present ideas, tools, and strategies that comprise an integrated lead discovery approach. The approach combines structural knowledge of target proteins and efforts to design conformationally constrained small peptides with combinatorial and parallel chemistry techniques.

Amino acid-derived peptide mimetics[9,10] corresponding to exposed protein epitopes, such as β-turns, loops, 3_{10}-helical and α-helical structures, can serve as valuable tools to probe potential ligand–receptor or ligand–enzyme interaction sites, and are discussed in Section II. Ligand–receptor interactions spanning large

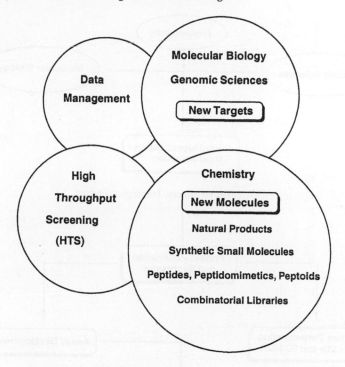

Figure 1. Four key elements of modern lead finding.

surface areas have been a challenge for drug discovery programs,[11] and so far it has been notoriously difficult to find small-molecule ligands as lead compounds (vide infra). In these cases, an integrated approach, combining structural knowledge derived from libraries of conformationally constrained small peptides with combinatorial chemistry, seems a very promising approach. In Section III we present ideas as to how such structural knowledge may be transferred to small non-peptidic molecules, again using parallel and combinatorial chemistry. This integrated approach of combining rational and random elements is schematically illustrated in Figure 2.

After a target protein has been identified and selected, it is expressed and purified using modern biochemical techniques. Once the material is available in pure form and sufficient quantity, the phases of structure determination and assay development usually start simultaneously. The assay is subsequently transferred into a format suitable for HTS, whereupon the process of screening a large number of structurally diverse compounds is initiated. The selection of these compound collections is usually performed in a random fashion, or by using 2D- and 3D-clustering techniques.[12] Once a ligand molecule has been identified in the screening

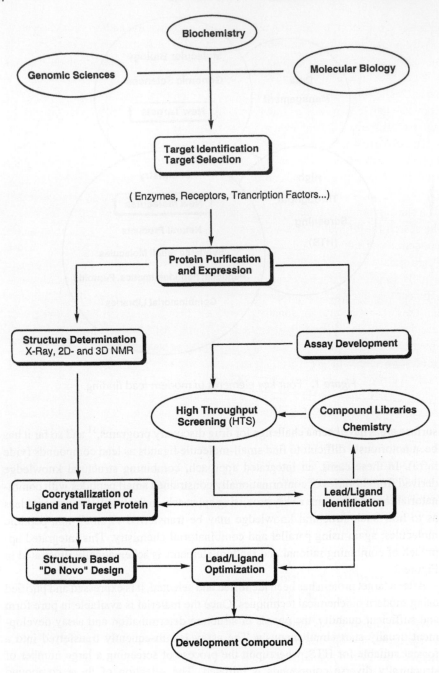

Figure 2. Key elements of modern drug discovery.

process, it can serve as a starting point for lead optimization and/or, in case crystals of the target protein are available, be cocrystallized with the protein. Based on structural information regarding the binding mode of a ligand molecule to an enzyme or receptor, an alternative approach involving "de novo" or rational design of an "ideal" ligand can be contemplated. In the ideal case, the two approaches converge into lead molecules showing common structural features which ultimately will be transferable into a development compound.

The different sources for novel lead compounds are schematically shown in Figure 3. Natural products isolated from microbial broths or plants are still very valuable sources of good lead compounds. Taxol,[13] avermectin,[14] mitomycin {AD-DIN},[15-17] and FK-506[18] are only a few examples of natural products that served as a starting point for intense optimization programs. Equally valuable sources in the search for interesting lead compounds are small molecular weight compound collections of industrial and academic origins, which have been amassed over the years or have been generated by parallel and combinatorial techniques. Due to their usually simpler structures, these small molecules can often be modified in a more efficient way than natural products. Based on the 3D structure of a target protein, libraries of small conformationally constrained peptides, incorporating amino acid templates [11,19] corresponding to exposed protein epitopes, can be designed, synthesized and tested, and such compound collections are potentially an additional interesting source of novel lead compounds. Section II of this account will especially focus on this aspect of lead discovery.

In the early phase of HTS, the compound collections will be rather large and of the random type (Figure 3). Based on first lead structures emerging from HTS, more tailor-made compound collections will be generated and screened. At that stage, rational design elements are usually of increasing importance. Lead compounds which originated from a compound collection generated by parallel and combinatorial chemistry are of particular value in this development phase since the optimization process is accelerated by the modular synthetic approaches used. In the lead optimization phase, focused libraries containing all the structural elements of the lead compounds play a greater role for fast development of clinical candidates, thereby helping to reduce the overall development costs.

B. Lead Finding Using Conformationally Constrained Peptides Mimicking Exposed Protein Epitopes

As the number of 3D structures of receptors that are relevant targets for drug design grow, new opportunities arise to apply design and combinatorial screening methods to identify small molecule peptide mimetics of the key "functional epitopes" involved in ligand recognition. We highlight below recent advances in protein chemistry, that provide a biochemical perspective to the theme of peptide and protein mimetics, discussed later in this chapter.

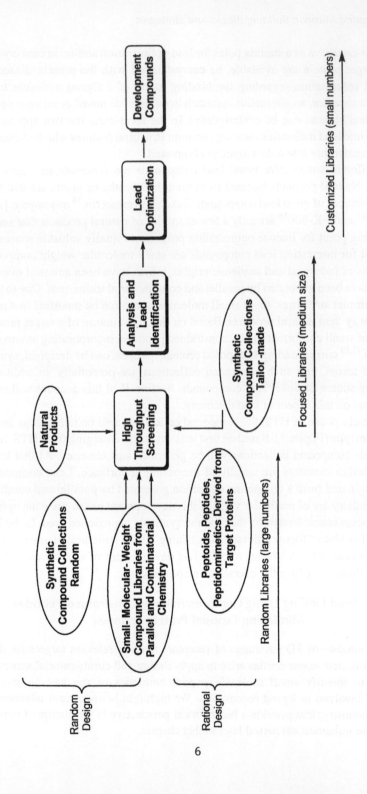

Figure 3. Sources for compound libraries in the drug development process.

We leave aside, here, the class of G protein-coupled membrane-bound receptors with seven transmembrane-spanning helices,[20] for which the discovery of novel small molecule drug leads is amply precedented. Instead we focus (for reasons of personal interest) on the many signaling pathways operating between mammalian cells, which are initiated by the binding of a *protein ligand* (e.g. integrins, cytokines, growth hormones, etc.) to a receptor displayed outside the cell. Docking of the protein ligand to its receptor triggers signal transduction across the membrane, which ultimately leads to changes in the levels of gene expression in the nucleus. Small molecules that block specific signaling pathways appear to have great potential for the treatment of a wide range of human disorders, yet up to the present have been difficult to identify.

The extracellular portion of receptors in the cytokine and hematopoietic receptor superfamily[21] typically comprise two or more immunoglobulin(Ig)-like domains[22–25] (Figure 4). A small transmembrane segment, typically sufficient in length for a membrane-spanning helix, leads to the cytoplasmic domain, which acts as the focal point for interaction with other cellular components. A key feature of the mechanism of signal transduction across the cell membrane involves the association of two or more receptor chains through complexation of their extracellular domains with a single copy of the ligand. In this way, the intracellular receptor domains and their associated kinases are also brought together, which is crucial for interactions with other cytoplasmic proteins.

From a structural viewpoint, perhaps the best studied ligand–receptor complex is that with human growth hormone (hGH).[26–28] Crystallographic studies have revealed a complex comprising a single growth hormone molecule bound to two extracellular receptor chains (Figure 5). The extracellular receptor consists of two Ig-like domains, each with a fibronectin-III fold. At each hGH–receptor (HGR) interface, intimate interactions between the largely helical ligand and the receptor, involving hydrogen bonding and van der Waals interactions, extend over an area of ca. 1300 Å^2. The two interfaces are not identical, showing that the same set of residues on the receptor monomers can bind to different parts of the hormone. In addition, there is a direct interaction over ca. 500 Å^2 between the two HGR molecules in their membrane-proximal Ig-like domains. Support for the dimerization mechanism of hGH–receptor activation has come from studies of monoclonal antireceptor antibodies that act as receptor agonists since they are able to dimerize the receptor, whereas monovalent fragments cannot and act as antagonists.[26]

At high concentrations, human growth hormone itself acts as an antagonist because of a difference in affinity for the receptor at its two receptor binding sites. This antagonist action has been accentuated by introducing mutations into the low-affinity site that further reduce affinity for the receptor, and by combining these with mutations that enhance binding at the second site.[27,29,30] The structure of one such antagonist mutant of human growth hormone (G120R) in complex with a single receptor chain has also been described.[31]

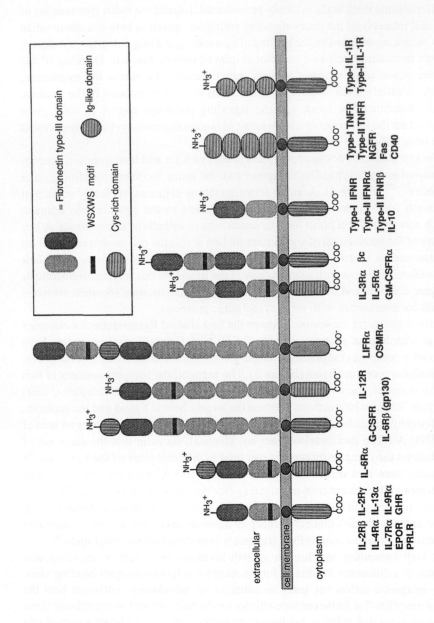

Figure 4. Some members of the hematopoietic cytokine receptor superfamily.

IL-2Rβ IL-2Rγ IL-6Rα G-CSFR IL-12R LIFRα IL-3Rα βc Type-I IFNR Type-I TNFR Type-I IL-1R
IL-4Rα IL-13α   IL-6Rβ (gp130) OSMRα IL-5Rα Type-II IFNRα Type-II TNFR Type-II IL-1R
IL-7Rα IL-9Rα   GM-CSFRα Type-II IFNRβ NGFR
EPOR GHR   IL-10 Fas
PRLR   CD40

extracellular

cell membrane

cytoplasm

= Fibronectin type-III domain

WSXWS motif

Cys-rich domain

Ig-like domain

8

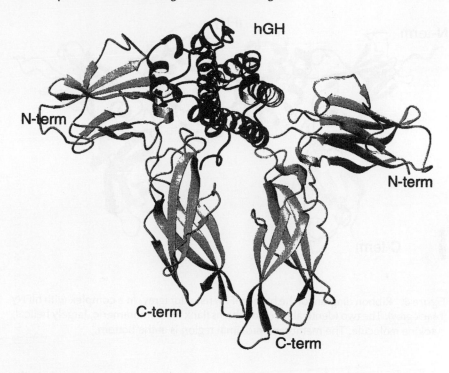

Figure 5. Ribbon diagram of the hGH receptor (grey) in a complex with growth hormone (black) (Brookhaven file 3HHR). The two receptor chains flank the largely helical growth hormone. The membrane proximal region is at the bottom.

A similar overall topology was found in the crystal structure of human interferon γ (IFNγ) bound to the IFNγ–receptor α-chain,[32] which also revealed a 2:1 receptor–ligand stoichiometry (Figure 6). In this case, the ligand is a homodimer (2×17 kDa), whereas each receptor chain again comprises two Ig-like domains. About 960 Å^2 of accessible surface is buried at each interface upon complex formation, but the two interfaces are essentially identical, and the receptor chains do not make direct contact with each other. In this system, however, signal transduction requires a second receptor chain, the IFNγR β-chain, which binds the 1:2 ligand–receptor complex.[33] Thereafter, on the cytoplasmic side of the membrane, STAT1, a transcriptional activator is recruited into a complex in which JAK-1 and JAK-2 kinases are associated with the intracellular portions of the receptor α- and β-chains, whereupon STAT1 is activated by phosphorylation. After release from the receptor, the phosphorylated STAT1 dimerizes, and with or without additional factors, migrates to the nucleus to activate transcription through GAS-related DNA response elements.[34]

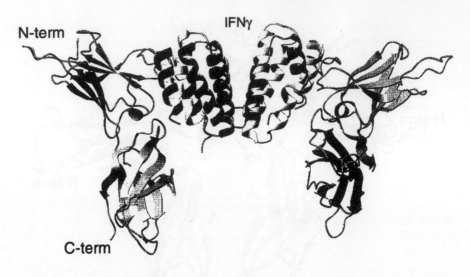

Figure 6. Ribbon diagram of the human IFNγ receptor (grey) in a complex with hIFNγ (black grey). The two identical receptor chains flank the homodimeric, largely helical, cytokine molecule. The membrane proximal region is at the bottom.

A quite different binding topology and stoichiometry was found recently in the crystal structure of IL-1β complexed with the type-I IL-1 receptor.[35] The extracellular IL-1R comprises three Ig-like domains, which wrap around the ligand in a manner quite distinct from that found in the GHR and IFNγR systems. In this case, a total of 2088 Å^2 of solvent accessible area is buried upon formation of the 1:1 ligand-receptor complex. IL-1 is the only cytokine for which a naturally occurring antagonist (IL-1ra) is known. The IL-1ra protein (152 residues; a six-stranded β-barrel similar to IL-1β) appears to be a physiologically important regulator of IL-1 activity, and shows about 30% sequence homology with IL-1β. Signaling through the IL-1 receptor requires the formation of a ternary complex between IL-1β, IL-1R, and another membrane receptor accessory protein. When IL-1ra binds the IL-1R the ternary complex with the accessory protein cannot be formed, and signaling does not occur. The crystal structure of IL-1ra bound to the IL-1R has also been solved to 2.5 Å resolution.[36] IL-1β and IL-1ra bind in the same concave elbow region of the IL-1R, although there are significant differences in the way the two ligands contact the receptor protein. In particular, IL-1β makes more extensive contacts with the third membrane proximal Ig-like domain of the receptor than does the antagonist IL-1ra.

Structural data is also available for tumor necrosis factor (TNF) bound to a soluble fragment of its receptor[37] in a 3:1 receptor:ligand stoichiometry, and for tissue factor bound to factor VIIA in a 1:1 complex.[38] Both of these complexes are characterized by the burial of a large area of solvent accessible surface upon complex formation.

The concept of rational drug design is based on the assumption that if the atomic structure of a ligand–receptor complex were known, it would become possible to design small molecules that block or mimic the action of the protein ligand, thus providing antagonists or agonists, respectively. However, these ligand–receptor crystal structures illustrate a major difficulty in the design of a small molecule antagonist; namely, the protein ligands bind their receptors by interacting at various points across a large surface area. Until recently, it was unclear whether a "small" molecule could be found that mimicked the same binding contacts to the receptor as the natural ligand.

With high-resolution structural data available, it becomes feasible to design and interpret mutagenesis studies which aim to identify the key interactions that contribute most to the stability of ligand–receptor association. In the hGH–receptor complex, residues buried in the ligand–receptor interface were systematically replaced by alanine and the effect on ligand–receptor binding was assayed.[27,39–42] Of about 30 side chains from each protein that make contact, on the receptor side a central hydrophobic region dominated by two tryptophan residues (W104 and W169) accounts for more than 75% of the binding free energy. A similar analysis on the hormone side showed that 8 out of 31 side chains in hGH accounted for about 85% of the binding energy. Moreover, the energetically important and unimportant residues on one molecule pack against those on the other. The buried surface thus appears to comprise a critical tightly packed hydrophobic core, surrounded by more polar groups involved in hydrogen bonding and charged interactions that make a much smaller contribution to the free energy of binding. In other words, the hGH–receptor complex appears to be stabilized by a few strong interactions near the center of the interface, a so-called energetic epitope or "hot-spot" of binding energy,[41] rather than through the accumulated effects of a network of weaker interactions spread more-or-less evenly across its surface. These appealing results suggest, of course, that a small molecule mimic of the central energetic or functional epitope might provide access to high-affinity inhibitors of the ligand–receptor interaction.[27]

An astonishing demonstration that small molecule cytokine receptor ligands can be found came recently with the discovery (through combinatorial methods) of a 20-amino acid disulfide bridged peptide that binds *and activates* the erythropoietin (EPO) receptor.[43,44] Human EPO is the principal regulator of red blood cell production, and comprises a 34 kDa glycoprotein of 166 amino acids. EPO is used to treat patients with renal anemia, and has the more general effect of stimulating erythropoiesis. EPO signaling is initiated when the ligand dimerizes the EPO receptor. By screening a random phage Cys-X_8-Cys peptide library a consensus, weak ($K_D \approx 10$ µM) EPOR binder was isolated. The affinity of this peptide for the EPOR was then improved by adding flanking residues and mutagenesis of the internal residues. This resulted in the isolation of a disulfide-bridged peptide

(GGTYSCHFGPLTWVCKPQGG) with a K_D of ≈ 200 nM (compared to ca. 200 pM for EPO). Even more remarkable, the peptide activates the EPO signaling pathway by dimerizing the EPOR.[44] How this is achieved was revealed in a crystal structure of the peptide–EPOR complex [43] (Figure 7). The extracellular part of the EPOR comprises two domains with high structural homology to the hGH–receptor complex. Two peptide molecules form a complex with two EPO–receptor chains, and each peptide takes up a β-hairpin conformation with a turn at Gly[9]-Pro[10]. The two β-hairpins in the dimer are arranged face-to-face at about right angles in the complex, such that each peptide interacts not only with its peptide partner but also with both EPO–receptor molecules. Hydrogen bonding and a small hydrophobic core apparently stabilize the complex.[44] It may be more than a coincidence that the dimeric peptide binds to the EPOR at a site analogous to the energetic hot spot identified on the hGH–receptor. This example illustrates the potential of phage-display methodology for ligand discovery, and importantly, establishes the feasibility of discovering small molecule mimetics of large polypeptide hormones.

Recently, peptide ligands of the IL-1 receptor[45] and ICAM-1[46] have also been isolated from phage-display libraries. We highlight later in this review how small molecule mimetics of loop epitopes are being sought by using synthetic methods. This raises the prospect of combining the biological and chemical approaches to ligand discovery, for example, by using phage display to screen large libraries of

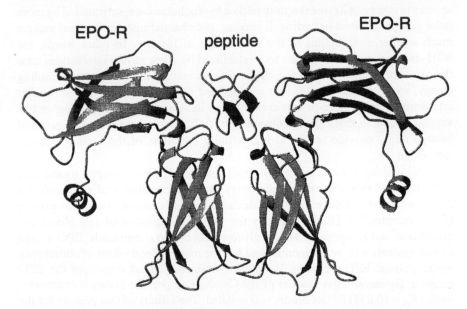

Figure 7. Ribbon diagram of the EPO-receptor bound to a disulfide-bridged agonist peptide, isolated by phage display methods (Brookhaven file 1EBP).

constrained peptides (10^6–10^8 members), and refining the hits using chemically based combinatorial and design methods.

Another interesting example reported recently is the peptide mimetic **1** (Figure 8), which is a small molecule inhibitor of the IL-2–IL-2Rα receptor interaction.[47]

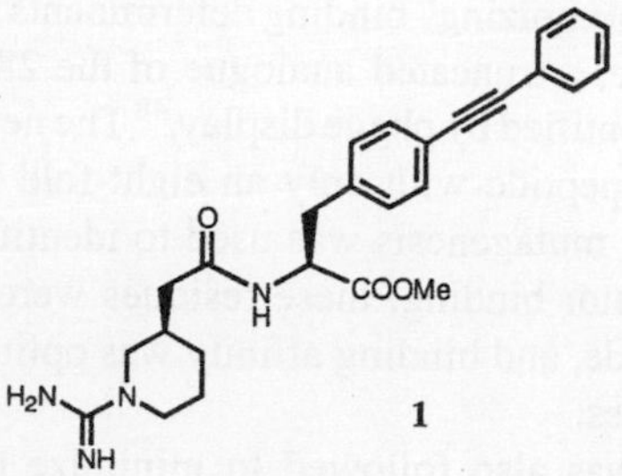

1

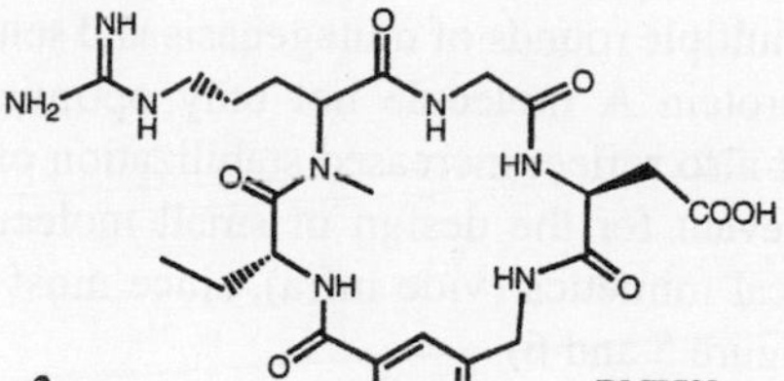

2 **DMP728**

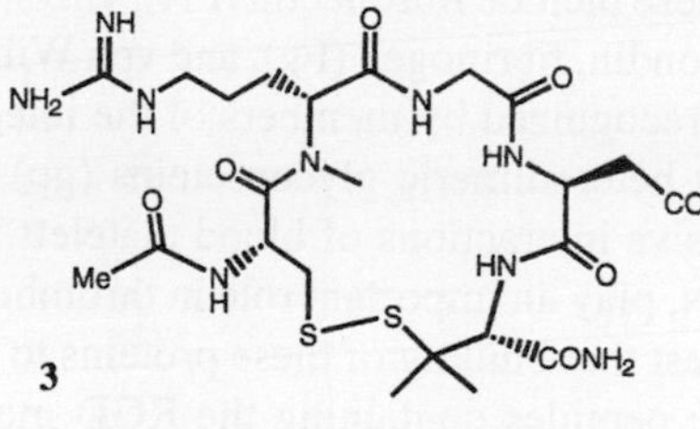

3

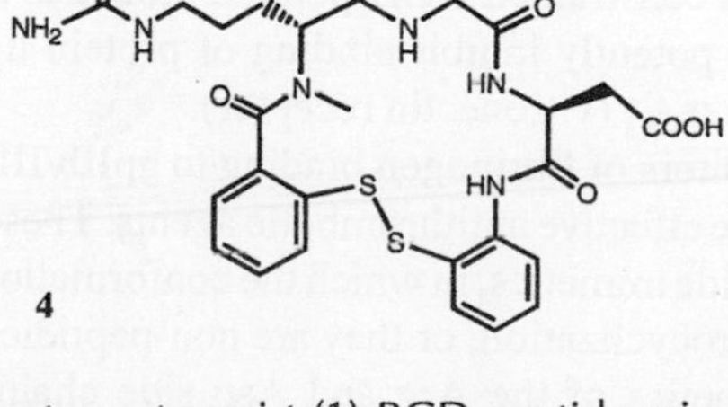

4

Figure 8. An IL-2 receptor antagonist (1) RGD peptide mimetic gpIIb/IIIa receptor antagonists.

IL-2 is a cytokine comprising four helices. Mutagenesis identified residues in the AB loop (K35, R38, T41, F42, K43, Y45) as being crucial for the binding of IL-2 to its receptor. Although the small molecule inhibitor 1 was intended to mimic the R38-F42 region of IL-2, and so might bind the receptor, NMR studies indicated that it actually binds to the *ligand*.

A related application of phage-display technology, equally relevant for peptide mimetic design, is in "minimizing" binding determinants in protein and polypeptide hormones. For example, a truncated analogue of the 28-amino-acid peptide hormone ANP has been identified by phage display.[48] The new analogue is a 15-amino-acid disulfide-bridged peptide with only an eight-fold lower biological potency. Here, alanine-scanning mutagenesis was used to identify residues in ANP which are important for receptor binding; these residues were combined into a smaller disulfide-bridged peptide, and binding affinity was optimized by randomly mutating nonessential residues.

A similar strategy was also followed to minimize the binding domain from protein A, a 59-residue 3-helix bundle protein that binds tightly to the Fc portion of an IgG_1 molecule.[49] In this case, a 2-helix derivative of just 33 residues was discovered following multiple rounds of mutagenesis and selection. The mutations selected in the mini-protein A molecule not only optimize the intermolecular contacts with IgG_1, but also reflect increased stabilization of the 2-helix scaffold. This work may be relevant for the design of small molecule cytokine receptor ligands, based on helical mimetics (vide infra), since most cytokines are helical bundle proteins (see Figure 5 and 6).

Many adhesive proteins in extracellular matrices and in the blood contain an Ig-like domain with an Arg-Gly-Asp (RGD) cell recognition motif located in an extended protein loop. These include fibronectin (FN), vitronectin (VN), osteopontin, collagens, thrombospondin, fibrinogen (Fg), and von Willebrand factor (vWF). The RGD sequences are recognized by members of the integrin receptor superfamily, which are typically heterodimeric glycoproteins (gp) with two membrane-spanning subunits. Adhesive interactions of blood platelets with plasma proteins, such as fibrinogen and FN, play an important role in thrombosis and hemostasis. It is therefore of great interest that binding of these proteins to integrin receptors can be inhibited by synthetic peptides containing the RGD motif. Even short linear peptides inhibit ligand binding to the fibrinogen receptor (gpIIb/IIIa) in the mid-micromolar range, but constrained RGD peptide mimetics have been discovered, which selectively and potently inhibit binding of protein ligands to the integrins gpIIb/IIIa ($\alpha_{IIb}\beta_3$) and $\alpha_v\beta_3$ (vitronectin receptor).

Several potent inhibitors of fibrinogen binding to gpIIb/IIIa are now known, and some have proven to be effective antithrombotic agents. These inhibitors tend either to be small cyclic peptide mimetics, in which the conformation of an RGD sequence is constrained by macrocyclization, or they are non-peptidic molecules containing functional group mimetics of the Arg and Asp side chains, held in the correct geometry for interaction with the receptor. Examples of both classes are shown in

Figures 8 and 9. Crucial to the success of these small molecule inhibitors is the fact that the key interactions needed for high-affinity binding to this receptor are localized in a short motif at the tip of one loop in the protein ligand.

Insights into the conformation(s) this loop might adopt in FN has come from structural studies by NMR and X-ray crystallography. FN is a dimeric glycoprotein with a subunit molecular weight of 220–250 kDa. It is a modular protein composed

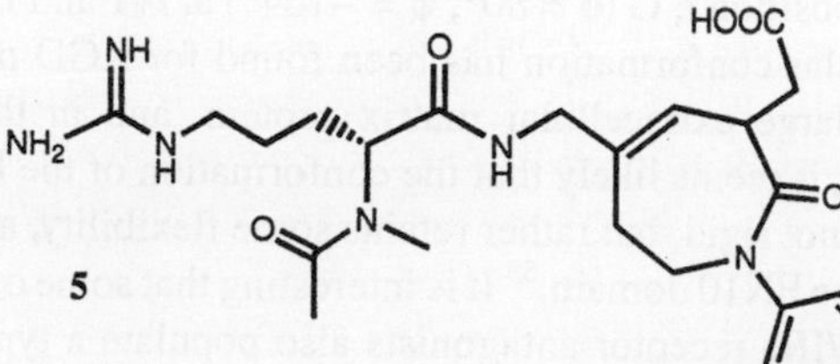

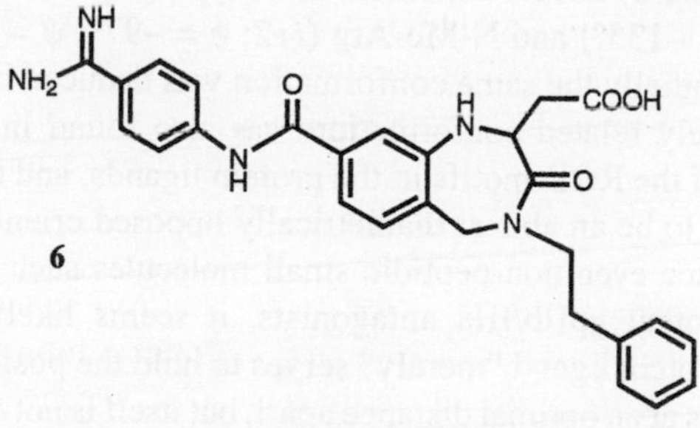

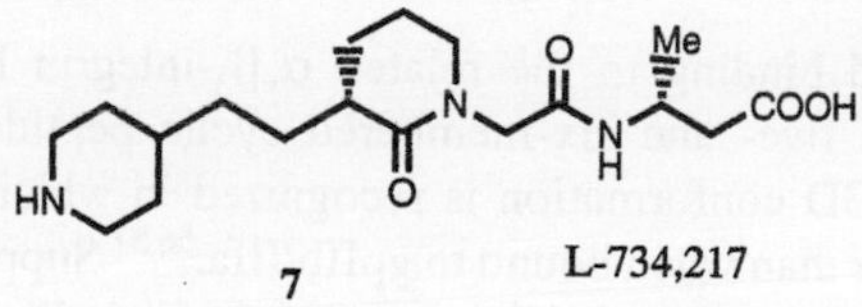

Figure 9. RGD non-peptide mimetic gpIIb/IIIa receptor antagonists.

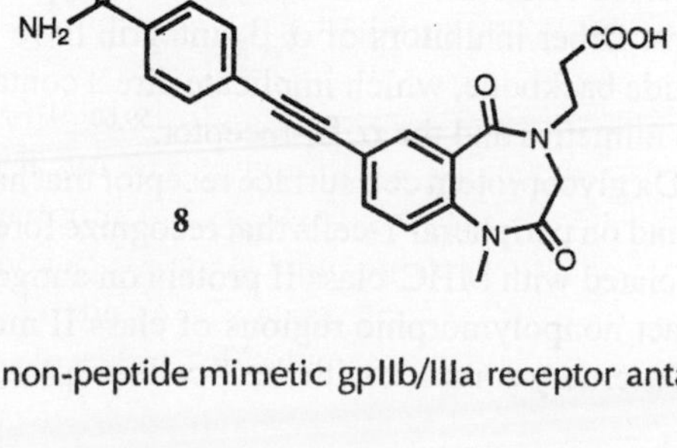

of homologous repeats of three prototypical domains known as types I, II, and III. Each monomer of FN contains a contiguous array of 15–17 of these ≈90 amino acid domains (depending on variations in RNA splicing). The RGD motif recognized by the $\alpha_{IIb}\beta_3$ integrin is found in the tenth type-III repeat. A crystal structure of the seventh through tenth type-III repeats revealed the RGDS sequence at the tip of a β hairpin-like loop that extends ≈10 Å away from the body of the FN7-10 molecule and makes no contacts with other regions of the FN molecule crystallography.[50] The RGDS motif has backbone torsion angles close to those for a type II′ β turn, with R at position i, G ($\phi = 80°$, $\psi = -164°$) at $i+1$ and D ($\phi = -75°$, $\psi = -24°$) at $i+2$. A similar conformation has been found for RGD motifs present in tenascin,[51] another large extracellular matrix protein, and in the leech protein decorsin.[52] However, it seems likely that the conformation of the RGD loop in the FN7-10 molecule is not rigid, but rather retains some flexibility, as suggested also by NMR studies of the FN10 domain.[53] It is interesting that some of the most potent RGD mimetic gpIIb/IIIa receptor antagonists also populate a type II′ β-turn, but not in the same register as seen in FN10, tenascin, or decorsin. For example DMP728 (**2**) (Figure 8) adopts an almost ideal type-II′ β-turn centered at D-Abu ($i+1$; $\phi = 60°$, $\psi = -135°$) and N-Me-Arg ($i+2$; $\phi = -97°$, $\psi = 13.5°$) in a crystal structure, and essentially the same conformation was deduced by NMR in DMSO solution.[54] A closely related conformation was also found in **4** (Figure 8).[55] A common feature of the RGD motifs in the protein ligands, and the small molecule mimetics, appears to be an almost diametrically opposed orientation of the R and D side chains. Since even non-peptidic small molecules such as those shown in Figure 9 make potent gpIIb/IIIa antagonists, it seems likely that the peptide backbone in the protein ligand "merely" serves to hold the positive and negatively charged side chains at an optimal distance apart, but itself is not directly recognized by this integrin.

Inhibitors of ligand binding to the related $\alpha_v\beta_3$-integrin have also attracted interest. Studies with five- and six-membered cyclic peptides have led to the hypothesis that an RGD conformation is recognized in which the R and D side chains are much closer than when bound to gpIIb/IIIa.[56,57] Support for this idea has come recently from studies of the RGD-mimetics shown in Figure 10,[58] which are closely related to DMP728 (**2**) (Figure 8), but selectively inhibit the $\alpha_v\beta_3$-integrin. Interestingly, the change in selectivity from gpIIb/IIIa to $\alpha_v\beta_3$ was accompanied by a change in preferred conformation from type-II′ to type-I β-turn centered on the Xaa-Arg residues. Other inhibitors of $\alpha_v\beta_3$-integrin have been found, based on a cyclic pentapeptide backbone, which implicate direct contacts between the peptide backbone of the mimetics and the $\alpha_v\beta_3$-receptor.[59,60]

CD4 is a 55-kDa glycoprotein cell surface receptor that has attracted great interest recently. It is found on peripheral T-cells that recognize foreign antigens in the form of peptides associated with MHC-class II protein on antigen presenting cells. CD4 appears to contact nonpolymorphic regions of class II molecules, which leads to the formation of a ternary complex with the T-cell receptor. Human CD4 also serves

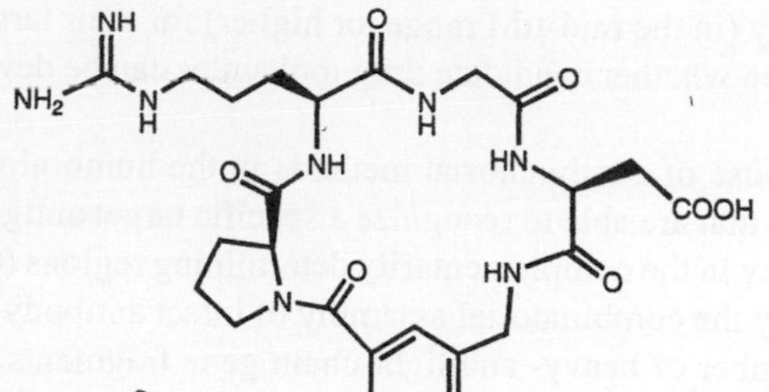

Figure 10. RGD peptide mimetic $\alpha_v\beta_3$ receptor antagonists.

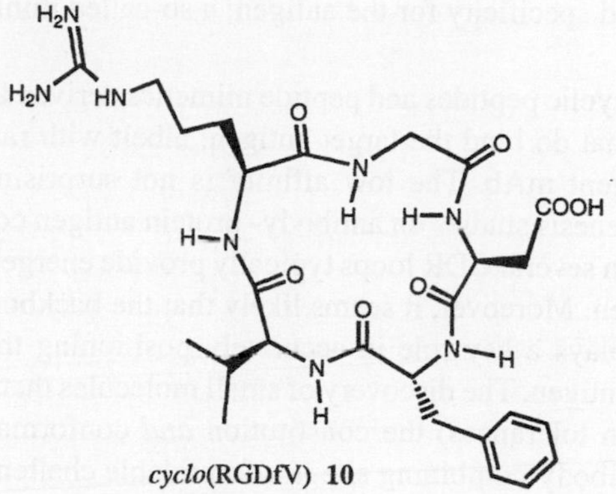

as a key component of the cellular receptor for the coat protein gp120 on the human immunodeficiency virus (HIV). The extracellular portion of CD4 comprises four Ig-like domains.[61] Cyclic peptides that mimic surface loops on CD4 involved in binding gp120 or MHC-class II are therefore of interest as inhibitors of HIV infection and CD4-dependent T-cell responses, respectively. Cyclic peptides and peptide mimetics derived from surface loops on CD4 have been found that inhibit, albeit weakly (in the mid µM range), *in vitro* CD4-dependent T-cell responses, including experimental allergic encephalomyelitis,[62] whereas others inhibit the association of CD4 with gp120 on HIV.[63–67]

Mimetics of several other Ig–receptor superfamily members, or their ligands have also been found in recent years [68] including, for example, residues in human granulocyte-macrophage colony stimulating factor (GM-CSF),[69,70] a loop region on tissue factor (TF),[71] a β-turn region of nerve growth factor,[72] a loop in intercellular adhesion molecule ICAM-1,[73] a loop in transforming growth factor-α,[74] and a loop on vascular adhesion molecule 1 (VCAM-1),[75] to name but a few. On the other hand, most of the molecules discovered display at best only a moderate inhibitory activity (in the mid-μM range, or higher) for their target receptor, and it remains to be seen whether candidate drug molecules can be developed from these leads.

Nature makes use of combinatorial methods in the humoral immune system to select antibodies that are able to recognize a specific target antigen. A high level of sequence diversity in the complementarity determining regions (CDRs) of antibodies is achieved by the combinatorial assembly of intact antibody-coding sequences from a large number of heavy- and light-chain gene fragments. It is an appealing idea, therefore, that monoclonal antibodies (mAbs) previously selected for their ability to bind a specific target might be used as a starting point for peptide mimetic design. The aim would be to produce mimetics of the CDR loops in an antibody that retain affinity and specificity for the antigen; a so-called minimal recognition unit.[76]

Several reports of cyclic peptides and peptide mimetics derived from mAb CDRs have appeared[77–86] that do bind the target antigen, albeit with rather low affinity compared to the parent mAb. The low affinity is not surprising since crystallographic and mutagenesis studies on antibody–protein antigen complexes[87] have shown that residues in several CDR loops typically provide energetically important contacts to the antigen. Moreover, it seems likely that the backbone conformation of each CDR loop plays a key role in accurately positioning these residues for interaction with the antigen. The discovery of small molecules that mimic (possibly within narrow design tolerances) the constitution *and* conformation of multiple CDR loops in an antibody-combining site is a formidable challenge. On the other hand, even if the full binding affinity of the intact mAb cannot be reached, useful applications for such mimetics may still be found. For example, a radiolabeled peptide from the CDR of an antitumor antibody has been used for specific tumor targeting, and may find application for breast cancer imaging and possibly therapy.[88]

The main purpose of this section has been to highlight the rapid progress over recent years in studies of ligand–receptor interactions at a molecular level, and the new opportunities that arise for peptide mimetic design. This also emphasizes the need for new approaches and molecular tools in the design of conformationally defined, small molecule peptide and protein mimetics, which is the main topic of the remaining sections in this article.

C.   Parallel and Combinatorial Chemistry Approaches

As mentioned earlier in the Introduction, combinatorial and parallel chemistry have emerged as novel technologies to provide libraries of diverse compounds. The basic idea of combinatorial chemistry is to synthesize starting e.g. from building blocks A, B, C^1, C^2, and C^3 (Figures 11–13) all possible combinations of products (P^1-P^3) as single compounds, or as mixtures on solid supports, or in solution. These compounds can be synthesized in a parallel or in a combinatorial fashion using the

Figure 11. Linear vs. convergent assembly strategies.

"split and mix" methodology.[89,90] Since compounds P^1-P^3 were created from similar building blocks using the same assembly strategy, they form an ensemble or a library of compounds having the same core structure with all possible combinations of appended substituents. Prerequisites for a successful combinatorial approach are sets of highly versatile and reactive building blocks of type A, B and C which can be readily assembled in a linear or in a convergent fashion (Figure 11). Not surprisingly, combinatorial chemistry has originated from linear assembly strategies employed in peptide and oligonucleotide chemistry,[91,92] since both building blocks and coupling chemistry have been extensively established in solution as well as on solid supports.

Using standard coupling procedures, a linear approach yields readily the six possible combinations of building blocks A^1-A^3 (assuming that for steric reasons not all three substituents can be adjacent to each other) (Figure 11). The convergent strategy is usually more demanding both in terms of having access to the reactive building blocks A, B, and C^1-C^3 ("reactophores") as well as in elaborating the corresponding assembly chemistry. Nevertheless, there is currently a significant shift observable from linear to more convergent strategies, probably due to the fact that convergent approaches are more likely than linear ones to generate small "drug-like" molecules.

In convergent approaches, the reactive components A, B, and C^1-C^3 (Figures 11-13) can be either combined in multicomponent one-pot reactions, where products P^1-P^3 are formed directly and no intermediates of type D can be isolated,

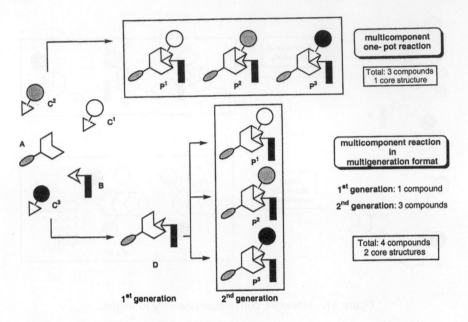

Figure 12. Multicomponent one-pot multigeneration reactions: sequence A.

or in multigeneration format where stable intermediates of type D can be isolated, thus forming the first generation products, which subsequently can be transformed into the same final products P^1-P^3. Among the multicomponent one-pot reactions, we would mention among many others the *Ugi* four-component reaction, the *Strecker* reaction, the different *Hantsch*-type condensations and the *Mannich* reaction.[93] As can be seen from Figures 12 and 13 there are in principle two multigeneration reactions possible starting from A: the sequence A in Figure 12, by combining A with B yields one intermediate D, which subsequently reacts with C^1-C^3 to form P^1-P^3. This sequence yields four compounds in total having two different core structures; the sequence B shown in Figure 13 combines first A with C^1-C^3 to yield intermediates D^1-D^3, which react further with B to yield P^1-P^3. This strategy produces six compounds in total having two different core structures.

This simple sketch illustrates clearly that convergent multicomponent reactions performed with a limited set of reactive building blocks (reactophores) in a multigeneration format offer a tremendous potential to produce diverse small-molecule compound collections, depending on the reaction sequence used (the "combinatorics of reactive building blocks"). The concept of combinatorics of reactive building blocks should ultimately lead to novel multicomponent reactions. In Section III we will focus on reactophores such as α-alkynyl ketones, which allow the construction of a wide variety of core structures.

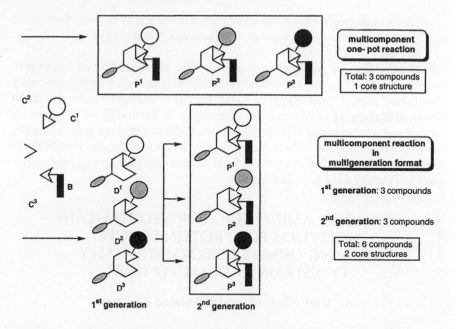

Figure 13. Multicomponent one-pot multigeneration reactions: sequence B.

In addition to using tailor-made reactive building blocks in multigeneration and/or multicomponent assembly strategies, the introduction of solid supports offers further interesting perspectives for efficient library generation. They can be employed as polymer-bound reagents[94,95] and polymer-bound trapping reagents, when reactions are performed in solution, or as supports for multigeneration reactions. Although there are some limitations in the reactions that can be employed on resins compared to solution chemistry, more and more solid-phase reaction procedures are currently being developed.[96]

Novel resin–cleavage strategies have emerged recently as valuable tools in combinatorial and parallel chemistry. The most promising strategies can be summarized as follows:

1. Cleavage of one specific functional group, such as a carboxylic ester, hydroxamate or amide. This strategy is especially useful to produce focused or targeted libraries.
2. Multidirectional cleavage strategies,[93,97,98] which offer the possibility to liberate several different functional groups or elements of diversity. "Safety-catch" linker strategies[97,99,100] have been of special interest in this context.
3. Cleavage procedures that take advantage of concomitant cyclizations have also proven to be interesting methods to generate novel core structures.[12]

This strategy often produces clean products since only those molecules are cleaved that have gone through the whole reaction sequence.

In Section II.C we will present novel tricyclic xanthene derived amino acid templates, which allow the construction of libraries of cyclic conformationally constrained peptide loop mimetics using the "split-and-mix" method without having to use tagging and deconvolution strategies. In Section III we will focus on parallel and combinatorial approaches devoted to the synthesis of small molecule, non-peptidic compound collections, which in addition offer the possibility to incorporate structural features derived from protein epitope mapping into conformationally constrained peptide mimetics.

II. NOVEL AMINO ACID-DERIVED TEMPLATE MOLECULES FOR PROTEIN EPITOPE MAPPING USING CONFORMATIONALLY CONSTRAINED SMALL PEPTIDES

A. Derivatives of α,α-Dialkylated Glycines

Introduction

The conformational flexibility of short peptides can be significantly reduced by replacing one coding amino acid by a sterically constrained unnatural amino acid analogue. As will be shown in this section, such a conformational constraint initiated by one unnatural amino acid is not limited to very short peptides but can extend into 9 to 12-mer peptides. The simple concept of replacing coding amino acids by sterically constrained analogues has been widely used in drug discovery programs and has led, for example, to potent inhibitors of renin,[101] HIV-protease,[102,103] and collagenase.[104] In the latter case, *tert*-butyl glycine was used to stabilize the β-sheet type conformation required for specific binding to the enzyme.

Among the growing number of noncoded synthetic amino acids, the open-chain and cyclic α,α-disubstituted amino acids of type **11–22** (Figure 14) have become important [105–107] due to their inherent propensities to stabilize rather well-defined conformations in short peptides, depending on the nature of the substituents at the α-position.[108–110] The α-methylated α-amino acids have been particularly well investigated due to their ability to stabilize 3_{10}- and α-helical as well as β-turn type conformations in peptides.[105] Much less was known about the conformational behavior of optically pure α,α-disubstituted glycines combining two side chains of proteinogenic amino acids ("α-chimeras") at the α-center.

In one example, (S)-α-methylproline (P^{Me}) has been used as a replacement for Pro to stabilize β-turn conformations in peptide antigens.[111,112] Proline is found frequently at the $i+1$ position of β-turns in protein crystal structures (vide infra), but the tendency for a turn to form in peptides is frequently enhanced upon

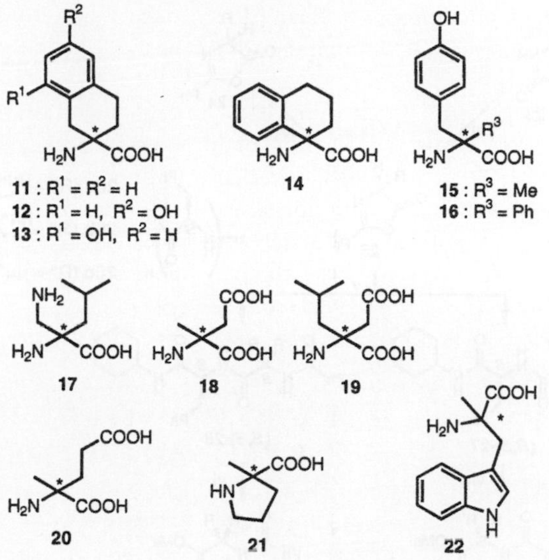

Figure 14. Optically pure α,α-disubstituted glycines (11–22).

substituting Pro by P^Me. For example, "nascent turns" in the interesting Asn-Pro-Asn-Ala (NPNA) motif, found as a tandemly repeated unit in the immunodominant site on the cicumsporozoite protein on the malaria parasite *Plasmodium falciparum*, were stabilized by substituting Pro by P^Me.[112] A 12-mer (NP^MeNA)₃ peptide was shown by NMR to populate extensively type-I β-turns, within each NP^MeNA motif. This methylated 12-mer peptide elicited antibodies that bind to the malaria parasite, thus providing some evidence for the biological relevance of the deduced β-turn conformations.[112] Molecular dynamics simulations with time-averaged distance restraints derived from NOE data also suggested how peptides containing multiple tandemly repeated NPNA motifs might fold to give a stem-like super-secondary structure with linked βI-turns.[113]

Recently, it was shown[108] that a large variety of novel and interesting open-chain and cyclic (*R*)- and (*S*)-α,α-disubstituted glycines could be synthesized in optically pure form using the strategy outlined in Scheme 1. Treatment of the 4,4-disubstituted-1,3-oxazol-5(4*H*)-ones **25**, which were obtained either from the hydantoins **23** via the classical Bucherer–Bergs reaction, or α-alkylation of the 4-monosubstituted-2-phenyl-1,3-oxazol-5(4*H*)-ones **24** (R′ = Ph) with an optically pure amine **26** derived from L-phenylalanine, yielded the diastereomeric peptides (*R,S*)-**27** and (*S,S*)-**28**, which were separated by crystallization and/or flash-chromatography (FC) on silica (Scheme 1). Selective amide cleavage using trifluoromethansulfonic acid (CF₃SO₃H) in methanol gave the optically pure esters (*R*)- and (*S*)-**29**. The corresponding amino acids could be obtained in high yield by treatment with 25%

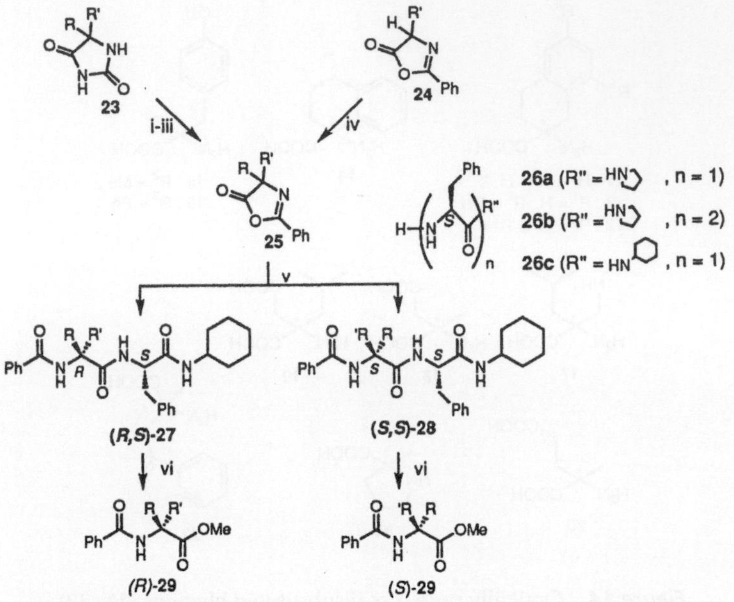

i: Ba(OH)$_2 \cdot$ 8 H$_2$O, Δ; ii: PhCOCl, NaOH; iii: N,N'-dicyclohexylcarbodiimide (DCC), CH$_2$Cl$_2$; iv: NaH, R-X, DMF, 0° --> r.t.; v: **26c**, N-methylpyrrolidone (NMP), 50-80°; vi: CF$_3$SO$_3$H, MeOH, 80°.

Scheme 1.

aqueous HCl in dioxane at 100 °C. We were able to show that the separation of the diastereomeric peptides (R,S)-**27** and (S,S)-**28** (Scheme 1) depended largely on the nature of R′ (Ph >> CH$_3$) and even more importantly on the amines **26a–c**. L-Phe-cyclohexylamide **26c** proved to be a particularly powerful chiral auxiliary for resolving a whole range of α,α-disubstituted glycines.

Many diasteromeric peptides of type (R,S)-**27** and (S,S)-**28** (Scheme 1) could be crystallized and their absolute configuration determined based on the known (S)-configuration of L-phenylalanine. Table 1 shows a compilation of the relevant torsional angles (Φ$_1$/Ψ$_1$) derived from crystal structures of the peptides shown.[108] It is interesting to note that the (S,S)-diastereomers show a high prevalence for β-turn type I conformations in the crystal state.

An interesting comparison can be made looking at the α- and β-tetralin derivatives entries—15/16 and —17 in Table 1 which can be regarded as cyclic conformationally constrained analogues of phenylglycine and phenylalanine. In an interesting study, 6-hydroxy-2-aminotetralin-2-carboxylic acid **12** (Hat) has been incorporated as a conformationally constrained tyrosine analogue into δ-opioid receptor selective tetrapeptides.[114,115] Whereas entry 15, the (S)-α-tetralin deriva-

Table 1. Conformations Found in Crystal Structures of Dipeptide Mimetics

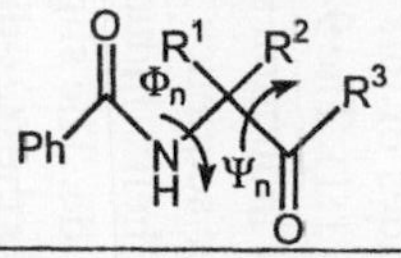

Diastereomer	*	R^1	R^2	R^3	Φ_1	Ψ_1	Φ_2	Ψ_2	d	Turn Type[1]
1	R/S	CH$_2$PhOMe	i-Pr	26c	-57	-27	-75	-10	2.92	Type I (ααα)
2	S/S	CH$_2$PhOMe	i-Pr	26c	-54	-27	-85	-8	2.96	Type I (ααα)
2[2]	S/S	CH$_2$PhOMe	i-Pr	26c	-57	-31	-81	-14	2.96	Type I (ααα)
3	R/S	CH$_2$PhOMe	Ph	26c	-63	-13	-82	-9	2.97	Type I (ααα)
3[2]	R/S	CH$_2$PhOMe	Ph	26c	-65	-15	-80	-6	2.97	Type I (ααα)
4	S/S	CH$_2$PhOMe	Ph	26c	+180	+175	-133	+118	—	Extended
5	S/S	CH$_2$NHCOPh	Me$_2$CHCH$_2$	26c	-60	-28	-74	-16	2.95	Type I (ααα)
6	R/S	CH$_2$NHCOPh	Me$_2$CHCH$_2$	26c	-43	-42	-84	+1	2.95	Type I (ααα)
7	R/S	CH$_2$NHCOPh	Me	26c	+64	+27	-77	+176	—	Extended
8	R/S	CH$_2$OCOPh	Me	26c	-48	-51	-111	+28	3.15	Type I (ααα)
9	R/S	Ph	Me	26c	+50	-130	-90	+4	2.87	Type II' («α)
10	R/S	i-Pr	Me	26c	+70	+40	-158	+117	—	Extended
11	S/S	CH$_2$Ph		26c	-60	-24	-90	-2	2.93	Type I (ααα)
12	R/S	CH$_2$OCOPh		26c	-55	-33	-81	-8	2.94	Type I (ααα)
13	S/S	CH$_2$CH$_2$COOtBu		26c	-44	-48	-81	+2	2.89	Type I (ααα)
14	R/S	CH$_2$CH$_2$COOtBu		26c	-53	-42	-103	+20	3.10	Type I (ααα)
15	S/S	(S)-14 (Fig. 14)		26b	+38	-130	-92	+2	2.85	Type II'
16	R/S	(R)-14 (Fig. 14)		26b	-41	+130	-153	+164	—	Extended
17	S/S	(S)-11 (Fig. 11)		26b	-57	-44	-109	+23	3.08	Type I (α-helical)

Notes: [1]see Table 5.
[2]Second of two molecules in the asymmetric unit.

25

tive, shows a β-turn type II′ (compare Section II.B) and its (*R*)-counterpart (entry 16) an extended conformation, the (*S*)-β-tetralin derivative (entry 17; Table 1) shows an almost perfect α-helical arrangement placing the *N*-benzoyl group in axial position and the carbamoyl group in a pseudo equatorial position. These observations prompted us to select the (*S*)-β-tetralin amino acid as our central template for α-helical stabilization of small peptides (see Section II).

Incorporation of α,α-Dialkylated Glycines into 9-Mer, 10-Mer, and 12-Mer Peptides

In order to evaluate the 3_{10}- and α-helix compatibility of novel optically pure unusual amino acid building blocks in short peptides, we designed the novel alanine-rich, hydrophobic peptides depicted in Figure 15.[116]

The number of residues in the peptides was chosen between 9 and 12, corresponding to a range of 2.5 to 3 α-helical turns, which is the peptide length where α- and 3_{10}-helices are known to coexist.[105] The helicities of the reference peptides (Xaa = Aib) measured by CD spectroscopy, were determined to be between 35 and 38% (assuming pure α-helicity). These values were derived by estimating the α-helicity at 222 nm. It has been shown,[109] however, that hydrophobic peptides of this type exist as a mixture of 3_{10}- and α-helical conformations in TFE/H_2O (1:1). The 9-mer reference peptides were intended to be used to calibrate an α-helix induction scale for the α,α-disubstituted glycines **11–22** (Figure 14). The 10-mer sequence with Xaa at position 1 was designed and synthesized to probe the helix induction potential of the unusual amino acids at the N-terminus of the peptide. The slightly longer 12-mer reference sequence was designed principally to assess the α,α-disubstituted glycines in position 5 of the peptide, where any unfavorable interactions should affect the conformational behavior of the model peptide.

To validate our approach, we synthesized numerous 9-mer, 10-mer, and 12-mer peptides[116] incorporating a series of optically pure cyclic- and open-chain α,α-

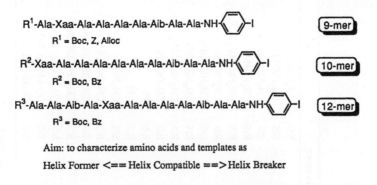

Figure 15. 9-,10-, 12-mer non-polar host peptides.

disubstituted glycines. For some of these peptides we were able to obtain crystals suitable for X-ray structural analyses. And for most of these peptides we measured CD-spectra in acidic, neutral, and basic media. While the residue ellipticities obtained at 222 nm ($\pi \rightarrow \pi^*$ amide transition) only reflect the sum of 3_{10} and α-helix compatibilities for Xaa in the different peptides, we successfully applied the convex constraint CD-analysis of Fasman et al. [117–119] for a selected set of 9-mer peptides, which allowed us to differentiate and determine the contents of 3_{10}- and α-helical conformations present. As one of the most striking results, we found that 9-mer peptide **31** (Table 2), incorporating Xaa = (*S*)-**11** (β-tetralin derivative) is predominantly β-helical (70%), whereas peptide Xaa = (*S*)-**14** (α-tetralin derivative) exhibits mainly a 3_{10}-helical conformation.

Interestingly, the high α-helix propensity of (*S*)-**11** has been observed in the X-ray structures of both tripeptides (see Table 1) and 9-mer peptides (Table 2). The molecular structures of the terminally blocked 9-mer peptides **31** and **32** were determined by X-ray diffraction and the observed Φ/Ψ angles are shown in Table 2. For comparison, the similar peptide containing Aib at position 2 (**30**) is also listed in Table 2. In all three peptides, the succession of similar pairs of Φ/Ψ values correspond to an α-helical conformation.

In peptide **32**, which has an (*R*)-α-tetralin derivative at position 2, we observed one 3_{10}-helical turn at the N-terminus with an exclusive H-bond between the Boc carbonyl group and the NH (i+3). The rest of the molecule adopts a α-helical backbone conformation (mean values: $\Phi = -62°$, $\Psi = -40°$). The amino group of the α-tetralin moiety is pseudo-equatorial. In contrast, the X-ray structure of peptide **31** shows almost two α-helical turns (mean values: $\Phi = -67°$, $\Psi = -37°$) starting from the Boc-carbonyl group of Ala(1) to Ala(5). Surprisingly, the carbonyl group of Ala(6) forms an exclusive H-bond to the NH(i+3), which leaves the C-terminal amide proton without a H-bonded interaction. This missing H-bond results in a fraying of the helix towards the C-terminus of the peptide which can be clearly seen from the Φ/Ψ angles (Table 2).

Further evidence for the strong 3_{10}-helix inducing properties of (*S*)-α-tetralin could be derived from the X-ray structure of a 10-mer peptide **33** containing (*S*)-**14** at the N-terminus. This peptide crystallized with two molecules in the asymmetric unit with almost identical conformations. The respective dihedral angles Φ and Ψ are listed in Table 3. The first four N-terminal residues form one 3_{10}-helical turn with Φ/Ψ-angles close to those of an ideal 3_{10}-helix (see Table 3). The residues from position 5 to 7 are characterized by Φ/Ψ dihedral angles of around $-60°$ and $-40°$, respectively. These values correspond to an α-helical conformation. For one of the 12-mer peptides it was possible to grow crystals from MeCN suitable for X-ray diffraction (Figure 16). Peptide **34** incorporates the (*S*)-6-hydroxy-β-tetralin **12** at the fifth position. The peptide makes one 3_{10}-helical turn at the first amino acid residue, as confirmed by its Φ and Ψ angles (see Table 4), and a hydrogen bond from the Boc-carbonyl (single acceptor) to the NH (i+3). From residue 2, the

Table 2. Backbone Dihedral Angles from X-ray Structures in Peptides of the 9-Mer Series[a]

Boc-Ala[1]-**Xaa**[2]-Ala[3]-Ala[4]-Ala[5]-Ala[6]-Aib[7]-Ala[8]-Ala[9]-NH-C_6H_4I

Peptide 30: Xaa = Aib

Peptide 31: Xaa = (S)-11

Peptide 32: Xaa = (R)-14

Dihedral Angle [°][b]	30	31 Molecule 1	31 Molecule 2	32
Φ_1/Ψ_1	−47.6/−41.8	−65.1/−45.6	−62.4/−49.7	−50.1/−43.7
Φ_2/Ψ_2	**−46.7/−45.5**	**−62.6/−46.4**	**−60.2/−47.5**	**−37.3/−49.7**
Φ_3/Ψ_3	−66.3/−33.4	−71.6/−27.6	−71.6/−27.1	−74.7/−34.7
Φ_4/Ψ_4	−61.9/−39.5	−65.9/−40.1	−65.9/−41.2	−64.6/−41.6
Φ_5/Ψ_5	−61.7/−35.5	−63.3/−44.9	−65.7/−41.6	−63.2/−41.8
Φ_6/Ψ_6	−64.8/−47.8	−66.4/−45.2	−69.8/−45.1	−63.4/−51.6
Φ_7/Ψ_7	**−50.7/−45.7**	**−56.2/−39.0**	**−57.1/−40.1**	**−55.1/−47.2**
Φ_8/Ψ_8	−65.9/−38.0	−74.6/−13.4	−75.9/−11.6	−68.3/−27.6
Φ_9/Ψ_9	−74.1/−25.4	−83.7/−28.9	−82.6/−29.4	−81.6/−20.4

Notes: [a] The dihedral angles of α,α-disubstituted amino acids are in bold.

[b] Mean dihedral angles for α-helix: $\Phi = -63$, $\Psi = -42$.

3_{10}-helix: $\Phi = -57$, $\Psi = -30$.

peptide is α-helical to the C-terminus (mean values: $\Phi = -61°$, $\Psi = -42°$). Table 4 gives the Φ/Ψ-angles for peptide **34**.

From this work we conclude that our novel nonpolar 9-, 10-, and 12-mer peptide sequences are interesting tools to assess the intrinsic propensities for 3_{10}- and α-helical conformations of natural and unnatural amino acids, such as cyclic and open-chain (R)- and (S)-α,α-disubstituted glycines **11–22** using CD-spectroscopy. As a highlight of this study, we found that the (R)- and (S)-β-tetralin derived amino acids **11** are significantly more α-helix-promoting than Ala and Aib. Such knowledge about the conformational behavior of these amino acid building blocks should be valuable for the design of novel peptide and protein mimetics.

B. β-Turn Mimetics

Turns in Peptides and Proteins

Turns are segments between secondary structural elements and are defined as sites in a polypeptide structure where the peptidic chain reverses its overall

Table 3. Backbone Dihedral Angles from X-ray Structures in Peptides of the
10-Mer Series[a]

Boc-Xaa^1-Ala^2-Ala^3-Ala^4-Ala^5Ala^6-Ala^7-Aib^8-Ala^9-Ala^{10}-NH-C_6H_4I

Peptide 33: Xaa = (S)-14

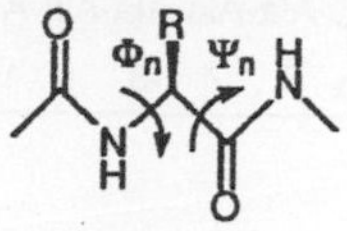

Dihedral Angle [°]	33 Molecule 1	33 Molecule 2
Φ_1/Ψ_1	**−60.4/−26.3**	**−61.5/−27.9**
Φ_2/Ψ_2	−63.6/−21.9	−61.1/−22.8
Φ_3/Ψ_3	−58.4/−30.6	−59.3/−25.3
Φ_4/Ψ_4	−64.5/−34.1	−61.3/−29.8
Φ_5/Ψ_5	−70.6/−42.5	−74.6/−35.4
Φ_6/Ψ_6	−59.6/−48.6	−62.9/−42.8
Φ_7/Ψ_7	−57.7/−49.8	−63.4/−43.9
Φ_8/Ψ_8	**−52.8/−44.7**	**−52.2/−43.9**
Φ_9/Ψ_9	−64.8/−27.4[a]	−72.5/−28.4[a]
Φ_{10}/Ψ_{10}	−84.9/−9.9[a]	−88.2/−18.9a

Note: [a] The dihedral angles of α,α-disubstituted amino acids are in bold.

direction. They are composed of four (β-turn) or three (γ-turn) amino acid residues. Compared to helices or sheets, turns are the only regular secondary structures which consist of nonrepeating backbone torsional angles. The turn structural element has been implicated as an important site for molecular recognition in biologically active peptides and globular proteins.[120–122] High-resolution crystal structures of several antibody–peptide complexes clearly showed turns as recognition motifs.[123,124] Furthermore, good correlations with turn conformations have been found in structure–activity studies of specific recognition sites in peptide hormones such as *bradykinin* and *somatostatin*.[125,126] β-Turns have also been hypothesized to be involved in posttranslational lysine hydroxylation,[127] processing of peptide hormones,[128,129] recognition of phosphotyrosine-containing peptides,[130,131] signal peptidase action,[132] receptor internalization signals,[133,134] and in glycosylation processes.[135] These findings together with the fact that turns occur mainly at the surface of proteins make them attractive targets as peptide mimetics.

Terminology of β-Turns

According to Venkatachalam β-turns are classified into conformational types depending on the values of four backbone torsional angles ($\Phi_1,\Psi_1,\Phi_2,\Psi_2$) (Table

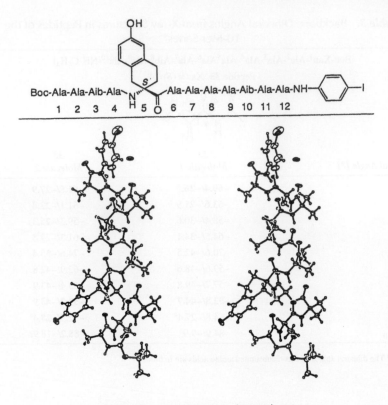

Boc-Ala-Ala-Aib-Ala—N—Ala-Ala-Ala-Ala-Aib-Ala-Ala-NH—◯—I
  1   2   3   4 H 5 O 6 7 8 9 10 11 12

Figure 16. ORTEP plot of peptide **34**.

5).[136] A second criterium is the $\alpha C_i \rightarrow \alpha C_{i+3}$ distance, introduced by Kabsch and Sander, which must be shorter than 7 Å[137] (Figure 17). Very typical for β-turn motifs is a H-bond between the carbonyl group at position i and the amide NH of residue $i+3$. To date, eight different types of β-turns have been classified and the torsional angles ($\Phi_1,\Psi_1,\Phi_2,\Psi_2$) of the most important naturally occurring β-turns are depicted in Table 5. Figure 18 shows the corresponding regions on the Ramachandran map.

The positional preferences of amino acids occurring in β-turns in protein crystal structures has been studied by Wilmot and Thornton,[138] and more recently by Hutchinson and Thornton.[139] In β-turns of type I, 48% have Asp, Asn, Ser, or Cys at position i, and 57% of Asp, Asn, and Ser side chains make a hydrogen bond with the NH at position $i+2$.

At the second position ($i+1$), Pro is the most common amino acid in type I and type II turns because of the restriction of Φ to about −60°. Ser also exhibits a reasonable preference for this position since its side chain oxygen can form

Table 4. Backbone Dihedral Angles from X-ray Structure of 12-Mer Peptide **34**[a]

Boc-Ala[1]-Ala[2]-Aib[3]-Ala[4]-**Xaa**[5]-Ala[6]-Ala[7]-Ala[8]-Ala[9]-Aib[10]-Ala[11]-Ala[12]-NH-C$_6$H$_4$I

Peptide **34**: Xaa = (S)-**12**

Dihedral Angle [°]	**34** Deg.
Φ_1/Ψ_1	−58/−29
Φ_2/Ψ_2	−57/−40
Φ_3/Ψ_3	**−59/−48**
Φ_4/Ψ_4	−66/−38
Φ_5/Ψ_5	**−57/−47**
Φ_6/Ψ_6	−65/−41
Φ_7/Ψ_7	−57/−45
Φ_8/Ψ_8	−64/−40
Φ_9/Ψ_9	−59/−46
Φ_{10}/Ψ_{10}	**−56/−50**
Φ_{11}/Ψ_{11}	−70/−38
Φ_{12}/Ψ_{12}	−67/−40

Note: [a] The dihedral angles of α,α-disubstituted amino acids are in bold.

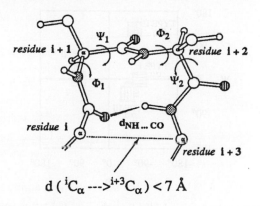

$$d\,(\,^iC_\alpha \dashrightarrow {}^{i+3}C_\alpha\,) < 7\ \text{Å}$$

Figure 17. Terminology of β-turns.

Table 5. Torsional Angle Values of the Most Frequent Types of β-Turns

Type[a]	Φ_1	Ψ_1	Φ_2	Ψ_2	Nomenclature[b]
I	−60	−30	−90	0	αα
I′	60	30	90	0	γγ
II	−60	120	80	0	βργ
II′	60	−120	−80	0	εα
III	−60	−30	−60	−30	αα
III′	60	30	60	30	γγ

Notes: [a] Turn nomenclature is according to Venkatalacham (136).

    [b] Turn nomenclature is according to Wilmot Thornton (138).

favorable electrostatic interactions with the NH at i+1. Somatostatin, LHRH (luteinizing hormone release hormone), bradykinin, enkephalin, gramicidin S, neurokinin, and angiotensin II are examples of biologically active peptides where β-turns have been linked with receptor recognition.[140–142]

Mimicking Strategies for β-Turns

Mimicking a β-turn principally means constraining correctly four torsional angles ($\Phi_1,\Phi_2,\Psi_1,\Psi_2$) and four bonds (bonds **a-d**, see Figure 19). Bonds **a** and **d** direct the entry and the exit of the peptide chain through the turn, respectively, whereas bonds **b** and **c** are responsible for the spatial disposition of the amino acid side chains at position i+1 and i+2 of a turn. The torsional angles determine the backbone geometry of the turn and consequently the shape of the turn hydrogen bond. Bonds **a** and **d**, describe the vectors of the C- and N-terminal peptide chain

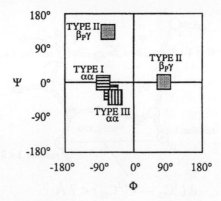

Figure 18. Ramachandran map indicating the predominant torsional angles for β-turns (see Table 5).

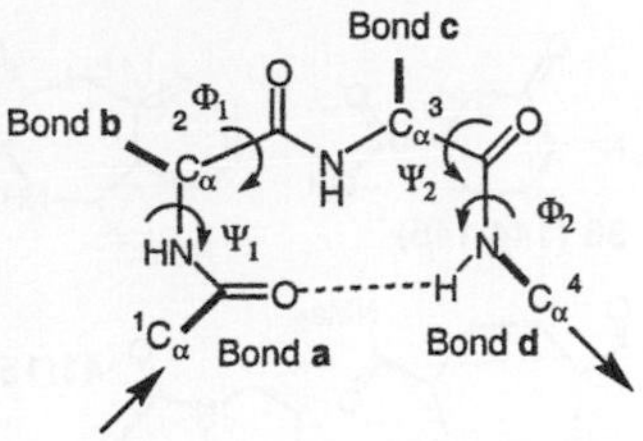

Figure 19. Vector angles determining the β-turn geometry.

entering and leaving the turn. The control of these vectors is essential for cases where additional amino acid residues are placed outside of the central amide unit.

Peptidic β-turn mimetics are generally based on cyclic backbone mimetics (e.g. replacing the turn hydrogen bond by a covalent bond) or by introducing one or several unusual amino acids, which constrain the backbone in β-turn conformations. Synthetic approaches to non-peptidic turn mimetics can be grouped into two classes: (1) external β-turn mimetics, and (2) internal β-turn mimetics. An external mimetic does not mimic per se the turn backbone but rather stabilizes β-turn geometries of an attached peptide. In an internal β-turn mimetic, the scaffold replaces the space in the turn which would normally be occupied by the peptide backbone. This methodology tries to correctly place key binding groups in 3D-space, while being less stringent with respect to the correct geometry of the turn backbone. Less effect on the backbone geometry can be expected, because only a small fragment of the peptide is replaced by the constraint. Furthermore, due to the small molecular size of the template, little steric hindrance with a receptor may be anticipated.

By far the greatest interest has been associated with lactam-derived mimetics, where early molecular modelling revealed an almost ideal overlap with β-turns of type II′ (Figure 20).[111,121,143–174] Lactams are easily prepared and the synthetic routes permit use of a variety of amino acids with defined configuration. Thus, many γ-, δ-, and ε-lactam derivatives have been synthesized, including mono- and bicyclic, fused and nonfused heterocycles, and many have been incorporated into receptor antagonists or agonists.[121]

Freidinger incorporated **35** (Figure 20) into LHRH to replace the Gly-Leu motif, which is believed to be part of a β-turn conformation.[143] The (S,S)-3-amino-2-piperidone-6-carboxylic acid **36** with the inherent *cis*-amide bond was used by Kemp and Sun as a constraint for a type VI β-turn.[144,145] However, after incorporation into peptide chains, NMR analyses revealed several possible conformations of the six-membered ring in **36**, thereby making a correlation of the preferred conformation(s) to the target turn conformation ambiguous. Lactam **37** might be envisaged as a Phe-Gly analogue having a covalent linkage between the *ortho*-position of the phenylalanine ring and the C_α of the Gly moiety.[146]

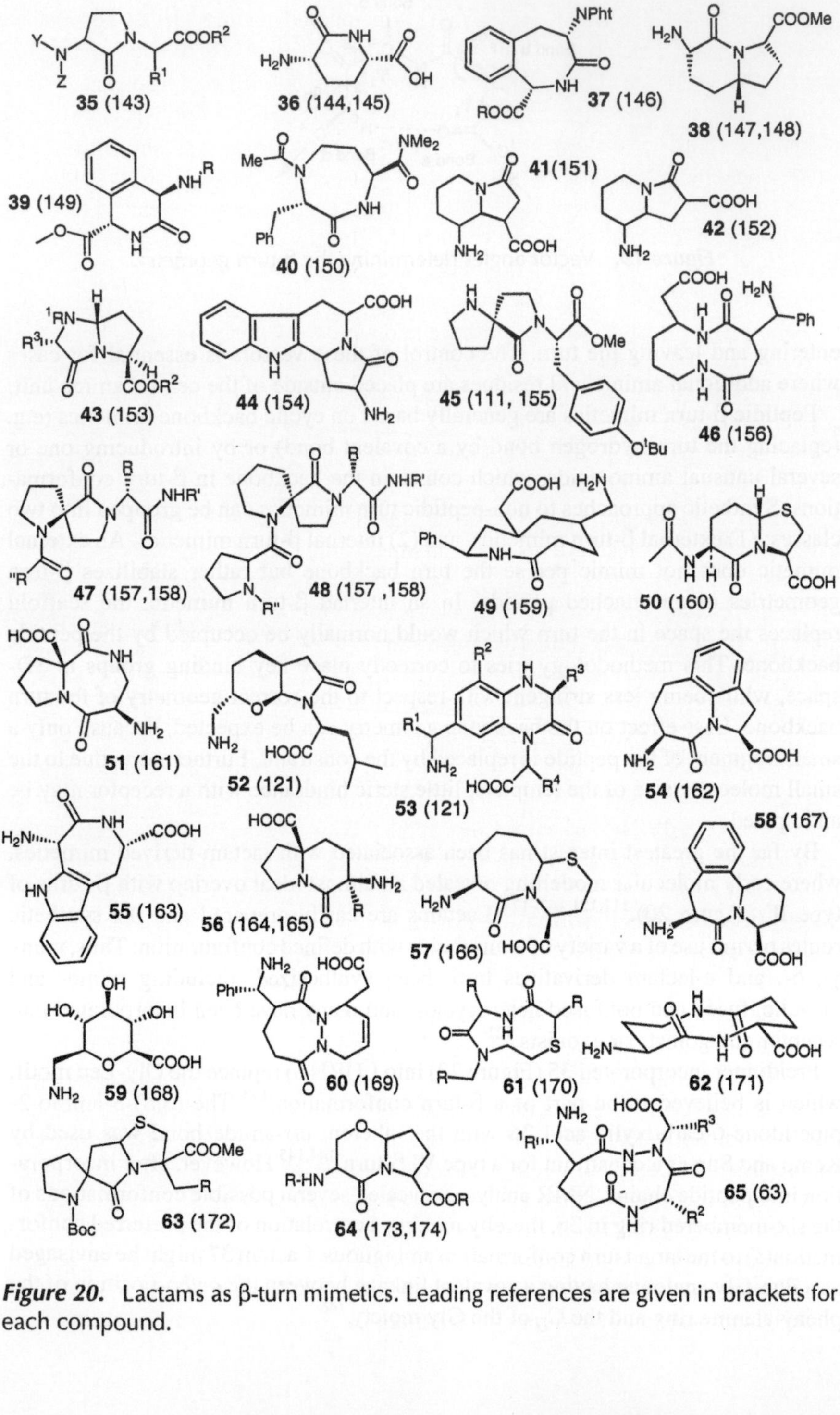

35 (143)   **36** (144,145)   **37** (146)   **38** (147,148)

39 (149)   **40** (150)   **41** (151)   **42** (152)

43 (153)   **44** (154)   **45** (111, 155)   **46** (156)

47 (157,158)   **48** (157, 158)   **49** (159)   **50** (160)

51 (161)   **52** (121)   **53** (121)   **54** (162)   **58** (167)

55 (163)   **56** (164,165)   **57** (166)

59 (168)   **60** (169)   **61** (170)   **62** (171)

63 (172)   **64** (173,174)   **65** (63)

Figure 20. Lactams as β-turn mimetics. Leading references are given in brackets for each compound.

Lactam **38**[147,148] and the more flexible 10-membered macrocycles **46** and **49** have been used as constraints for β-turns of type II and I.[156,159] The bicyclic lactam **39** can be regarded as an unnatural phenylglycine analogue with the carboxy and amino termini in *trans* orientation.[149] Baldwin et al. described a bicyclic γ-lactam of type **43** as a restrained dipeptide mimetic.[153] Furthermore, it could be shown by NMR studies that the indolizinoindole derivative **44** (a *Trp*-analogue) adopts conformations similar to a β-turn of type II'.[154] A different approach was undertaken by Genin & Johnson in designing spiro[4.4]- and [5.4]lactam systems **47** and **48** as turn mimetics.[157,158] Depending on the chirality at position *i*+1 they could mimic β-turns of type II or II', respectively. The same spirocyclic unit, as in **45**, had been developed earlier as a conformationally locked analogue of a natural peptide replacing a Pro-Tyr unit within a β-turn conformation.[111,155] β-turn-dipeptide (BTD) **50** was used by Nagai in the synthesis of analogues of biologically active peptides such as LHRH, somatostatin or gramidicin S.[160] The bicyclic lactam **56** has been synthesized as a mimic for a type-VI β-turn, with the characteristic *cis* Xaa-Pro peptide bond.[164,165]

Among the bi- and tricyclic mimetics, the phenoxathiine derivative **66** (Figure 21),[11,54,175-183] designed by Feigel, is an interesting template where C(2) and C(7) mimic the C_α-positions *i*+1 and *i*+2 of a β-turn.[176] It was incorporated into a series of cyclic peptides, and NMR investigations revealed the correct hydrogen-bonding pattern expected for a β-turn.[121] Using a similar approach Mueller et al. developed xanthene-, phenoxazin-, and phenothiazin-derived tricyclic systems of type **67** as external β-turn and loop-stabilizing templates[11] (see Section II.C). Analyses of low-energy conformations of macro-heterocyclic compounds led to molecules such as benzodiazepines **69** and **70** (Figure 21). Benzodiazepines showed a good correlation with a selected set of β-turn conformations.[177] Finally, succinimide analogues of type **71**, derived from α-alkylated aspartic acids, were described as efficient type II and II' β-turn mimetics by Obrecht et al.[178,184] (see Section II.B).

Synthesis of Conformationally Constrained α-Substituted Serine and Aspartic Acid Derivatives

α-Methyl(alkyl) aspartic acid derivatives and their corresponding succinimide analogues show a high preference for β-turn type I, II, and II'-conformations[184] (Figure 22). As shown by X-ray and NMR solution structures, a shift from β-turn type I to type II or II' occurs upon cyclization of the open-chain Asp to the cyclic succinimide derivatives, depending on the absolute configuration of the noncoded amino acid.[141] This β-turn conformation was also observed when the Asp-imide moiety was incorporated into a 10-mer model peptide (Figure 23).

Since serine shows a high occurrence at the exposed positions (*i*+1) and (*i*+2) of β-turns in protein crystal structures,[138,185] it was decided to investigate whether or not the (*R*)- and (*S*)-"α-chimeras" of Asp-Ser would be interesting β-turn mimetics at the (*i*+1)-position of β-turns. Furthermore, it was of interest to see whether or

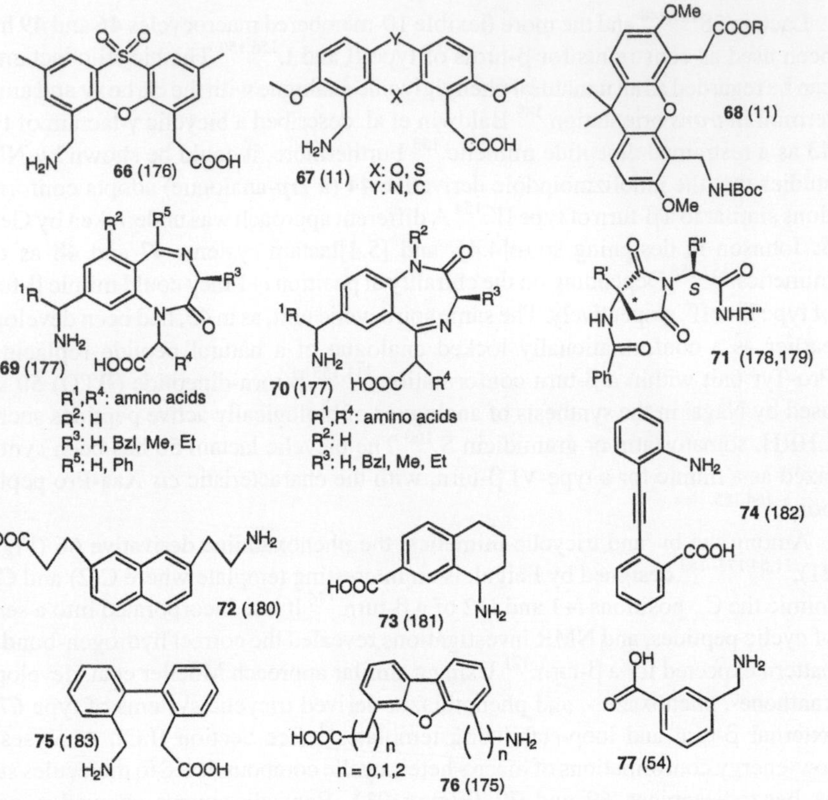

Figure 21. Other β-turn mimetics. Leading references are given in brackets for each compound.

not the corresponding succinimide derivatives of Asp-Ser would also show this β-turn switch from type I to type II or type II′, depending on the absolute configuration.

Using a novel azlactone/oxazoline interconversion reaction (**I ⇔ II**, Figure 24), the synthesis of optically pure (*R*)- and (*S*)-α-methyl(alkyl)-serine derivatives is particularly straightforward and requires a minimum number of protective groups. This synthetic route provides a rapid and easy access to α-chimeras combining the side chains of serine and functionalized amino acids like Asp, Glu, Ser,[179,186] Lys,[109] and tyrosine (Tyr).[187] Using Asp side chain cyclization to the corresponding succinimide analogues in combination with the novel azlactone/oxazoline interconversion reaction allowed the synthesis of the conformationally constrained spiro-cyclic Asp-Ser analogues (*R/S*)-**86**, (*S/S*)-**86**, and corresponding (*R/S*)-, (*S/S*)-Asp-Glu analogues (Scheme 2).

(R/S)-78

β-turn type I [7]

(R/S)-79

β-turn type II [7]

(S/S)-80

β-turn type I [7]

(S/S)-81

β-turn type II' [7]

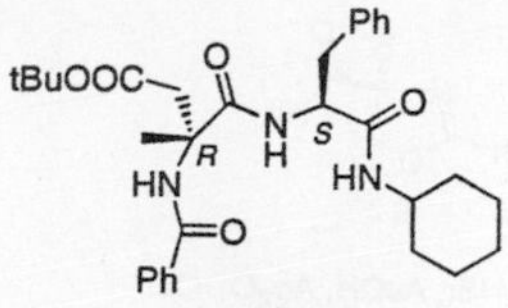

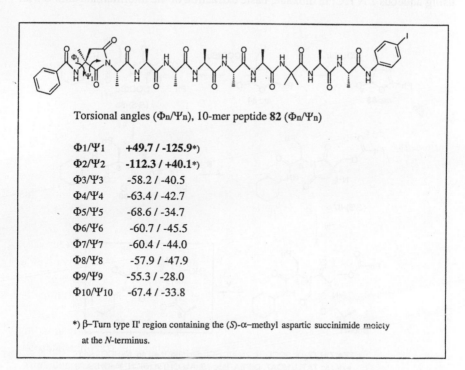

Figure 22. α-Methyl(alkyl) aspartic acid analogues and corresponding succinimides as turn mimetics.

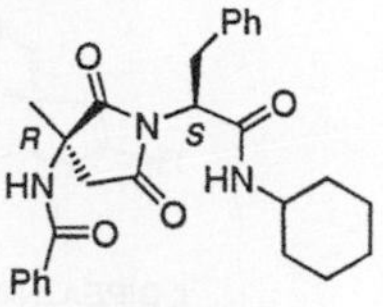

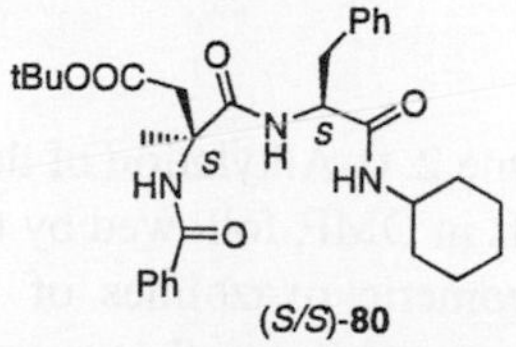

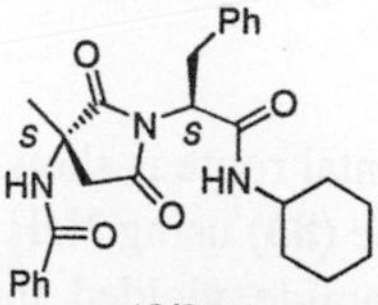

Torsional angles (Φn/Ψn), 10-mer peptide **82** (Φn/Ψn)

Φ1/Ψ1	**+49.7 / -125.9*)**
Φ2/Ψ2	**-112.3 / +40.1*)**
Φ3/Ψ3	-58.2 / -40.5
Φ4/Ψ4	-63.4 / -42.7
Φ5/Ψ5	-68.6 / -34.7
Φ6/Ψ6	-60.7 / -45.5
Φ7/Ψ7	-60.4 / -44.0
Φ8/Ψ8	-57.9 / -47.9
Φ9/Ψ9	-55.3 / -28.0
Φ10/Ψ10	-67.4 / -33.8

***) β–Turn type II' region containing the (*S*)-α–methyl aspartic succinimide moiety at the *N*-terminus.**

Figure 23. 10-mer model peptide **82** its backbone torsional angles derived from the X-ray structure.

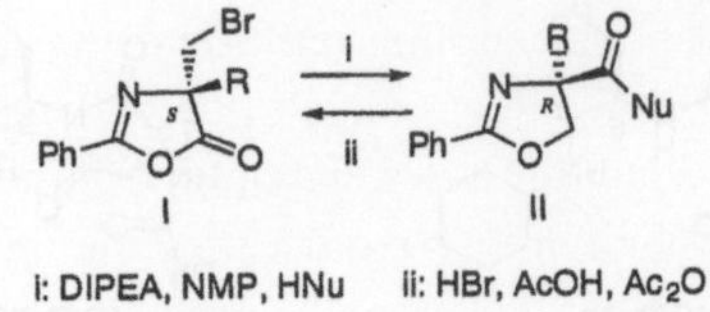
Br
R
N
s
Ph
O
O
I
i
ii
B
O
N
R
Nu
Ph
O
II
i: DIPEA, NMP, HNu
ii: HBr, AcOH, Ac₂O

Figure 24. Azlactone/oxazoline-interconversion.

The experimental route is shown in Scheme 2. α-Alkylation of the 4-monosubstituted azlactone (**83**) using NaH and CH_2I_2 in DMF, followed by treatment with Phe-cyclohexylamide yielded the diastereomeric oxazolines of type (*S/S*)-**85**, which could be easily separated. (*S/S*)-**85** was subsequently treated with TFA in CH_2Cl_2 at 0°C and the intermediate acid cyclized in high yields to the spiro derivative (*S/S*)-**86** using $SOCl_2$. Hydrolysis of the spiro-compound (*S/S*)-**86** with aqueous 2 N HCl in dioxane gave the free amine (*S/S*)-**87**. (*S/S*)-**87** was finally incorporated into the tripeptide derivative (*S/S/S*)-**89** via (*S/S/S*)-**88** by hydrolysis using aqueous 2 N HCl in dioxane, basic extraction of the intermediate amine with

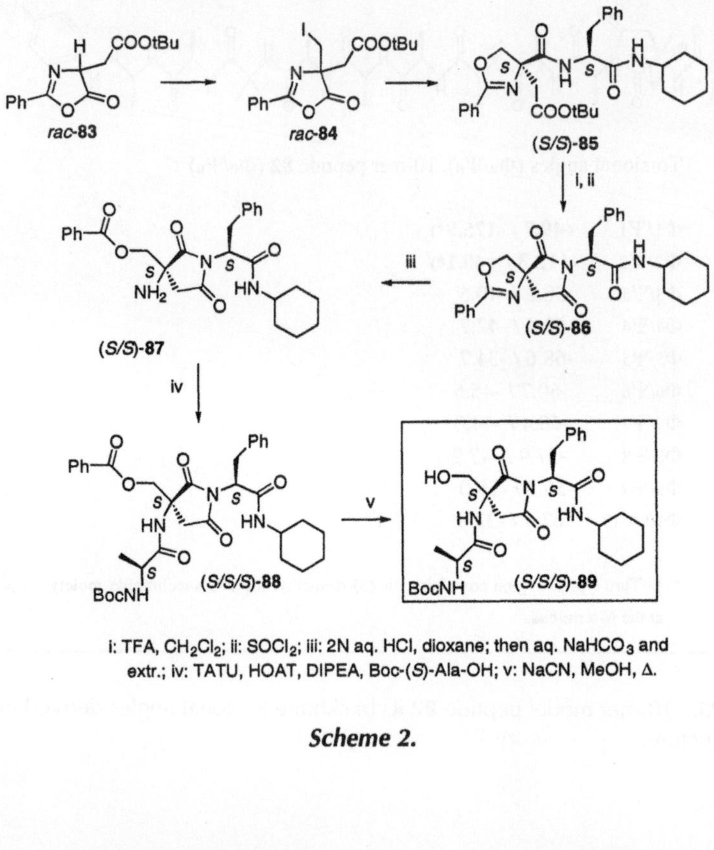
rac-83
rac-84
(S/S)-85
i, ii
iii
(S/S)-86
(S/S)-87
iv
(S/S/S)-88
v
(S/S/S)-89
COOtBu
Ph
BocNH
HN
HO
i: TFA, CH₂Cl₂; ii: SOCl₂; iii: 2N aq. HCl, dioxane; then aq. NaHCO₃ and extr.; iv: TATU, HOAT, DIPEA, Boc-(S)-Ala-OH; v: NaCN, MeOH, Δ.

Scheme 2.

$CHCl_3$, and subsequent peptide coupling reaction using TATU, HOAT,[188] DIPEA and Boc-(S)-Phe-OH in DMF, and cleavage of the O-benzoyl group using NaCN in MeOH. This reaction sequence shows that the succinimide ring is fully compatible with the mild reaction conditions required for the incorporation of this Asp-Ser-"α-chimera" into a peptide sequence. CD measurements confirmed the β-turn type II conformation in solution.

C. Loop Mimetics

Beside the classical secondary structure elements occurring at the surface of proteins such as γ- and β-turns, 3_{10}- and α-helices, and β-sheets, loops of variable size seem to play a crucial role in many protein–protein interactions (see Section I.A). Several amino acid-derived templates have been proposed to stabilize β-turns as external templates (see Section II.B) and a compilation of loop-stabilizing templates is shown in Figure 25.[189–194]

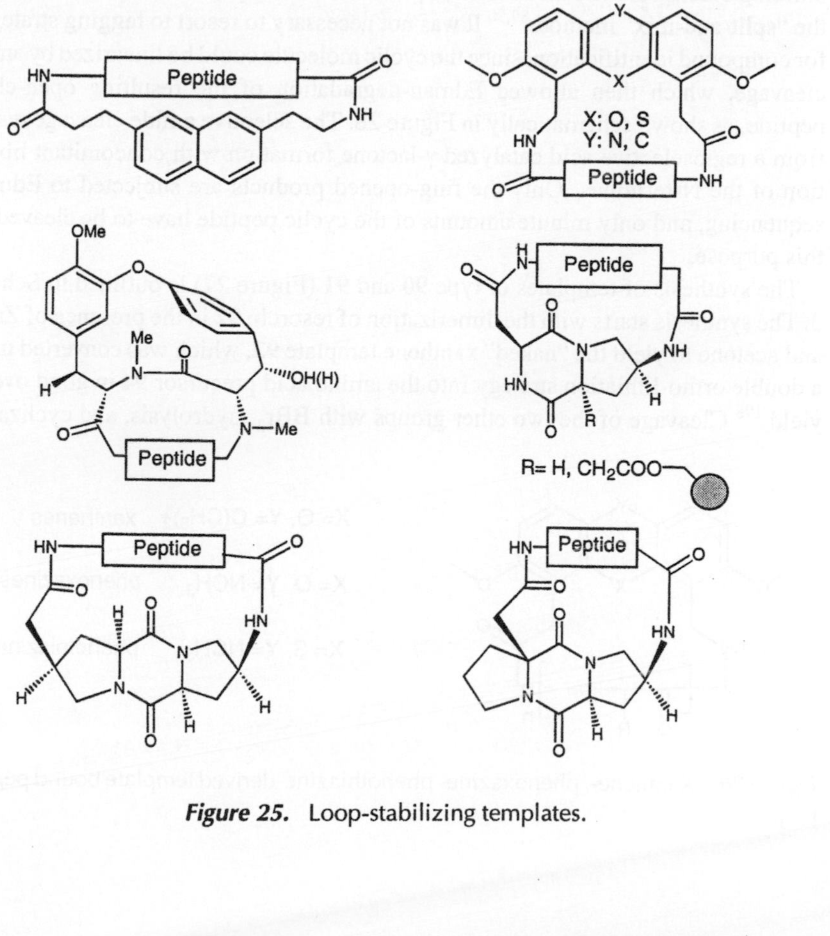

Figure 25. Loop-stabilizing templates.

Figure 25 shows a series of tricyclic xanthene-, phenoxazine-, and phenothiazine-derived templates that have been shown to stabilize β-turn conformations in attached peptides for $n = 2$ and 4.[11] These templates were also useful for the synthesis of larger loop mimetics (n up to 14). Although the larger loops are more flexible and the conformational influence enforced by the template is smaller, the templates usually increased significantly the cyclization yields. As a validation of this concept for loop stabilization, cyclic peptides incorporating xanthene-derived templates corresponding to the sequence of the exposed loop structure in the snake toxin flavoridin[195,196] were synthesized on a solid support, and tested for binding to the gpIIb/IIIa receptor, and for inhibiting platelet aggregation.[11]. The xanthene-derived loop incorporating the sequence Ile-Ala-Arg-Gly-Glu-Phe-Pro showed an IC_{50}-value of 15 nM and inhibited significantly platelet aggregation. In contrast, the open-chain and disulfide-bridged peptides were markedly less potent.

In order to use this template approach to synthesize combinatorial libraries of mimetics of exposed protein loops, we synthesized the unsymmetrically substituted xanthene-derived templates 90 and 91 (Figure 27).[197] These proved to be useful building blocks for the construction of β-turn stabilized peptide loop libraries using the "split-and-mix" method.[89,90] It was not necessary to resort to tagging strategies for compound identification, since the cyclic molecule could be linearized by amide cleavage, which then allowed Edman-degradation of the resulting open-chain peptide, as shown schematically in Figure 28. The selective amide cleavage results from a regioselective acid catalyzed γ-lactone formation with concomitant liberation of the N-terminus. Only the ring-opened products are subjected to Edman-sequencing, and only minute amounts of the cyclic peptide have to be cleaved for this purpose.

The synthesis of templates of type 90 and 91 (Figure 27) is outlined in Scheme 3. The synthesis starts with the dimerization of resorcin 92 in the presence of $ZnCl_2$ and acetone to yield the "naked" xanthene template 93, which was converted using a double ortho-lithiation strategy into the amino acid precursor 94 in good overall yield.[198] Cleavage of the two ether groups with BBr_3, hydrolysis, and cyclization

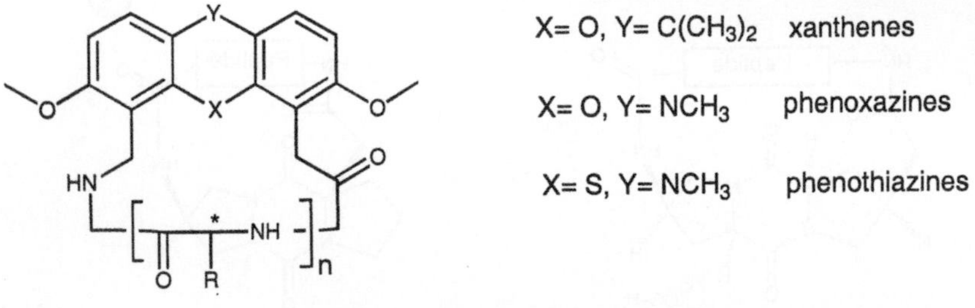

X= O, Y= C(CH₃)₂   xanthenes

X= O, Y= NCH₃   phenoxazines

X= S, Y= NCH₃   phenothiazines

Figure 26. Xanthene-, phenoxazine- phenothiazine-derived template bound peptide loops.

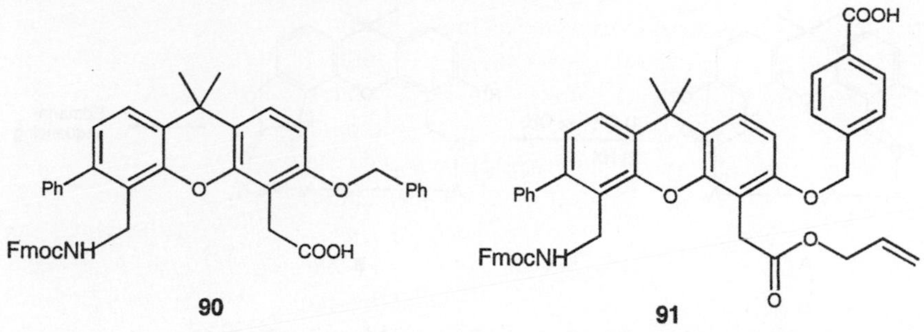

Figure 27. Unsymmetrically substituted xanthene derived templates.

with $SOCl_2$ afforded the key intermediate **96**. The synthesis takes advantage of anchimeric participation with intermediate formation of a γ-lactone to differentiate between the pseudo-symmetrical phenolic groups present e.g. in **95**. Intermediate **95** was converted efficiently into the starting templates **90** and **91**.[197]

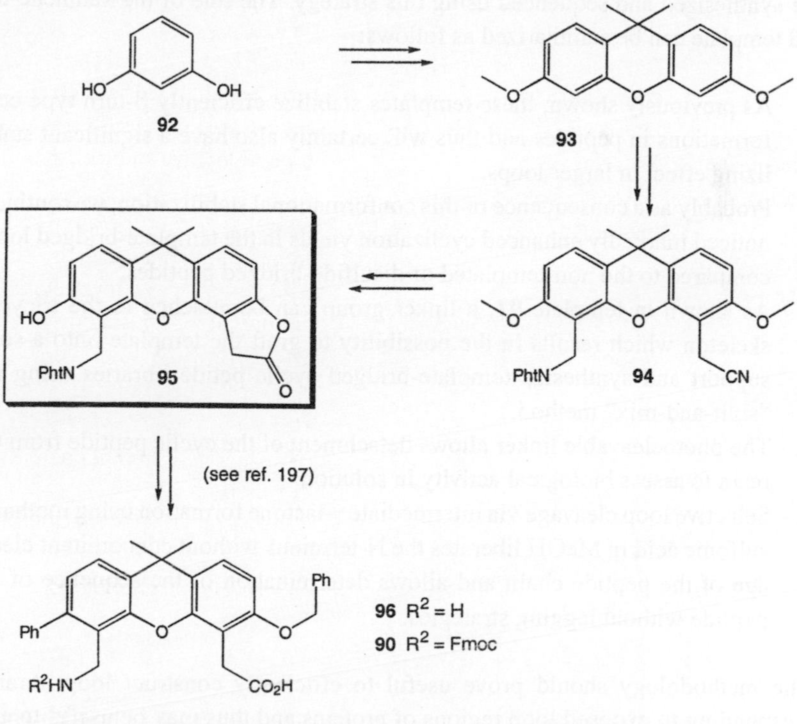

Scheme 3.

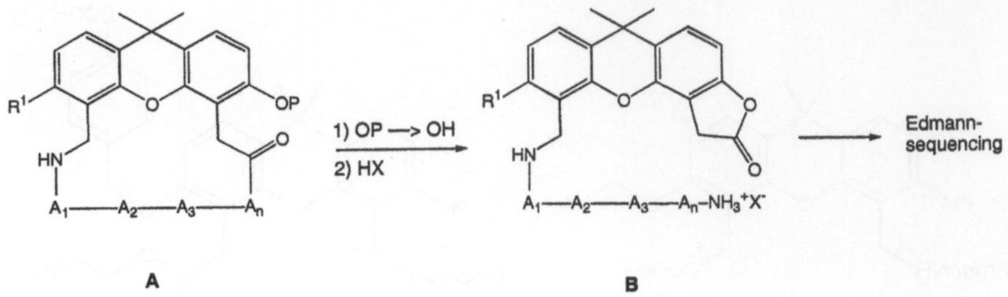

Figure 28. Strategy for loop identification.

In order to validate this concept, the template bridged cyclic peptides **97** and **100** containing the sequences Ala-Lys-Glu-Phe and Asp-Pro-Phe-Asp-Gly-Arg-Ala-Ile were synthesized by standard protocols,[198] ring-opened using methanesulfonic acid in MeOH at 50–60 °C, and sequenced by Edman-degradation (Figure 29). Using the solid-phase-compatible template **91**, which was attached via the photocleavable PPOA-linker [199,200] to a polystyrene resin, template-bridged cyclic peptide libraries were synthesized and sequenced using this strategy. The role of the xanthene-derived template can be summarized as follows:

1. As previously shown, these templates stabilize efficiently β-turn type conformations in peptides and thus will certainly also have a significant stabilizing effect in larger loops.
2. Probably as a consequence of this conformational stabilization, we routinely noticed markedly enhanced cyclization yields in the template-bridged loops compared to the non-templated or disulfide-bridged peptides.
3. As shown in template **91**, a linker group can be attached to the tricyclic skeleton which results in the possibility to graft the template onto a solid support and synthesize template-bridged cyclic petide libraries using the "split-and-mix" method.
4. The photocleavable linker allows detachment of the cyclic peptide from the resin to assess biological activity in solution.
5. Selective loop cleavage via intermediate γ-lactone formation using methanesulfonic acid in MeOH liberates the N-terminus without concomitant cleavage of the peptide chain and allows determination of the sequence of the peptide without tagging strategies.

The methodology should prove useful to efficiently construct loop-libraries corresponding to exposed loop regions of proteins and thus may be useful to gain new insights into protein-protein interactions.

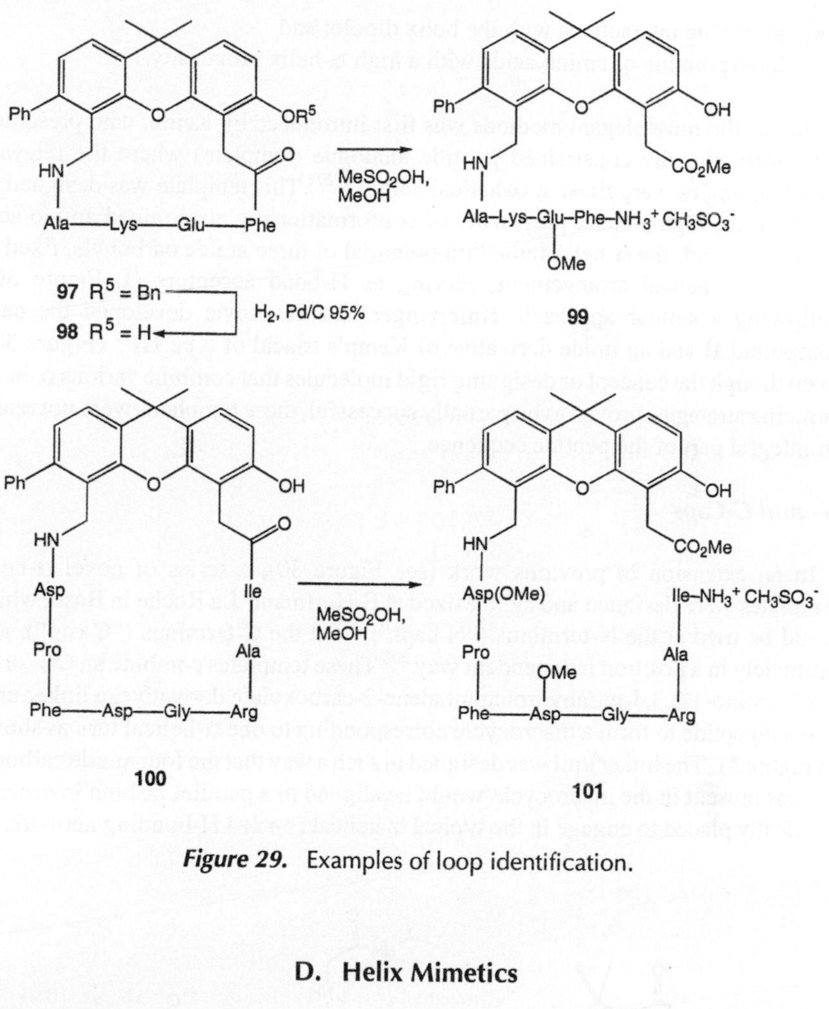

Figure 29. Examples of loop identification.

D. Helix Mimetics

The α-helix is the most abundant secondary structural element, determining the functional properties of proteins as diverse as α-keratin, hemoglobin and the transcription factor GCN4. The average length of an α-helix in proteins is approximately 17 Å, corresponding to 11 amino acid residues or three α-helical turns. In short peptides, the conformational transition from random coil to α-helix is usually entropically disfavored. Nevertheless, several methods are known to induce and stabilize α-helical conformations in short peptides, including:

- side-chain to side-chain cross-linking;
- directed formation of a hydrophobic core (helix bundles);
- induction of an α-helical H-bond network;

- favorable interactions with the helix dipole; and
- incorporation of amino acids with a high α-helix propensity.

One of the most elegant methods was first introduced by Kemp, who presented a conformationally constrained peptide analogue (template) where the relevant torsional angles were fixed at α-helical values.[9,201] This template was designed to combine the high α-helix propensity of conformationally constrained amino acid derivatives with the α-helix induction potential of three amide carbonyls, fixed in the correct helical arrangement, serving as H-bond acceptors (**I**, Figure 30). Following a similar approach, Knierzinger and Stankovic developed the cage compound **II** and an imide derivative of Kemp's triacid of type **III**[11] (Figure 30). Even though the concept of designing rigid molecules that combine various α-helix inducing strategies proved to be partially successful, these templates were not really an integral part of the peptide sequence.

N- and C-Caps

In an extension of previous work (see Figure 30), a series of novel α-helix templates were designed and synthesized at F. Hoffmann-La Roche in Basel which could be used at the N-terminus ("N-caps"),[202] at the C-terminus ("C-cap"), and ultimately in a position independent way.[203] These templates combine an (*R*)- or an (*S*)-2-amino-1,2,3,4-tetrahydronaphthalene-2-carboxylate derivative, a linker unit, and a dipeptide to form a macrocycle corresponding to one α-helical turn as shown in Figure 31. The linker unit was designed in such a way that the four amide carbonyl groups present in the macrocycle would be aligned in a parallel fashion in order to be ideally placed to engage in the typical α-helical *i* → *i*+4 H-bonding network. At

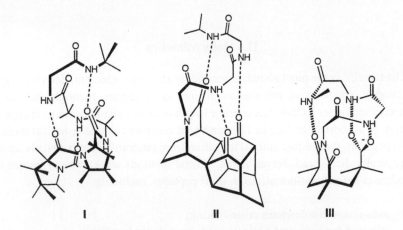

Figure 30. α-helix inducing templates **I–III**.

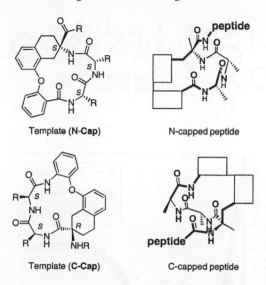

Template (N-Cap)   N-capped peptide

Template (C-Cap)   C-capped peptide

Figure 31. N-cap and C-cap templates and the corresponding conformationally constrained N- and C-capped peptides.

the peptide attachment site was placed a conformationally constrained amino acid, which had previously been shown (see Sections II.A and B) to inherently induce and stabilize α-helical conformations in peptides. It was anticipated that these templates would not only efficiently induce and stabilize α-helical conformations in peptides, but at the same time they would be mimetics of one α-helical turn and thus be an integral part of the α-helix.

The conformational properties of peptides linked to these capping templates were investigated by CD spectroscopy, NMR-measurements, and X-ray diffraction analysis.[204] Figure 32 shows the CD spectra of the two template-linked model peptides **102, 103**, and an uncapped 12-mer reference peptide **104** with comparable amino acid sequence (Boc-Ala$_2$-Aib-Ala$_6$-Aib-Ala$_2$-pIa). The CD spectra exhibit the negative maxima at about 207 and 222 nm, characteristic for α-helical peptides. Based on the ellipticity values at 222 nm (amide $n \rightarrow \pi^*$ transition) the N-capped peptide was calculated to be approximately 90% α-helical, whereas the helicity of the reference peptide is approximately 50% in TFE/H$_2$O (1:1).

These findings agree well with the results obtained from 2D-NMR measurements. NOE measurements revealed that not only the template itself, but also the second helical turn adopted a well defined α-helical conformation. In the third turn, the conformation was more flexible but α-helical conformations still predominate. Furthermore, the X-ray structure of the same N-capped model peptide revealed that the peptide is fully α-helical in the solid state (see Figure 33). Figure 34 schematically shows a retrosynthetic analysis of the N-caps.

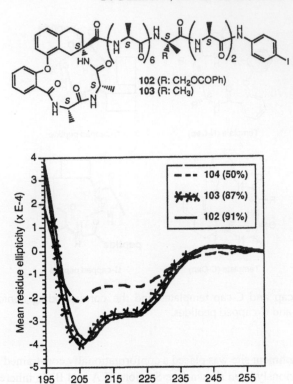

102 (R: CH₂OCOPh)
103 (R: CH₃)

Figure 32. CD-spectra of template linked model peptide **102** and **103** 12-mer reference peptide (Boc-Ala₂-Aib-Ala₆-Aib-Ala₂-pla) **104**.

As mentioned above, these three subunits were designed to form a first α-helical turn in which the four amide groups are ideally oriented to participate in the H-bond network of an α-helix. Peptides that are linked to these templates adopt an α-helical conformation due to the conformational constraint induced by the high inherent α-helix propensity of the template and the H-bond network that is initiated by the proper orientation of the carbonyl groups. Thus, helix induction is achieved by the combination of steric and electronic properties of the three subunits.

The β-tetralin amino acid induces the α-helical conformation by fixing the torsional angles along the peptide backbone at about −60° (φ) and −50° (ψ).[109] β-Tetralin amino acids may be regarded as cyclic-constrained phenylalanine analogues. As shown in Section II.A, this class of unnatural amino acids is known to stabilize distinct conformations in peptides since the two substituents at the α-center restrict the available conformational space. Cyclic α,α-dialkylated glycines and α-substituted alanines preferentially adopt α-helical conformations.[205]

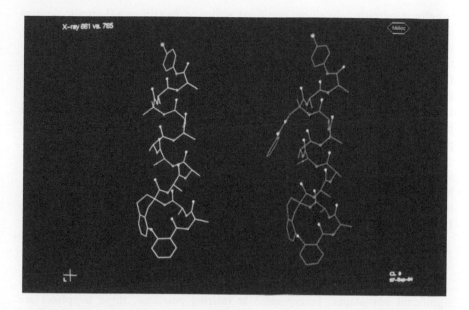

Figure 33. X-ray structures of two N-capped 12-mer peptides **102** (right) **103** (left).

A careful analysis of conformational energy maps (Ramachandran plots) revealed that both enantiomers of β-tetralin amino acids were compatible with the right handed α-helical conformation.[109] This was an important prerequisite for the development of N- and C-caps since the (S)-configuration of the β-tetralin amino acid is needed for N-terminal helix induction, whereas the (R)-enantiomer was used in the C-cap series. The fact that both configurations were compatible with α-helical conformations made these amino acids our first choice as building blocks for

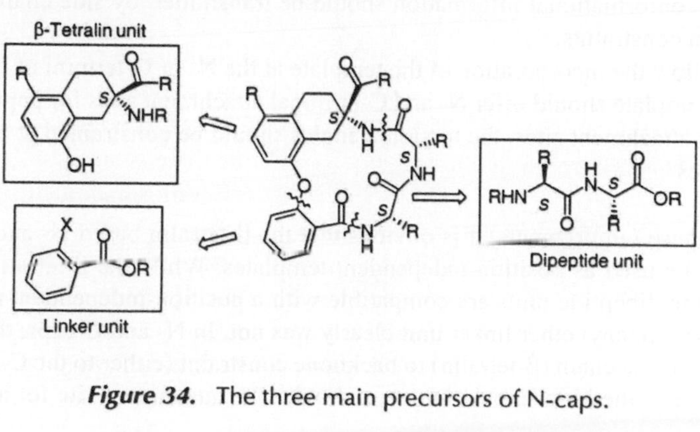

Figure 34. The three main precursors of N-caps.

α-helix stabilization. In all but one X-ray structure containing a β-tetralin derived amino acid building block, the amino group was found in the pseudo-axial position, which is consistent with molecular modeling calculations that favor this conformation by about 0.7 kcal/mol over the pseudo-equatorial one.[109] This conformational flexibility was needed to use these building blocks at both ends of the α-helix. N- and C-caps should alter the physicochemical properties of attached peptides as little as possible. In this approach, the only unnatural amino acid in the N- and C-capped peptides is the α-tetralin amino acid that is linked to the dipeptide unit via a phenoxy group.

The fact that both N- and C-terminal α-helix induction was achieved using the same concept raised the question of whether it would be possible to combine the N- and C-cap concepts, by generating a position-independent template that stabilizes β-helical conformations not only from N-terminal or C-terminal ends but also from internal positions.

Position-Independent α-Helix-Inducing Templates

In comparison to N- and C-caps, position-independent templates potentially offer significant advantages. Incorporated at internal positions, these templates are supposed to transmit the conformational information along the helix axis towards both termini of the peptide. If the helix induction potential is comparable to that observed for N- and C-caps, position-independent templates at internal positions might stabilize up to five consecutive α-helical turns, or peptides with 18 amino acids.[202]

A position-independent template had to exhibit structural and electronic properties that met the requirements for N-caps, C-caps as well as internal templates:

• The template has to bear H-bond acceptors for N-terminal—as well as H-bond donors for C-terminal helix induction.
• The template should be uncharged and amphiphilic.
• The conformational information should be transmitted by side chain to side chain constraints.
• To allow the incorporation of the template at the N- or C-termini of peptides, the template should offer N- and C-terminal attachment sites for peptides. At both attachment sites, the torsional angles should be constrained at α-helical values.

Given these requirements, it is obvious that the β-tetralin based N- and C-caps could not be used as position-independent templates. While the β-tetralin amino acid and the dipeptide units are compatible with a position-independent template design, the biphenyl ether linker unit clearly was not. In N- and C-caps, the linker unit forms a side chain (β-tetralin) to backbone constraint (either to the C-terminal carboxylate or the N-terminal amine), and only one attachment site for a peptide chain is available. Both shortcomings could be overcome by introducing a second peptide attachment site within the linker unit. By using an (*R*)-serine-derived linker group, the following requirements could be met:

- Side chain to side chain constraint is allowed.
- A second attachment site for peptides is provided.
- The template stabilizes an α-helical conformation.
- The ring size of the template is maintained.

The three different linker units used in N-caps, C-caps, and position-independent templates are shown in Figure 35. Figure 36 shows the structure of our first position-independent α-helix-inducing template **105** (HIT-1). In order to induce the α-helical conformation in peptides in a position-independent way, the template has to be able to adopt an α-helical conformation. This does not necessarily mean that the template is α-helical on its own. NMR measurements have revealed that our N-caps and the proline-based templates proposed by Kemp do not preferentially adopt the nucleating α-helical conformation unless they are linked to peptides.[206-210]

The conformational preferences of HIT-1 were estimated by means of molecular modeling calculations. Using the MOLOC force field,[211] a library with more than 700 conformations of the protected HIT-1 derivative Ac-(HIT-1)-NHMe was generated. The ideal α-helical conformation was found to be at least 2.1 kcal/mol more stable than any other conformation (Figure 37).

In Figure 38 a retrosynthetic analysis of HIT is shown schematically giving rise to three main building blocks, namely a suitably protected (*S*)-2-amino-1,2,3,4-tetrahydronaphthalen-2-carboxylate derivative, a protected serine derived building block as linker unit, and a suitably protected dipeptide unit. A synthesis of HIT-1 is shown in Scheme 4. Coupling an excess of tosylate (*S*)-**108** with the (*S*)-β-tetralin derivative in the presence of NaH gave **109** and **110** in good yields, respectively. Transketalization, using 1,3-propanediol and camphorsulfonic acid in methanol at 30 °C gave the Z-protected alcohols **111** and **112**. These conditions were selected after several trials and proved to be critical to obtain good yields. Subsequent oxidation of the alcohols using Pt, $NaHCO_3$, O_2 (2 bar) in dioxane/H_2O (4:1) [212,213] gave very cleanly the corresponding acids in excellent yields. This method was

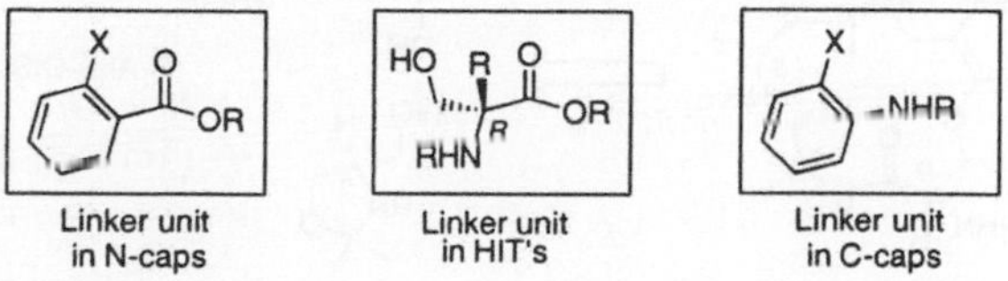

Linker unit in N-caps

Linker unit in HIT's

Linker unit in C-caps

Figure 35. Linker units in N-caps, C-caps, position-independent templates.

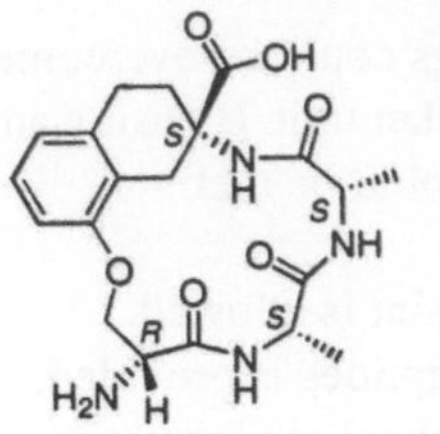

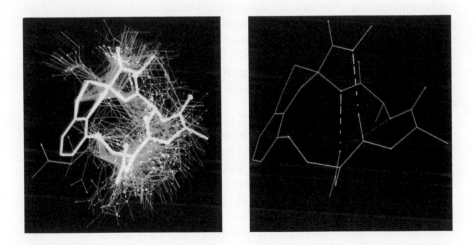

105

HIT-1

Figure 36. HIT-1, a position-independent α-helix inducing template.

Figure 37. Superposition of the 79 most stable conformations of Ac-(HIT-1)-NHMe (left). The α-helical conformation was found to be the most stable conformation (right).

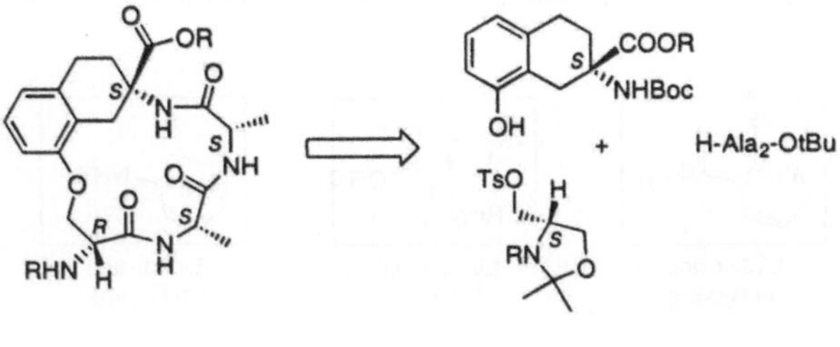

Figure 38. A retrosynthetic analysis of HIT-1.

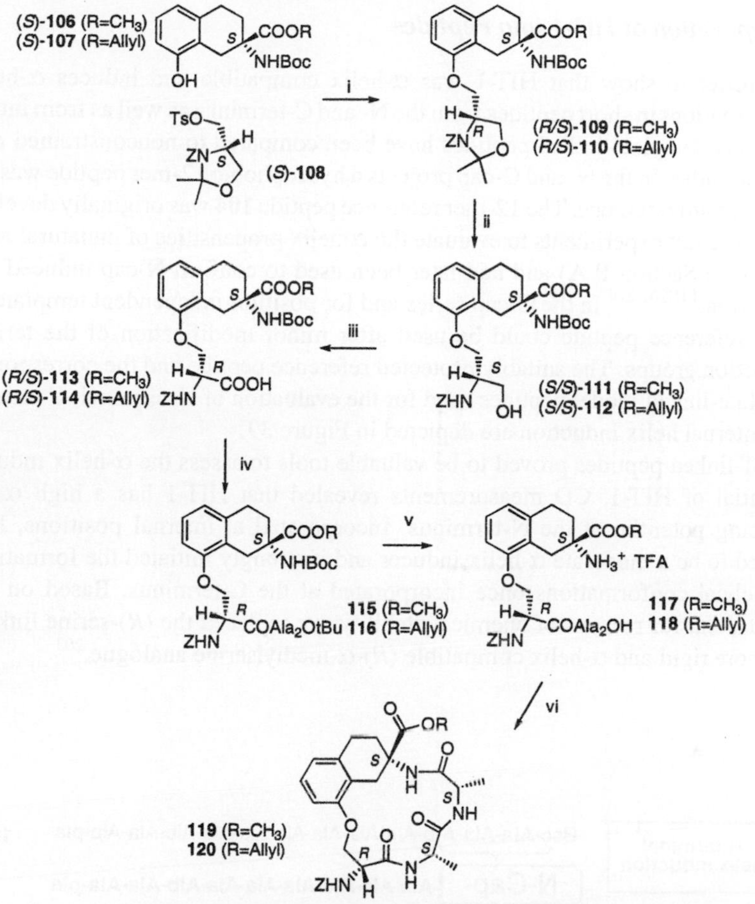

i: NaH, DMF, TDA-1; ii: CSA, 1,3-propanediol, MeOH; iii: Pt, NaHCO₃, O₂ (2 bar),
H₂O/dioxane (4:1); iv: (S/S)-H-Ala-Ala-OtBu, HBTU, HOBT, NaHCO₃, DMF;
v: TFA, CH₂Cl₂; vi: TATU, HOAT, NaHCO₃, DMF.

Scheme 4.

clearly superior to other oxidation methods, such as pyridinium dichromate (PDC), Jones reagent, pyridinium dichromate/Al₂O₃, or NaIO₄/RuCl₃. The acids **113** and **114** were coupled to H-Ala₂-OtBu, followed by cleavage of the *t* butyl ester and the Boc protecting group using TFA in CH₂Cl₂. The final cyclization was accomplished using TATU, HOAT, and NaHCO₃ in DMF in good yield. By using NaHCO₃ as a mild base, β-elimination was largely suppressed and dimerization was avoided by running the reaction at high dilution. A wide range of alternative cyclization strategies failed to give good yields of HIT-1.

Incorporation of HIT-1 into Peptides

In order to show that HIT-1 was α-helix compatible and induces α-helical conformations in short peptides from the N- and C-terminus as well as from internal positions, template-linked peptides have been compared to nonconstrained reference peptides. In the N- and C-cap projects a hydrophobic 12-mer peptide was used as a standard reference. The 12-mer reference peptide **104** was originally developed for host guest experiments to evaluate the α-helix propensities of unnatural amino acids (see Section II.A) and had later been used to confirm N-cap induced helix formation.[141,202,214] In the C-cap series and for position-independent templates the same reference peptide could be used after minor modification of the terminal protection groups. The suitably protected reference peptide and the corresponding template-linked model peptides used for the evaluation of N-terminal, C-terminal, and internal helix induction are depicted in Figure 39.

HIT-linked peptides proved to be valuable tools to assess the α-helix induction potential of HIT-1. CD measurements revealed that HIT-1 has a high α-helix inducing potential at the N-terminus. Incorporated at internal positions, HIT-1 proved to be a moderate α-helix inducer and it strongly initiated the formation of 3_{10}-helical conformations once incorporated at the C-terminus. Based on these results, and for reasons of chemical stability, we replaced the (R)-serine linker by the more rigid and α-helix compatible (R)-α-methylserine analogue.[203]

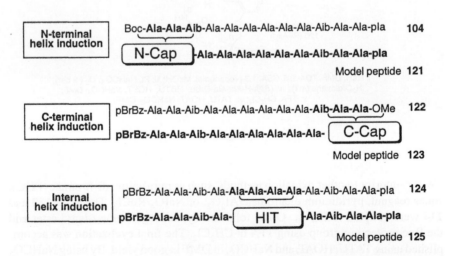

Figure 39. Reference peptides (first row) template-linked model peptides (second row) used for the evaluation of N-terminal, C-terminal internal α-helix induction.

III. NOVEL STRATEGIES FOR THE COMBINATORIAL SYNTHESIS OF LOW-MOLECULAR-WEIGHT COMPOUND LIBRARIES

A. General

As mentioned in Section I.C, we focus our attention primarily on convergent assembly strategies (Figure 11) performed in solution or preferentially on solid supports which should, more likely than linear strategies, give access to low-molecular-weight "drug-like" molecules. We selected primarily those reactions, assembly strategies, and reactive building blocks which would allow for high flexibility in the synthesis, both in terms of arriving at a large number of pharmacologically relevant core structures and of easy subsequent derivatizations.

As shown in Figures 12 and 13, we focused essentially on the principle of combining tailor-made reactive building blocks to assemble interesting core structures (the "combinatorics of reactophores") in a multigeneration format performed in a one-pot or cascade manner.

B. General Reactive Building Blocks ("Reactophores")

In order to classify the reactive building blocks, we grouped them into mono- and bidentate nucleophiles (e.g. amines, alcohols, thioureas, isothioureas, amidines, amidrazones), into their mono- and bis-acceptor counter parts (e.g. alkyl halides, α-bromomethyl ketones, α-alkinyl ketones), and into donor–acceptor species (e.g. isocyanates, isothiocyanates, 2-azido benzoic acids[215]), as depicted in Figure 40.

For the synthesis of optically pure building blocks we mainly focused on the synthesis of protected noncoded (R)- and (S)-amino acids, as they can be synthesized reliably in enantiomerically pure form with a large variety of side chains using asymmetric hydrogenation of α-amino-α,β-didehydroamino acids using cationic diphosphine rhodium catalysts.[216,217] As a typical example of a reactophore we present α-alkynyl ketones, which is a representative bis-acceptor molecule. In Scheme 5 are depicted some of the many synthetic applications of acetylenic ketones in heterocyclic synthesis, which have great potential for combinatorial and parallel organic synthesis.

α-Alkynyl ketones can be easily synthesized with a whole variety of orthogonally protected functional groups and have been used to synthesize various five-membered heterocycles such as 3-halofuranes[218] (Scheme 5, entry a), 3-halothiophenes[219] (Scheme 5, entry b), and 3-halopyrroles[220] (Scheme 5, entry c). Condensation of methyl thioglycolate with substituted acetylenic ketones followed by intramolecular Knoevenagel type cyclization yielded thiophenes (Scheme 5, entry d), which may be employed as highly versatile building blocks.[221] Acetylenic ketones have also been employed successfully as electron deficient dienophiles in

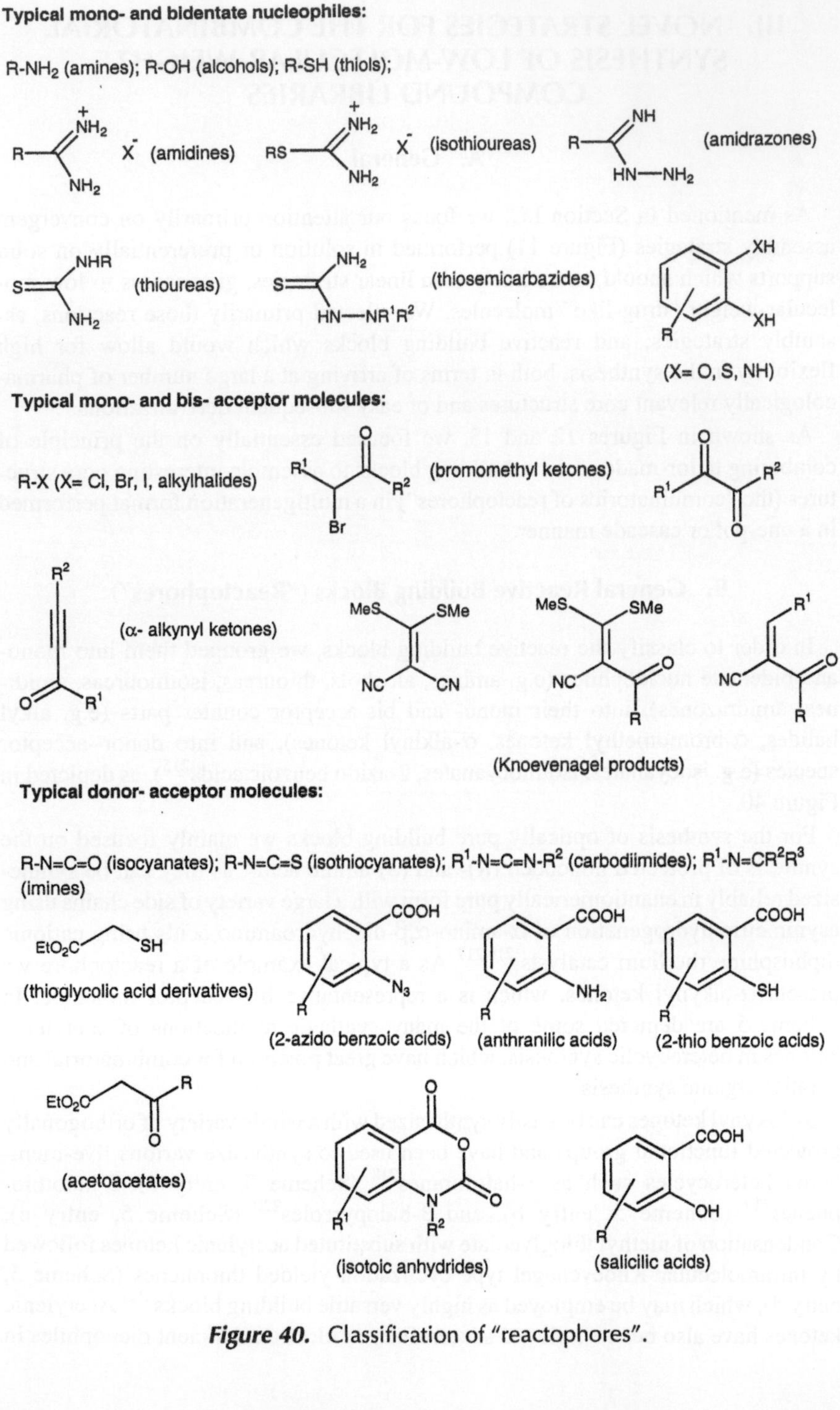

Typical mono- and bidentate nucleophiles:

R-NH$_2$ (amines); R-OH (alcohols); R-SH (thiols);

(amidines)   (isothioureas)   (amidrazones)

(thioureas)   (thiosemicarbazides)

(X= O, S, NH)

Typical mono- and bis- acceptor molecules:

R-X (X= Cl, Br, I, alkylhalides)   (bromomethyl ketones)

(α- alkynyl ketones)

(Knoevenagel products)

Typical donor- acceptor molecules:

R-N=C=O (isocyanates); R-N=C=S (isothiocyanates); R^1-N=C=N-R^2 (carbodiimides); R^1-N=CR^{2}R^3 (imines)

(thioglycolic acid derivatives)

(2-azido benzoic acids)   (anthranilic acids)   (2-thio benzoic acids)

(acetoacetates)

(isotoic anhydrides)   (salicilic acids)

Figure 40. Classification of "reactophores".

carbonyl-alkyne exchange reactions[222] to yield highly substituted phenols with excellent degrees of regioselectivity (Scheme 5, entry e). Finally, acetylenic ketones have been used as building blocks for the construction of 2,4-substituted quinolines via the intermediate formation of thiazepines followed by sulfur extrusion[223] (Scheme 5, entry f). As shown in Section III.F they have also been employed in versatile solid-phase syntheses of pyrimidines.

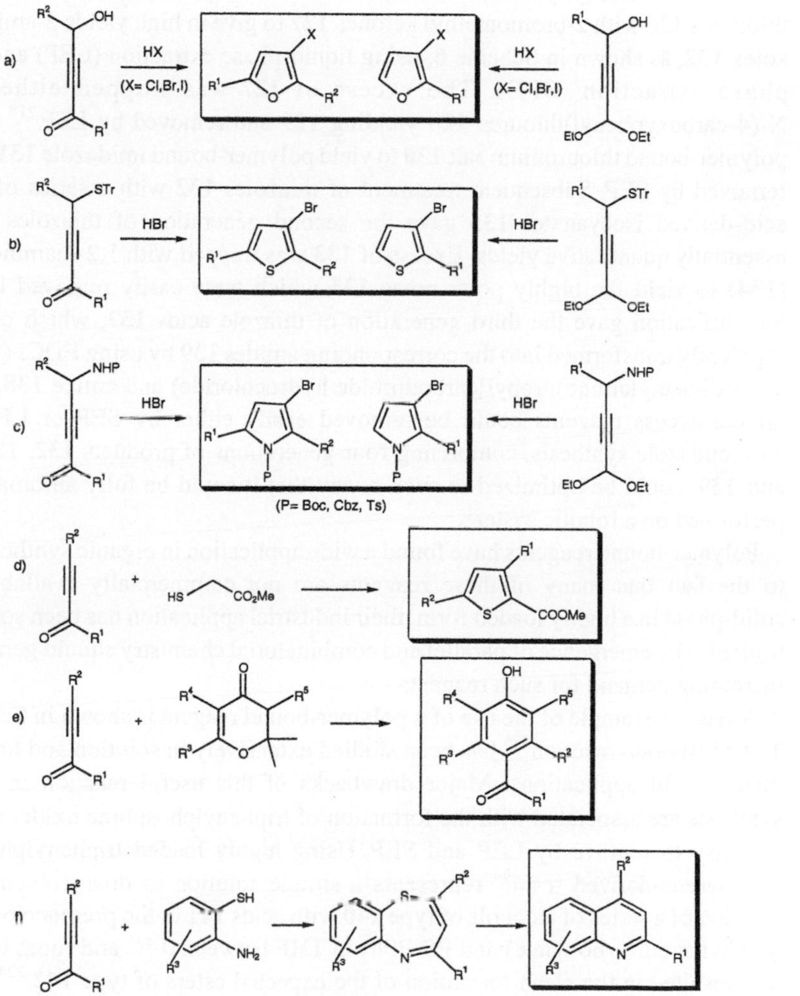

Scheme 5.

C. Parallel Solution Strategies Using Polymer-Bound Reagents

One major advantage in the use of solution chemistry is certainly the fact that many well-established reaction procedures are available, and subsequently less effort has to be investigated into the development of methods that are compatible with solid supports. To overcome the complex and time-consuming extraction and purification procedures associated with solution chemistry, more and more polymer-bound reagents and polymer-supported scavenging reagents are being successfully employed. As a first example, we present a solution-phase multigeneration strategy towards highly substituted thiazoles based on the Hantsch condensation of thioureas **126** with 2-bromomethyl ketones **127** to give in high yields 2-aminothiazoles **132**, as shown in Scheme 6, using liquid-phase extraction (LEP) and solid-phase extraction (SPE). The excess of **127** was trapped either with N-(4-carboxyphenyl)thiourea **128** yielding **129** and removed by LEP,[215] or with polymer-bound thiouronium salt **130** to yield polymer-bound imidazole **131** [224] and removed by SEP. Subsequent treatment of thiazoles **132** with a series of amino acid-derived isocyanates **133** gave the second generation of thiazoles **136** in essentially quantitative yields. Excess of **133** was trapped with 1,2-diaminoethane (**134**) to yield the highly polar ureas **135** which were easily removed by SEP. Saponification gave the third generation of thiazole acids **137**, which could be efficiently transformed into the corresponding amides **139** by using EDCI (1-ethyl-3-[3-(dimethylamino)propyl]carbodiimide hydrochloride) and amine **138**. Again, all the excess reagents could be removed easily either by SEP or LEP. This aminothiazole synthesis, comprising four generations of products **132**, **136**, **137**, and **139**, could be optimized in such a way that it could be fully automated and performed on a robotic system.

Polymer-bound reagents have found a wide application in organic synthesis. Due to the fact that many of these reagents are not commercially available on a solid-phase in a highly loaded form, their industrial application has been somewhat limited. The emergence of parallel and combinatorial chemistry should generate an increasing demand for such reagents.

A typical example of the use of a polymer-bound reagent is shown in Scheme 7. The Mitsunobu reaction[225] has been studied extensively in solution and has found many useful applications. Major drawbacks of this useful reaction in parallel synthesis are associated with the formation of triphenylphosphine oxide, which is not easy to remove by LEP and SEP. Using highly loaded triphenylphosphine polystyrene-derived resin[94] represents a simple solution to this problem. Thus, reaction of a series of alcohols of type **140** with acids **141** in the presence of DEAD (diethyl azodicarboxylate) and PS-PPh$_2$ in THF between 0 °C and room temperature resulted in the clean formation of the expected esters of type **142**.[224] Simple filtration through a plug of Al$_2$O$_3$ gave the essentially pure products in high yields (70–90%).

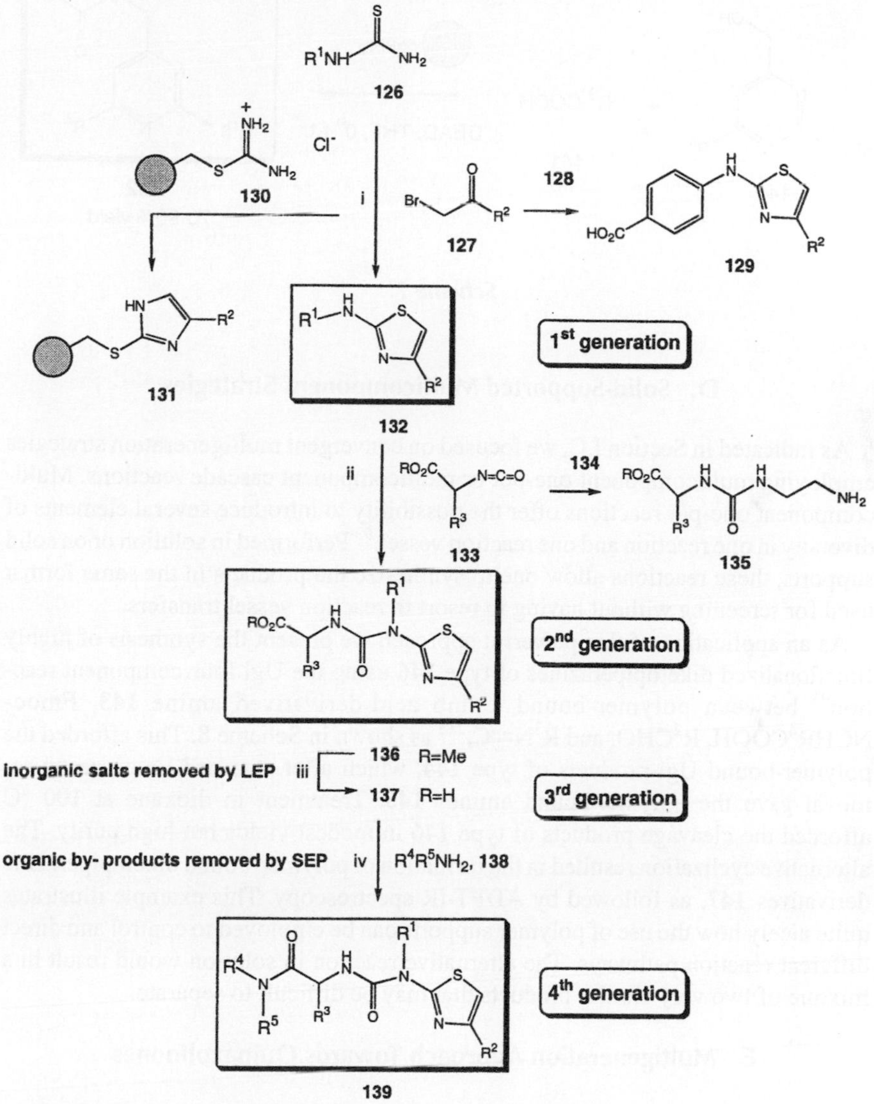

i: EtOH, dioxane, 60°, followed by **128** or **130**; ii: toluene, 70°, followed by **134**; iii:LiOH, dioxane, H_2O,r.t.;
iv: EDCI, cat. DMAP, CH_2Cl_2, r.t.
128= N-(4-carboxyphenyl)thiourea; **134**= 1,2-diaminoethane

Scheme 6.

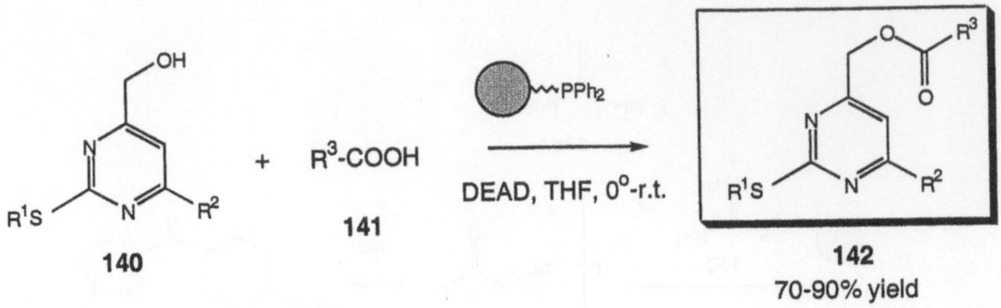

Scheme 7.

D. Solid-Supported Multicomponent Strategies

As indicated in Section I.C, we focused on convergent multigeneration strategies employing multicomponent one-pot or multicomponent cascade reactions. Multicomponent one-pot reactions offer the possibility to introduce several elements of diversity in one reaction and one reaction vessel.[93] Performed in solution or on solid supports, these reactions allow one to synthesize the products in the same format used for screening without having to resort to reaction vessel transfers.

As an application of this powerful approach we present the synthesis of highly functionalized diketopiperazines of type **146** using the Ugi four-component reaction[93] between polymer-bound amino acid-derivatived amine **143**, Fmoc-NCHR^{4}COOH, R^{2}CHO, and R^{3}N=C,[215] as shown in Scheme 8. This afforded the polymer-bound Ugi-products of type **144**, which after standard Fmoc-group removal gave the polymer-bound amines **145**. Treatment in dioxane at 100 °C afforded the cleavage products of type **146** in modest yields but high purity. The alternative cyclization resulted in the formation of polymer-bound diketopiperazine derivatives **147**, as followed by AT/FT-IR spectroscopy. This example illustrates quite nicely how the use of polymer supports can be employed to control and direct different reaction pathways. The alternative reaction in solution would result in a mixture of two very similar products that may be difficult to separate.

E. Multigeneration Approach Towards Quinazolinones

Another key strategy in solid-phase organic synthesis is to cleave molecules from the resin by means of intramolecular cyclization as the last step. This strategy normally leads to very pure compounds since only those molecules will be cleaved from the resin that have gone through the whole reaction sequence. This concept was used for a versatile synthesis of quinazolinones of type **151** and **152** as shown in Scheme 9.[215]

Coupling of *ortho*-azido benzoic acids of type **148** to chloromethylated Merrifield resin, gave high yields of esters of type **149**. After treatment with triphenyl-

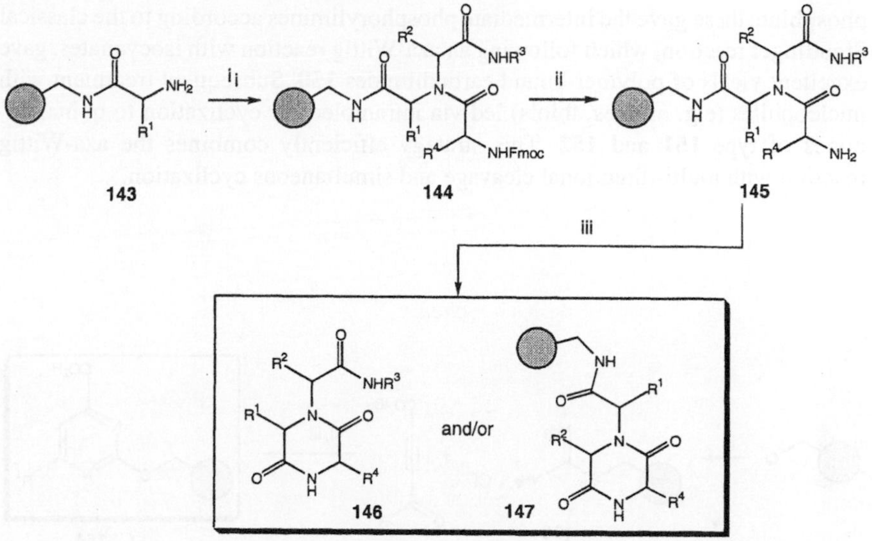

i: R^2CHO, $R^3N=C$, FmocNHCHR^{4}COOH, dioxane/CH$_2$Cl$_2$/MeOH(4:4:1); ii: 20% piperidine/DMF, r.t.;
iii: dioxane, 100°

Scheme 8.

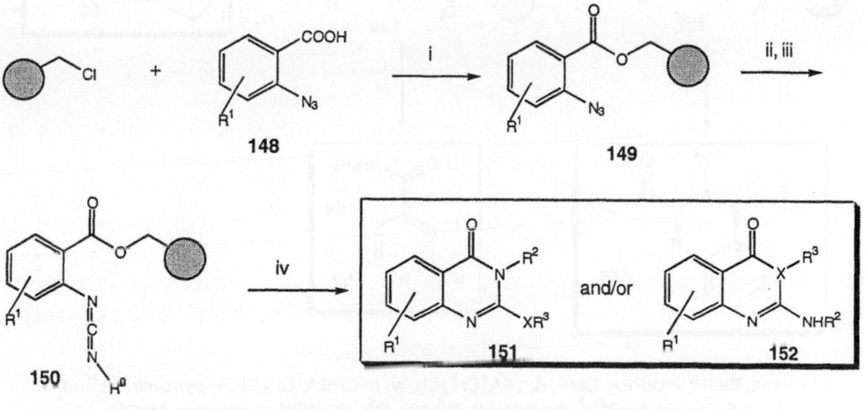

i: Cs$_2$CO$_3$, DMF, KI; ii: PPh$_3$, r.t.; iii: R$_2$N=C=O; iv: R$_3$-XH (X= O, S, NH), THF, 50°

Scheme 9.

phosphine, these gave the intermediate phosphorylimines according to the classical Staudinger reaction, which following an aza-Wittig reaction with isocyanates, gave excellent yields of polymer-bound carbodiimides **150**. Subsequent treatment with nucleophiles (e.g. amines, thiols) led via intramolecular cyclization to quinazolinones of type **151** and **152**. This strategy efficiently combines the aza-Wittig reaction with multi-directional cleavage and simultaneous cyclization.

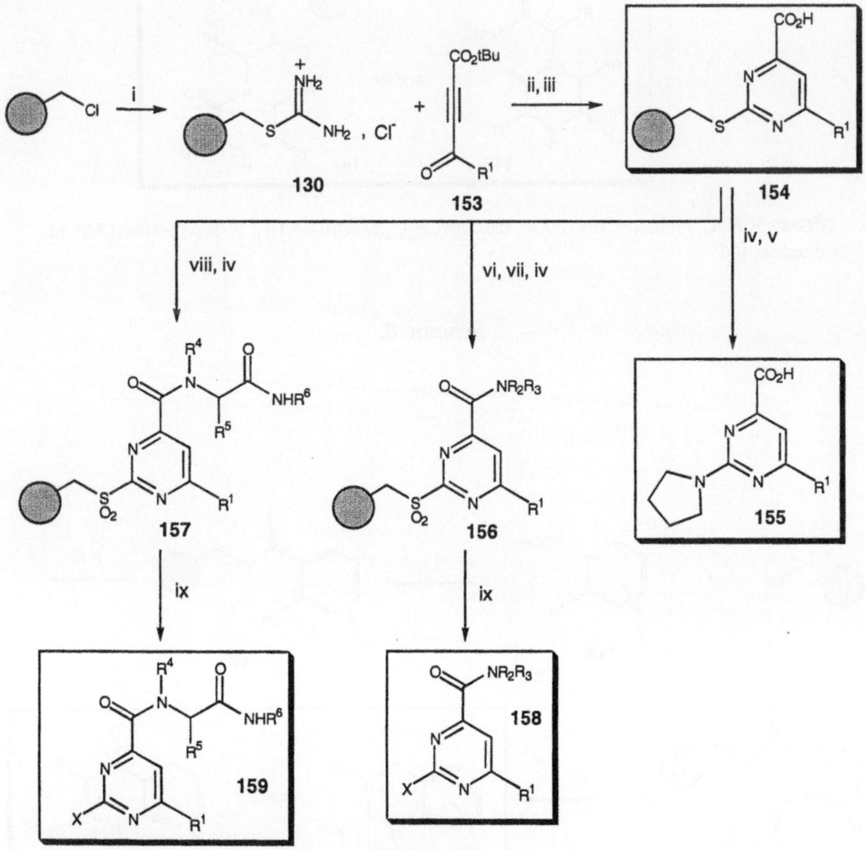

i: thiourea, EtOH; ii: DIPEA, DMF; iii: TFA, CH$_2$Cl$_2$; iv: m-CPBA, CH$_2$Cl$_2$; v: pyrrolidine, dioxane; vi: DCC, C$_6$F$_5$OH; vii: NHR^{2}R^3, dioxane; viii: R^{4}NH2, R^{5}CHO, R^{6}N=C, dioxane, MeOH; ix: XH, dioxane (X: amines, alcoholates, thiolates)

Scheme 10.

F. Multigeneration Solid-Phase Approach to Pyrimidines

Next, we focused on a multigeneration strategy towards the pharmacologically interesting pyrimidine derivatives of type **155**, **158**, and **159**, as shown in Scheme 10. This strategy features a cyclo-condensation reaction of polymer-bound thiouronium salt **130** with the acetylenic ketones of type **153** in the presence of diisopropylethylamine, followed by cleavage of the *tert*-butyl ester with TFA to give excellent yields of polymer-bound acids **154**.[226] All these reactions were monitored by on-bead AT/FT-IR spectroscopy.[227] Oxidation of **3** with meta-chloroperbenzoic acid resulted in clean formation of the intermediate sulfones and treatment with pyrrolidine gave the cleaved pyrimidine acids of type **155** in high yields and purity.

The oxidative activation of the pyrimidine–sulfide linkage, followed by reaction with nucleophiles (e.g. alcoholates, thiolates, amines) and simultaneous cleavage[98,226] constitutes a novel example of the "safety-catch" linker principle.[100] Having demonstrated the utility of this novel traceless linker, the resin-bound acids **154** were partitioned and treated either with amines to give amides of type **156**, or reacted in an Ugi reaction with amines, aldehydes, and isonitriles to give amides **157**. Oxidation to the sulfones and cleavage with nucleophiles such as amines, alcoholates and thiolates gave high yields of pyrimidines of type **158** and **159** (Scheme 10). It should be noted, that cleavage of the intermediate polymer-bound sulfones results directly in the formation of pharmacologically important 2-aminopyrimidines.

This multigeneration strategy for the synthesis of pyrimidines combines efficiently a novel cyclocondensation reaction using the highly reactive acetylenic ketones **153**[215,218] to build the pyrimidine skeleton, with a multicomponent reaction, and a multidirectional cleavage procedure. The Ugi four component reaction is especially useful in the context of building peptidomimetic-derived combinatorial libraries as it affords directly dipeptide analogues of type **159**.

ACKNOWLEDGMENTS

The authors would like to thank Drs. U. Bohdal, R. Kimmich, R. Ruffieux, P. Waldmeier, and Prof. K. Müller of F. Hoffmann-La Roche, Basel.

REFERENCES

1. Oefner, C.; D'Arci, A.; Winkler, F. K.; Eggimann, B.; Hosang, M. *EMBO J.* **1992**, *11*, 3921.
2. Ross, R.; Raines, E. W.; Bowden-Pope, D. F. *Cell* **1986**, *46*, 155–159.
3. Ibanez, C. F.; Ebendahl, T.; Barbany, G.; Murray-Rust, J.; Blundell, T.; Perrson, H. *Cell* **1992**, *69*, 320–341.
4. Kramer, R. A.; Schaber, M. D.; Skalka, A. M.; Ganguly, K.; Wong-Staal, F.; Reddy, E. P. *Science* **1986**, *231*, 1580.
5. Davidson, A. H.; Drummond, A. H.; Galloway, W.; Whittaker, M. *Chem. Ind.* **1997**, *7*, 258–261.

6. McDonald, N. Q.; Murray-Rust, J.; Blundell, T. *Structure* **1995**, *3*, 1–6.
7. Hubbard, S. R.; Wei, L.; Ellis, L.; Hendrickson, W. A. *Nature* **1994**, *372*, 746–754.
8. Griffith, J. P.; Kim, J. L.; Kim, E. E.; Sintchak, M. D.; Thompson, J. A.; Fitzgibbon, M. J.; Fleming, M. A.; Caron, P. R.; Hsiao, K.; Navia, M. A. *Cell* **1995**, *82*, 507–522.
9. Kemp, D. S. *TIBTECH* **1990**, *8*, 249–255.
10. Schneider, J. P.; Kelly, J. W. *Chem. Rev.* **1995**, *95*, 2169–2187.
11. Müller, K.; Obrecht, D.; Knierzinger, A.; Stankovic, C.; Spiegler, C.; Bannwarth, W.; Trzeciak, A.; Englert, G.; Labhardt, A. M.; Schönholzer, P. *Perspectives in Medicinal Chemistry, Helv. Chim. Acta*: **1993**, 513–531.
12. Balkenhohl, F.; von dem Bussche-Huennefeld, C.; Lansky, A.; Zechel, C. *Angew. Chem.* **1996**, *108*, 2436–2488.
13. Nicolaou, K. C.; Guy, R. K. *Angew. Chem.* **1995**, *107*, 2247–2259.
14. Albers-Schoenberg, G.; Arison, B. H.; Chabala, Y.; Douglas, A. W.; Eskola, P.; Fisher, M. H.; Lusi, A.; Mrozik, H.; Smith, J. L.; Tolman, R. L. *J. Am. Chem. Soc.* **1981**, *103*, 4216.
15. Hata, T.; Sano, Y.; Sugwara, R.; Matsume, A.; Kanamori, K.; Shima, T.; Hoshi, T. *J. Antibiot. Ser. A* **1956**, *9*, 141.
16. Fukuyama, T.; Nakasubo, F.; Coccuzza, J.; Kishi, J. *Tetrahedron Lett.* **1977**, *14*, 4295.
17. Kishi, J. *J. Nat. Prod.* **1979**, *42*, 549.
18. Bierer, B. E.; Matila, B. S.; Standaert, R.; Herzenberg, L. A.; Burakoff, S. J.; Grabtree, G.; Schreiber, S. L. *Proc. Natl. Acad. Sci. USA* **1990**, *87*, 9231.
19. Mutter, M. *Angew. Chem.* **1985**, *97*, 639–654.
20. Beck-Sickinger, A. G. *Drug Discovery Today* **1996**, *1*, 502–513.
21. Silvennoinen, O.; Ihle, J. N. *Signaling by the Hematopoietic Cytokine Receptors*; R. G. Landes; Austin, TX, 1996.
22. Heldin, C.-H. *Cell* **1995**, *80*, 213–223.
23. Ihle, J. N. *Nature* **1995**, *377*, 591–594.
24. Taga, T.; Kishimoto, T. *Curr. Opin. Immunol.* **1995**, *7*, 17–23.
25. Watowich, S. S.; Wu, H.; Socolovsky, M.; Klingmuller, U.; Constantinescu, S. N.; Lodish, H. F. *Annu. Rev. Cell Dev. Biol.* **1996**, *12*, 91–128.
26. Wells, J. A. *Curr. Opin. Cell. Biol.* **1994**, *6*, 163–173.
27. Wells, J. A. *Proc. Natl. Acad. Sci. USA* **1996**, *93*, 1–6.
28. Kossiakoff, A.; Somers, W.; Ultsch, M.; Andow, K.; Muller, Y. A.; De Vos, A. M. *Protein Sci.* **1994**, *3*, 1697–1705.
29. Fuh, G.; Colosi, P.; Wood, W. I.; Wells, J. A. *J. Biol. Chem.* **1993**, *268*, 5376–5381.
30. Fuh, G.; Cunningham, B. C.; Fukunaga, R.; Nagata, S.; Goeddel, D. V.; Wells, J. A. *Science* **1992**, *256*, 1677–1680.
31. Sundstrom, M.; Lundqvist, T.; Rodin, J.; Giebel, L. B.; Milligan, D.; Norstedt, G. *J. Biol. Chem.* **1996**, *271*, 32197–32203.
32. Walter, M. R.; Windsor, W.; Nagabhushan, T. L.; Lundell, D. J.; Lunn, C. A.; Zauodny, P. J.; Narula, S. K. *Nature* **1995**, *376*, 230–235.
33. Marsters, S. A.; Pennica, D.; Bach, E.; Schreiber, R. D.; Ashkenazi, A. *Proc. Natl. Acad. Sci. USA* **1995**, *92*, 5401–5405.
34. Briscoe, J.; Guschin, D.; Rogers, N. C.; Watling, D.; Muller, M.; Horn, F.; Heinrich, P.; Stark, G. R.; Kerr, I. M. *Philos. Trans. Roy. Soc. Lond. [Biol]* **1996**, *351*, 167–171.
35. Vigers, G. P. A.; Anderson, L. J.; Caffes, P.; Brandhuber, B. J. *Nature* **1997**, *386*, 190–194.
36. Schreuder, H.; Tardif, C.; Trump-Kallmeyer, S.; Soffientini, A.; Sarubbi, E.; Akeson, A.; Bowlin, T.; Yanofsky, S.; Barrett, R. W. *Nature* **1997**, *386*, 194–200.
37. Banner, D. W.; D'Arcy, A.; Janes, W.; Gentz, R.; Schoenfeld, H. J.; Broger, C.; Loetscher, H.; Lesslauer, W. *Cell* **1993**, *73*, 431–445.
38. Banner, D. W.; D'Arcy, A.; Chène, C.; Winkler, F. K. *Nature* **1996**, *380*, 41–46.
39. Cunningham, B. C.; Wells, J. A. *Science* **1989**, *244*, 1081–1085.

40. Bass, S. H.; Mulkerrin, M. G.; Wells, J. A. *Proc. Natl. Acad. Sci. USA* **1991**, *88*, 4498–4502.
41. Clackson, T.; Wells, J. A. *Science* **1995**, *267*, 383–386.
42. Pearce, K. H.; Ultsch, M. H.; Kelley, R. F.; Devos, A. M.; Wells, J. A. *Biochemistry* **1996**, *35*, 10300–10307.
43. Livnah, O.; Stura, E. A.; Johnson, D. L.; Middleton, S. A.; Mulcahy, L. S.; Wrighton, N. C.; Dower, W. J.; Jolliffe, L. K.; Wilson, I. A. *Science* **1996**, *273*, 464–471.
44. Wrighton, N. C.; Farrell, F. X.; Chang, R.; Kashyap, A. K.; Barbone, F. P.; Mulcahy, L. S.; Johnson, D. L.; Barrett, R. W.; Jolliffe, L. K.; Dower, W. J. *Science* **1996**, *273*, 458–463.
45. Akeson, A. L.; Woods, C. W.; Hsieh, L. C.; Bohnke, R. A.; Ackermann, B. L.; Chan, K. Y.; Robinson, J. L.; Yanofsky, S. D.; Jacobs, J. W.; Barrett, R. W.; Bowlin, T. L. *J. Biol. Chem.* **1996**, *271*, 30517–30523.
46. Welply, J. K.; Steininger, C. N.; Caperon, M.; Michener, M. L.; Howard, S. C.; Pegg, L. E.; Meyer, D. M.; De Ciechi, P. A.; Devine, C. S.; Casperson, G. F. *Proteins: Struct. Funct. Genet.* **1996**, *26*, 262–270.
47. Tilley, J. W.; Chen, L.; Fry, D. C.; Emerson, S. D.; Powers, G. D.; Biondi, D.; Varnell, T.; Trilles, R.; Guthrie, R.; Mennona, F.; Kaplan, G.; LeMahieu, R. A.; Carson, M.; Han, R.-J.; Liu, C.-M.; Palermo, R.; Ju, G. *J. Am. Chem. Soc.* **1997**, *119*, 7589–7590.
48. Li, B.; Tom, J. Y. K.; Oare, D.; Yen, R.; Fairbrother, W. J.; Wells, J. A.; Cunningham, B. C. *Science* **1995**, *270*, 1657–1660.
49. Braisted, A. C.; Wells, J. A. *Proc. Natl. Acad. Sci. USA* **1996**, *93*, 5688–5692.
50. Leahy, D. J.; Aukhil, I.; Erickson, H. P. *Cell* **1996**, *84*, 155–164.
51. Leahy, D. J.; Hendrickson, W. A.; Aukhil, I.; Erickson, H. P. *Science* **1992**, *258*, 987–991.
52. Krezel, A. M.; Wagner, G.; Seymour-Ulmer, J.; Lazarus, R. A. *Science* **1994**, *264*, 1944–1947.
53. Main, A. L.; Harvey, T. S.; Baron, M.; Boyd, J.; Campbell, I. D. *Cell* **1992**, *71*, 671–678.
54. Bach, A. C.; Eyermann, C. J.; Gross, J. D.; Bower, M. J.; Harlow, R. L.; Weber, P. C.; DeGrado, W. F. *J. Am. Chem. Soc.* **1994**, *116*, 3207–3219.
55. Kopple, K. D.; Baures, P. W.; Bean, J. W.; D'Ambrosio, C. A.; Hughes, J. L.; Peishoff, C. E.; Eggleston, D. S. *J. Am. Chem. Soc.* **1992**, *114*, 9615–9623.
56. Gurrath, M.; Müller, G.; Kessler, H.; Aumailley, M.; Timpl, R. *Eur. J. Biochem.* **1992**, *210*, 911–921.
57. Pfaff, M.; Tangemann, K.; Müller, B.; Gurrath, M.; Müller, G.; Kessler, H.; Timpl, R.; Engel, J. *J. Biol. Chem.* **1994**, *269*, 20233–20238.
58. Bach, A. C.; Espina, J. R.; Jackson, S. A.; Stouten, P. F. W.; Duke, J. L.; Mousa, S. A.; Degrado, W. F. *J. Am. Chem. Soc.* **1996**, *118*, 293–294.
59. Haubner, R.; Gratias, R.; Diefenbach, B.; Goodman, S. L.; Jonczyk, A.; Kessler, H. *J. Am. Chem. Soc.* **1996**, *118*, 7461–7472.
60. Wermuth, J.; Goodman, S. L.; Jonczyk, A.; Kessler, H. *J. Am. Chem. Soc.* **1997**, *119*, 1328–1335.
61. Wu, H.; Kwong, P. D.; Hendrickson, W. A. *Nature* **1997**, *387*, 527–530.
62. Jameson, B. A.; McDonnell, J. M.; Marini, J. C.; Korngold, R. *Nature* **1994**, *368*, 744–746.
63. Chen, S.; Chruschiel, R. A.; Nakanishi, H.; Raktabutr, A.; Johnson, M. E.; Sato, A.; Weiner, D.; Hoxie, J.; Saragovi, H. U.; Greene, M. I.; Kahn, M. *Proc. Natl. Acad. Sci. USA* **1992**, *89*, 5872–5876.
64. Kumagai, K.; Tokunaga, K.; Tsutsumi, M.; Ikuta, K. *FEBS Lett.* **1993**, *330*, 117–121.
65. Ma, S.; McGregor, M. J.; Cohen, F. E.; Pallai, P. V. *Biopolymers* **1994**, *34*, 987–1000.
66. Zhang, X.; Piatier-Tonneau, D.; Auffray, C.; Murali, R.; Mahapatra, A.; Zhang, F.; Maier, C. C.; Saragovi, U.; Greene, M. I. *Nature Biotech.* **1996**, *14*, 472–475.
67. Zhang, X.; Gaubin, M.; Briant, L.; Srikantan, V.; Murali, R.; Saragovi, U.; Weiner, D.; Devaux, C.; Autiero, M.; Piatier-Tonneau, D.; Greene, M. I. *Nature Biotech.* **1997**, *15*, 150–154.
68. Nakanishi, H.; Ramurthy, S.; Raktabutr, A.; Shen, R.; Kahn, M. *Gene* **1993**, *137*, 51–56.
69. Monfardini, C.; Kieberemmons, T.; Voet, D.; Godillot, A. P.; Weiner, D. B.; Williams, W. V. *J. Biol. Chem.* **1996**, *271*, 2966–2971.

70. VonFeldt, J. M.; Monfardini, C.; Kieber-Emmons, T.; Voet, D.; Weiner, D. B.; Williams, W. V. *Immunol. Res.* **1994**, *13*, 96–109.

71. Paborsky, L. R.; Law, V. S.; Mao, C. T.; Leung, L. L. K.; Gibbs, C. S. *Biochemistry* **1995**, *34*, 15328–15333.

72. LeSauteur, L.; Wei, L.; Gibbs, B. F.; Saragovi, H. U. *J. Biol. Chem.* **1995**, *270*, 6564–6569.

73. Michne, W. F.; Schroeder, J. D. *Int. J. Pept. Protein Res.* **1996**, *47*, 2–8.

74. Chamberlin, S. G.; Sargood, K. J.; Richter, A.; Mellor, J. M.; Anderson, D. W.; Richards, N. G. J.; Turner, D. L.; Sharma, R. P.; Alexander, P.; Davies, D. E. *J. Biol. Chem.* **1995**, *270*, 21062–21067.

75. Wang, J.-H.; Pepinsky, R. B.; Stehle, T.; Liu, J.-H.; Karpusas, M.; Browning, B.; Osborn, L. *Proc. Natl. Acad. Sci. USA* **1995**, *92*, 5714–5718.

76. Winter, G.; Milstein, C. *Nature* **1991**, *349*, 293–299.

77. Williams, W. V.; Moss, D. A.; Kieber-Emmons, T.; Cohen, J. A.; Myers, J. N.; Weiner, D. B.; Greene, M. I. *Proc. Natl. Acad. Sci. USA* **1989**, *86*, 5537–5541.

78. Williams, W. V.; Kieber-Emmons, T.; Weiner, D. B.; Rubin, D. H.; Greene, M. I. *J. Biol. Chem.* **1991**, *266*, 9241–9250.

79. Williams, W. V.; Kieber-Emmons, T.; Von Feldt, J.; Greene, M. I.; Weiner, D. B. *J. Biol. Chem.* **1991**, *266*, 5182–5190.

80. Saragovi, H. U.; Fitzpatrick, D.; Raktabutr, A.; Nakanishi, H.; Kahn, M.; Greene, M. I. *Science* **1991**, *253*, 792–795.

81. Saragovi, H. U.; Greene, M. I.; Chrusciel, R. A.; Kahn, M. *Bio/Technology* **1992**, *10*, 773–778.

82. Taub, R.; Hsu, J.-C.; Garsky, V. M.; Hill, B. L.; Erlanger, B. F.; Kohn, L. D. *J. Biol. Chem.* **1992**, *267*, 5977–5984.

83. Smythe, M. L.; von Itzstein, M. *J. Am. Chem. Soc.* **1994**, *116*, 2725–2733.

84. Monfardini, C.; Kieberemmons, T.; Vonfeldt, J. M.; Omalley, B.; Rosenbaum, H.; Godillot, A. P.; Kaushansky, K.; Brown, C. B.; Voet, D.; Mccallus, D. E.; Weiner, D. B.; Williams, W. V. *J. Biol. Chem.* **1995**, *270*, 6628–6638.

85. Levi, M.; Sällberg, M.; Rudén, U.; Herlyn, D.; Maruyama, H.; Wigzell, H.; Marks, J.; Wahren, B. *Proc. Natl. Acad. Sci. USA* **1993**, *90*, 4374–4378.

86. Doring, E.; Stigler, R.; Grutz, G.; Vonbaehr, R.; Schneidermergener, J. *Mol. Immunol.* **1994**, *31*, 1059–1067.

87. Davies, D. R.; Cohen, G. H. *Proc. Natl. Acad. Sci. USA* **1996**, *93*, 7–12.

88. Sivolapenko, G. B.; Douli, V.; Pectasides, D.; Skarlos, D.; Sirmalis, G.; Hussain, R.; Cook, J.; Courtenayluck, N. S.; Merkouri, E.; Konstantinides, K.; Epenetos, A. A. *Lancet* **1995**, *346*, 1662–1666.

89. Furka, A.; Sebestyen, F.; Asgedom, M.; Dibo, G. *Int. J. Pept. Protein Res.* **1991**, *37*, 487–493.

90. Sebestyen, F.; Dibo, G.; Kovacs, A.; Furka, A. *Bioorg. Med. Chem. Lett.* **1993**, *3*, 413–418.

91. Geysen, H. M.; Meloen, R. H.; Parteling, S. J. *Proc. Natl. Acad. Sci. USA* **1984**, *81*, 3998–4002.

92. Gordon, E. M.; Barrett, R. W.; Dower, W. J.; Fodor, S. P.; Gallop, M. A. *J. Med. Chem.* **1994**, *37*, 1385.

93. Armstrong, R. W.; Combs, A. P.; Tempest, P. A.; Brown, S. D.; Keating, T. A. *Acc. Chem. Res.* **1996**, *29*, 123–131.

94. Frechet, J. M. J. *Tetrahedron* **1981**, *37*, 663–683.

95. Leznoff, C. C. *Acc. Chem. Res.* **1978**, *11*, 327–333.

96. Hermkens, P. H. H.; Ottenheijm, H. C. J.; Rees, D. *Tetrahedron* **1996**, *52*, 4527–4554.

97. Obrecht, D.; Abrecht, C.; Grieder, A.; Villalgordo, J.-M. *Helv. Chim. Acta* **1997**, *80*, 65–72.

98. Gayo, L. M.; Suto, M. J. *Tetrahedron Lett.* **1997**, *38*, 211–214.

99. Ellman, J. A. *Acc. Chem. Res.* **1996**, *29*, 132–143.

100. Kenner, G. W.; Mc Dermott, J. R.; Sheppard, R. C. *J. Chem. Soc., Chem. Commun.* **1971**, 636.

101. Smith III, A. B.; Akaishi, R.; Jones, D. R.; Keenan, T. P.; Guzman, M. C.; Holcomb, R. C.; Sprengeler, P. A.; Wood, J. L.; Hirshmann, R. *Biopolymers* **1995**, *37*, 29–53.

102. Goehring, W.; Gokhale, S.; Hilpert, H.; Roessler, F.; Schlageter, M.; Vogt, P. *Chimia* **1996**, *11*, 532–537.
103. Roberts, N. A.; Martin, J. A.; Kinchington, D.; Broadhurst, A. V.; Craig, J. C.; Duncan, J. B.; Calpin, S. A.; Handa, B. K.; Kay, J.; Krohn, A.; Lambert, M. B.; Merrett, J. H.; Mills, J. S.; Parkes, K. E. B.; Redshaw, S.; Ritchie, A. J.; Taylor, D. L.; Thomas, G. J.; Machin, P. J. *Science* **1990**, *248*, 358.
104. Davidson, A. H.; Drummond, A. H.; Galloway, W.; Whittaker, M. *Chem. Ind.* **1997**, *7*, 258–261.
105. Karle, I. L.; Balaram, P. *Biochemistry* **1990**, *29*, 6747–6761.
106. Prasad, B. V. V.; Balaram, P. *Crit. Rev. Biochem.* **1984**, *16*, 307–348.
107. Di Blasio, B.; Pavone, V.; Pedone, C.; Lombardi, A.; Benedetti, E. *Biopolymers* **1993**, *33*, 1037–1049.
108. Obrecht, D.; Spiegler, C.; Schoenholzer, P.; Mueller, K.; Heimgartner, H.; Stierli, F. *Helv. Chim. Acta.* **1992**, *75*, 1666–1696.
109. Spiegler, C. E. Ph.D. Thesis, University of Zürich, 1993.
110. Wipf, P. Ph.D Thesis, University of Zurich, 1987.
111. Hinds, M.; Welsh, J. H.; Brennand, D. M.; Fisher, J.; Glennie, M. J.; Richards, N. G. J.; Turner, D. L.; Robinson, J. A. *J. Med. Chem.* **1991**, *34*, 1777–1789.
112. Bisang, C.; Weber, C.; Inglis, J.; Schiffer, C. A.; van Gunsteren, W. F.; Jelesarov, I.; Bosshard, H. R.; Robinson, J. A. *J. Am. Chem. Soc.* **1995**, *117*, 7904–7915.
113. Nanzer, D.; Torda, A. E.; Bisang, C.; Weber, C.; Robinson, J. A.; van Gunsteren, W. F. *J. Mol. Biol.* **1997**, *267*, 1011–1024.
114. Mosberg, H. I.; Lomize, A. L.; Wang, C.; Kroona, H.; Heyl, D. L.; Sobczyk-Kojiro, K.; Ma, W.; Mousigian, C.; Porreca, F. *J. Med. Chem.* **1994**, *37*, 4371–4383.
115. Mosberg, H. I.; Lomize, A. L.; Wang, C.; Kroona, H.; Heyl, D. L.; Sobczyk-Kojiro, K.; Ma, W.; Mousigian, C.; Porrecca, F. *J. Med. Chem.* **1994**, *37*, 4371–4383.
116. Obrecht, D.; Altorfer, M.; Bohdal, U.; Daly, J.; Huber, W.; Labhardt, A.; Lehmann, C.; Müller, K.; Ruffieux, R.; Schönholzer, P.; Spiegler, C.; Zumbrunn, C. *Biopolymers* **1997**, in press.
117. Perczel, A.; Park, K.; Fasman, G. D. *Proteins: Struct. Funct. Genet.* **1992**, *13*, 57–69.
118. Perczel, A.; Hollósi, M.; Tusnády, G.; Fasman, G. D. *Protein Eng.* **1991**, *4*, 669–679.
119. Fasman, G. D. *Biopolymers* **1995**, *37*, 339–362.
120. Hoelzemann, G. *Kontakte(Merck, Darmstadt)* **1991**, 55.
121. Hoelzemann, G. *Kontakte (Merck, Darmstadt)* **1991**, 3–12.
122. Rose, G. D.; Gierasch, L. M.; Smith, J. A. *Adv. Protein Chem.* **1985**, *37*, 1–109.
123. Wilson, I. A.; Stanfield, R. L. *Curr. Opin. Struct. Biol.* **1994**, *4*, 857–867.
124. Wilson, I. A.; Ghiara, J. B.; Stanfield, R. L. *Res. Immunol.* **1994**, *145*, 73–78.
125. Veber, D. F. In *Peptides, Synthesis, Structure Function; Proceedings of the Seventh American Peptide Symposium*; Rich, D. H.; Gross, V. J., Eds.; Pierce Chemical Company: Rockford, IL, 1981, Vol. , pp 685–694.
126. Kyle, D. J.; Chakravarty, S.; Sinsko, J. A.; Storman, T. M. *J. Med. Chem.* **1994**, *37*, 1347–1354.
127. Jiang, P.; Ananthanarayanan, V. S. *J. Biol. Chem.* **1991**, *266*, 22960–22967.
128. Brakch, N.; Rholam, M.; Boussetta, H.; Cohen, P. *Biochemistry* **1993**, *32*, 4925–4930.
129. Paolillo, L.; Simonetti, M.; Brakch, N.; D'Auria, G.; Saviano, M.; Dettin, M.; Rholam, M.; Scatturin, A.; DiBello, C.; Cohen, P. *EMBO J.* **1992**, *11*, 2399–2405.
130. Mandiyan, V.; Obrien, R.; Zhou, M.; Margolis, B.; Lemmon, M. A.; Sturtevant, J. M.; Schlessinger, J. *J. Biol. Chem.* **1996**, *271*, 4770–4775.
131. Trüb, T.; Choi, W. E.; Wolf, G.; Ottinger, E.; Chen, Y.; Weiss, M.; Shoelson, S. E. *J. Biol. Chem.* **1995**, *270*, 18205–18208.
132. Barkocy-Gallagher, G. A.; Cannon, J. G.; Bassford, P. J. *J. Biol. Chem.* **1994**, *269*, 13609–13613.
133. Bansal, A.; Gierasch, L. *Cell* **1991**, *67*, 1195–1201.
134. Eberle, W.; Sander, C.; Klaus, W.; Schmidt, B.; von Figura, K.; Peters, C. *Cell* **1991**, *67*, 1203–1209.

135. Imperiali, B.; Spencer, J. R.; Struthers, M. D. *J. Am. Chem. Soc.* **1994**, *116*, 8424–8425.
136. Venkatachalam, C. M. *Biopolymers* **1968**, *6*, 1425–1434.
137. Kabsch, W.; Sander, C. *Biopolymers* **1983**, *22*, 2577.
138. Wilmot, C. M.; Thornton, J. M. *J. Mol. Biol.* **1988**, *203*, 221–232.
139. Hutchinson, E. G.; Thornton, J. M. *Protein Sci.* **1994**, *3*, 2207–2216.
140. Brady, S. F.; Paleweda, W. J.; Arison, B. H.; Saperstein, R.; Brady, E. J.; Raynor, K.; Reisin, T.; Veber, D. F.; Freidinger, A. M. *Tetrahedron* **1993**, *49*, 3449.
141. Bohdal, U. W. Ph.D. Thesis, University of Zürich, **1996**.
142. Freidinger, R. M. *Peptides: Synthesis, Structure and Function*; Pierce Chem. Co.: Rockford, IL, **1981**, pp 673–683.
143. Freidinger, R. M.; Perlow, D. S.; Veber, D. F. *J. Org. Chem.* **1982**, *47*, 104.
144. Kemp, D. S.; McNamara, P. E. *Tetrahedron Lett.* **1982**, *23*, 3761.
145. Kemp, D. S.; McNamara, P. E. *J. Org. Chem.* **1985**, *50*, 5834.
146. Flynn, G. A.; Burkholder, T. P.; Huber, E. W.; Bey, P. *Bioorg. Med. Chem. Lett.* **1991**, *1*, 309.
147. Mueller, R.; Revesz, L. *Tetrahedron Lett.* **1994**, *35*, 4091–4092.
148. Colombo, L.; DiGiacomo, M.; Scolastico, C.; Manzoni, L.; Belvisi, L.; Molteni, V. *Tetrahedron Lett.* **1995**, *36*, 625–628.
149. Ben-Ishai, D.; McMurray, A. R. *Tetrahedron* **1993**, *49*, 6399.
150. Pellegrini, M.; Weitz, I. S.; Chorev, M.; Mierke, D. F. *J. Am. Chem. Soc.* **1997**, *119*, 2430–2436.
151. Monterrey, I. M. G.; González-Muñiz, R.; Herranz, R.; García-López, M. T. *Tetrahedron Lett.* **1995**, *51*, 2729–2736.
152. González-Muñiz, R.; Domínguez, M. J.; García-López, M. T. *Tetrahedron* **1992**, *48*, 5191–5198.
153. Baldwin, J. E.; Hulme, C.; Edwards, A. J.; Schofield, C. J. *Tetrahedron Lett.* **1993**, *34*, 1665–1668.
154. Figuera, N.; Alkorta, I.; Garcia-Lopez, T.; Herranz, R.; Gonzalez-Muniz, R. *Tetrahedron* **1995**, *51*, 7841.
155. Hinds, M. G.; Richards, N. G.; Robinson, J. A. *J. Chem. Soc., Chem. Commun.* **1447–1449**.
156. Kahn, M.; Wilke, S.; Chen, B.; Fujita, K.; Lee, Y.-H.; Johnson, M. E. *J. Mol. Recogn.* **1988**, *1*, 75–79.
157. Genin, M. J.; Ojala, W. H.; Gleason, W. B.; Johnson, R. L. *J. Org. Chem.* **1993**, *58*, 2334.
158. Genin, M. J.; Gleason, W. B.; Johnson, R. L. *J. Org. Chem.* **1993**, *58*, 860.
159. Olson, G. L.; Voss, M. E.; Hill, D. E.; Kahn, M.; Madison, V. S.; Cook, C. M. *J. Am. Chem. Soc.* **1990**, *112*, 323–333.
160. Nagai, U.; Sato, K. *Tetrahedron Lett.* **1985**, *26*, 647.
161. Curran, T. P.; McEnaney, P. M. *Tetrahedron Lett.* **1995**, *36*, 191–194.
162. DeLombaert, S.; Blanchard, L.; Stamford, L. B.; Sperbeck, D. M.; Grim, M. D.; Jenson, T. M.; Rodriguez, H. R. *Tetrahedron Lett.* **1994**, *35*, 7513–7516.
163. Esser, F.; Carpy, A.; Briem, H.; Köppen, H.; Pook, K.-H. *Int. J. Pept. Protein Res.* **1995**, *45*, 540–546.
164. Gramberg, D.; Weber, C.; Beeli, R.; Inglis, J.; Bruns, C.; Robinson, J. A. *Helv. Chim. Acta* **1995**, *78*, 1588–1606.
165. Kim, K. H.; Dumas, J. P.; Germanas, J. P. *J. Org. Chem.* **1996**, *61*, 3138–3144.
166. Slusarchyk, W. A.; Robl, J. A.; Taunk, P. C.; Asaad, M. M.; Bird, J. E.; DiMarco, J.; Pan, Y. *Bioorg. Med. Chem. Lett.* **1995**, *5*, 753–758.
167. Robl, J. A.; Karanewsky, D. S.; Asaad, M. M. *Tetrahedron Lett.* **1995**, *36*, 1593–1596.
168. vonRoedern, E. G.; Kessler, H. *Angew. Chem., Int. Ed. Engl.* **1994**, *33*, 687–689.
169. Nakanishi, H.; Chrusciel, R. A.; Shen, R.; Bertenshaw, S.; Johnson, M. E.; Rydel, T. J.; Tulinsky, A.; Kahn, M. *Proc. Natl. Acad. Sci. USA* **1992**, *89*, 1705–1709.
170. Virgilio, A. A.; Bray, A. A.; Zhang, W.; Trinh, L.; Snyder, M.; Morrissey, M. M.; Ellman, J. A. *Tetrahedron* **1997**, *53*, 6635–6644.
171. Kemp, D. S.; Stites, W. E. *Tetrahedron Lett.* **1988**, *29*, 5057–5060.
172. Subasinghe, N. L.; Khalil, E. M.; Johnson, R. L. *Tetrahedron Lett.* **1997**, *38*, 1317–1320.

173. Slomczynska, U.; Chalmers, D. K.; Cornille, F.; Smythe, M. L.; Beusen, D. D.; Moeller, K. D.; Marshall, G. R. *J. Org. Chem.* **1996**, *61*, 1198–1204.
174. Baldwin, J. E.; Hulme, C.; Schofield, C. J.; Edwards, A. J. *J. Chem. Soc., Chem. Comm.* **1993**, 935–936.
175. Tsang, K. Y.; Diaz, H.; Graciani, N.; Kelly, J. W. *J. Am. Chem. Soc.* **1994**, *116*, 3988–4005.
176. Feigel, M. *J. Am. Chem. Soc.* **1986**, *108*, 181.
177. Ripka, W. C.; DeLucca, G. V.; Bach, A. C.; Pottorf, R. S.; Blany, J. M. *Tetrahedron* **1993**, *49*, 3609.
178. Obrecht, D.; Abrecht, C.; Altorfer, M.; Bohdal, U.; Grieder, A. M. K.; Pfyffer, P.; Mueller, K. *Helv. Chim. Acta* **1996**, *79*, 1315–1337.
179. Obrecht, D.; Altorfer, M.; Lehmann, C.; Schoenholzer, P.; Mueller, K. *J. Org. Chem.* **1996**, *61*, 4080–4086.
180. Olson, G. L.; Bolin, D. R.; Bonner, M. P.; Bös, M.; Cook, C. M.; Fry, D. C.; Graves, B. J.; Hatada, M.; Hill, D. E.; Kahn, M.; Madison, V. S.; Rusiecki, V. K.; Sarabu, R.; Sepinwall, J.; Vincent, G. P.; Voss, M. E. *J. Med. Chem.* **1993**, *36*, 3039–3049.
181. Ernest, I.; Kalvoda, J.; Rihs, G.; Mutter, M. *Tetrahedron Lett.* **1990**, *31*, 4011–4014.
182. Kemp, D. S.; Li, Z. Q. *Tetrahedron Lett.* **1995**, *36*, 4175–4178.
183. Brandmeier, V.; Feigel, M. *Tetrahedron* **1989**, *45*, 1365–1376.
184. Obrecht, D.; Bohdal, U.; Daly, J.; Lehmann, C.; Schönholzer, P.; Müller, K. *Tetrahedron* **1995**, *51*, 10883–10900.
185. Wilmot, C. M.; Thornton, J. M. *Protein Eng.* **1990**, *3*, 479.
186. Obrecht, D.; Altorfer, M.; Lehmann, C.; Schönholzer, P.; Müller, K. *J. Org. Chem.* **1995**.
187. Obrecht, D.; Lehmann, C.; Ruffieux, R.; Schönholzer, P.; Müller, K. *Helv. Chim. Acta* **1995**, *78*, 1567–1587.
188. Carpino, L. A. *J. Am. Chem. Soc.* **1993**, *115*, 4397–4398.
189. Sarabu, R.; Lovey, K.; Madison, V.; Fry, D. C.; Greeley, D. N.; Cook, C. M.; Olson, G. L. *Tetrahedron* **1993**, *49*, 3629–3640.
190. Boger, D. L.; Patane, M. A.; Zhou, J. *J. Am. Chem. Soc.* **1995**, *117*, 7357–7363.
191. Beeli, R.; Steger, M.; Linden, A.; Robinson, J. A. *Helv. Chim. Acta* **1996**, *79*, 2235–2248.
192. Bisang, C.; Weber, C.; Robinson, J. A. *Helv. Chim. Acta* **1996**, *79*, 1825–1842.
193. Emery, F.; Bisang, C.; Favre, M.; Jiang, L.; Robinson, J. A. *J. Chem. Soc., Chem. Comm.* 2155–2156.
194. Pfeifer, M.; Linden, A.; Robinson, J. A. *Helv. Chim. Acta* **1997**, *80*, 1513–1527.
195. Klaus, W.; Broger, C.; Gerber, P.; Senn, H. *J. Mol. Biol.* **1993**, *232*, 897–906.
196. Senn, H.; Klaus, W. *J. Mol. Biol.* **1993**, *232*, 907–925.
197. Waldmeier, P. Thesis, Zurich, 1997.
198. Bannwart, W.; Gerber, F.; Grieder, A.; Knierzinger, A.; Muller, K.; Obrecht, D.; Trzceciak, A. *Can. Pat. Appl. CA2101599* 131 pages.
199. Abraham, N. A. (1991) *Tetrahedron Lett.* **1991**, *32*, 577.
200. Bellof, D.; Mutter, M. *Chimia* **1985**, *39*, 10.
201. Kemp, D. S.; Curran, T. P. *Tetrahedron Lett.* **1988**, *29*, 4935–4938.
202. Abrecht, C.; Obrecht, D.; Mueller, K.; Trzeciak, A. EP 0640618A1, 1995.
203. Altorfer, M. Ph.D. Thesis, University of Zürich, 1996.
204. Kimmich, R. Ph. D. Thesis, University of Zürich, 1997.
205. Toniolo, C.; Benedetti, E. *ISI Atlas Sci.: Biochem.* **1988**, 255.
206. Kemp, D. S.; Curran, T. P.; Davies, W. M.; Boyd, J. G.; Muendel, C. (1991) *J. Org. Chem.* **1991**, *56*, 6672–6682.
207. Kemp, D. S.; Curran, T. P.; Boyd, J. G.; Allen, T. J. *J. Org. Chem.* **1991**, *56*, 6683–6697.
208. Kemp, D. S.; Rothman, J. H. *Tetrahedron Lett.* **1995**, *36*, 4023–4026.
209. Kemp, D. S.; Rothman, J. H. *Tetrahedron Lett.* **1995**, *36*, 4019–4022.
210. Kemp, D. S.; Allen, T. J.; Oslick, S. L. *J. Am. Chem. Soc.* **1995**, *117*, 6641–6657.

211. Gerber, P. R.; Müller, K. *J. Computer-Aided Mol. Design* **1995**, *9*, 251–268.
212. Kogen, H.; Nishi, T. *J. Chem. Soc., Chem. Commun.* 311–312.
213. Paulsen, H.; Koebernick, W.; Autschbach, E. *Chem. Ber.* **1972**, *105*, 1524–1531.
214. Ruffieux, R. S. Ph.D. Thesis, University of Zürich, 1995.
215. Chucholowski, A.; Masquelin, T.; Obrecht, D.; Stadlwieser, J.; Villalgordo, J. M. *Chimia* **1996**, 530–532.
216. Masquelin, T.; Broger, E.; Müller, K.; Schmid, R.; Obrecht, D. *Helv. Chim. Acta* **1994**, *77*, 1395–1411.
217. Burk, M. *J. Am. Chem. Soc.* **1991**, *113*, 8518.
218. Obrecht, D. *Helv. Chim. Acta* **1989**, *72*, 447.
219. Masquelin, T.; Obrecht, D. *Tetrahedron Lett.* **1994**, *35*, 9387.
220. Masquelin, T.; Obrecht, D. *Synthesis* **1995**, 276–284.
221. Obrecht, D.; Gerber, F.; Sprenger, D.; Masquelin, T. *Helv. Chim. Acta* **1997**, *80*, 531.
222. Obrecht, D. *Helv. Chim. Acta* **1991**, *74*, 27.
223. Masquelin, T.; Obrecht, D. *Tetrahedron* **1997**, *53*, 641–646.
224. Obrecht, D., 1996. Unpublished results.
225. Mitsunobu, O.; Wada, M.; Sano, T. *J. Am. Chem. Soc.* **1972**, *94*, 679–680.
226. Obrecht, D.; Abrecht, C.; Grieder, A.; Villalgordo, J.-M. *Helv. Chim. Acta* **1997**, *80*, 65–72.
227. Yan, B. *J. Org. Chem.* **1995**, *60*, 5736.

RECENT ADVANCES IN THE MEDICINAL CHEMISTRY OF TAXOID ANTICANCER AGENTS

Iwao Ojima, Scott D. Kuduk, and Subrata Chakravarty

Advances in Medicinal Chemistry
Volume 4, pages 69–124.
Copyright © 1999 by JAI Press Inc.
All rights of reproduction in any form reserved.
ISBN: 0-7623-0064-7

ABSTRACT

Taxol® (paclitaxel) and Taxotère® (docetaxel) are currently considered to be the most promising leads in cancer chemotherapy. Both paclitaxel and docetaxel exhibit significant antitumor activity against various cancers, especially breast and ovarian cancers, which have not been effectively treated by existing chemotherapeutic drugs. The anticancer activity of these drugs is ascribed to their unique mechanism of action, i.e. causing mitotic arrest in cancer cells leading to apoptosis through inhibition of the depolymerization of microtubules. Although both paclitaxel and docetaxel possess potent antitumor activity, treatment with these drugs often results in a number of undesired side effects as well as multidrug resistance (MDR). Therefore, it has become essential to develop new anticancer agents with fewer side effects, superior pharmacological properties, and improved activity against various classes of tumors. This chapter describes the accounts of our research on the chemistry of paclitaxel and taxoid anticancer agents at the biomedical interface including: (i) the development of a highly efficient method for the semisynthesis of paclitaxel and a variety of taxoids by means of the β-Lactam Synthon Method (β-LSM), (ii) the structure–activity relationship (SAR) study of taxoids for their activities against human cancer cell lines, (iii) the discovery and development of "second-generation" taxoid anticancer agents that possess exceptional activities against drug-resistant cancer cells expressing the MDR phenotype as well as solid tumors (human cancer xenografts in mice), (iv) the development of fluorine-containing taxoids as a series of the second-generation taxoid anticancer agents and as excellent probes for the identification of bioactive conformation(s) of paclitaxel and taxoids by means of ^{19}F NMR in solution as well as in solid state for the microtubule–taxoid complex, (v) the development of radiolabeled photoreactive analogues of paclitaxel for photoaffinity labeling and mapping of the drug-binding domain on microtubules as well as P-glycoprotein that is responsible for MDR, and (vi) an SAR study of taxoids on their activities for inducing NO and tumor necrosis factor (TNF) through macrophage activation, which may be operative as an alternative mechanism of action. Thus, this chapter covers a wide range of issues associated with these powerful taxoid anticancer agents, discussing current status and future prospects.

I. INTRODUCTION

Taxol® (1, paclitaxel), a highly functionalized naturally occurring diterpenoid, is currently considered one of the most important drugs in cancer chemotherapy.[1-4] A tremendous amount of research focusing on the science and applications of paclitaxel has been performed since its initial isolation in 1966 from the bark of the Pacific yew tree (*Taxus brevifolia*) and subsequent structural determination in 1971.[5] Paclitaxel was approved by the FDA for the treatment of advanced ovarian cancer (December 1992) and metastatic breast cancer (April 1994). A semisynthetic analogue of paclitaxel, Taxotere® (2, docetaxel),[6] was also approved by the FDA in 1996 for the treatment of advanced breast cancer. These two taxane anticancer drugs are currently undergoing Phase II and III clinical trials worldwide for a variety of other cancers as well as for combination therapy with other agents.

Paclitaxel and docetaxel have been shown to act as spindle poisons, causing cell division cycle arrest, based on a unique mechanism of action.[7-10] These drugs bind to the β-subunit of the tubulin heterodimer, the key constituent protein of cellular microtubules (spindles). The binding of these drugs accelerates the tubulin polymerization, but at the same time stabilizes the resultant microtubules, thereby inhibiting their depolymerization. The inhibition of microtubule depolymerization between the prophase and anaphase of mitosis results in the arrest of the cell division cycle, which eventually leads to the apoptosis of cancer cells.

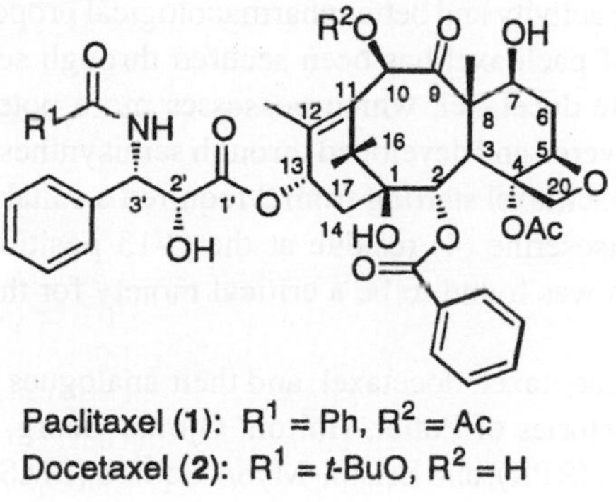

Paclitaxel (1): R¹ = Ph, R² = Ac
Docetaxel (2): R¹ = *t*-BuO, R² = H

In this chapter, we describe an account of our research on the chemistry and biology of paclitaxel and taxoid anticancer agents (taxoid = taxol-like compound). The topics covered in this chapter include (i) the development of a practical and efficient method for the semisynthesis of paclitaxel and docetaxel using chiral 3-hydroxy-β-lactams as synthetic intermediates, (ii) structure–activity relationship (SAR) studies of various taxoids that led to the discovery of the extremely potent "second-generation" taxoids, and (iii) biological and conformational studies with the use of fluorine-containing taxoids as "probes."

II. HIGHLY EFFICIENT SEMISYNTHESIS OF PACLITAXEL AND DOCETAXEL BY MEANS OF THE β-LACTAM SYNTHON METHOD

Clinical trials and the subsequent FDA approval of paclitaxel brought about a serious problem in the supply of this drug. Extraction of paclitaxel from its natural source, i.e. the bark of Pacific yew trees, was a cumbersome and low-yielding process. Furthermore, the removal of the bark requires cutting down the yew trees which are not so abundant. Accordingly, a long-term use of this drug cannot be secured once the supply of the yew trees becomes depleted. It was estimated in the late 1980s that the supply of the Pacific yew trees would be exhausted within 5 years at the rate of use of paclitaxel for clinical trials. Therefore, it was urgent and essential to explore all possibilities for securing the supply of paclitaxel. Although six total syntheses of paclitaxel have been reported in the past 4 years,[11–17] none of them are practical in terms of large-scale preparation, although they represent tremendous academic achievements. Fortunately, French scientists found that 10-deacetylbaccatin III (3, DAB), a diterpenoid which comprises the complex tetracyclic core of paclitaxel, could be isolated from the leaves of the European yew, *Taxus baccata*, in good yield (1 g per 1 kg of fresh leaves).[18] This was an extremely important discovery by Potier and coworkers since the yew leaves are readily renewable sources. Thus, an efficient and practical semisynthesis using 3 could ensure the long-term supply of paclitaxel and make it possible to develop novel taxoids with superior activity and better pharmacological properties than paclitaxel. In fact, the supply of paclitaxel has been secured through semisynthesis starting from DAB (3), while docetaxel, which possesses more potent cytotoxicity than paclitaxel, was discovered and developed through semisynthesis based on 3 as well. A semisynthesis of paclitaxel starting from 3 required a suitable preparation of the N-benzoyl-3-phenylisoserine (4) residue at the C-13 position with the (2′R,3′S) configuration, which was found to be a critical moiety for the cytotoxicity of the drug.[3,19–21]

Semisynthesis of paclitaxel, docetaxel, and their analogues has been extensively studied by the laboratories of Potier, Holton, Ojima, Georg, Swindell, Kingston, Rhône-Poulenc Rorer (RPR), and Bristol-Myers Squibb (BMS). Except for the very

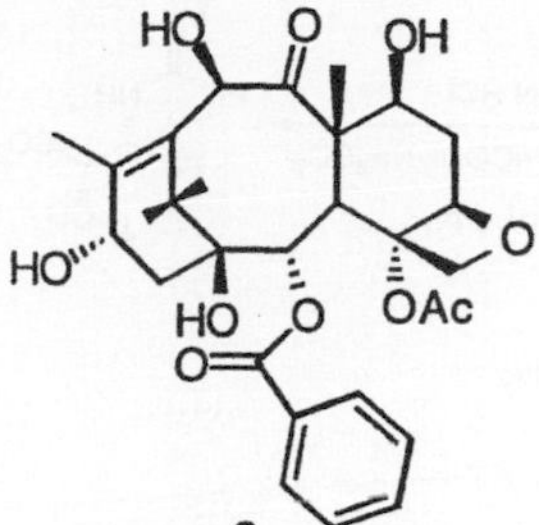

3

10-Deacetylbaccatin III (DAB)

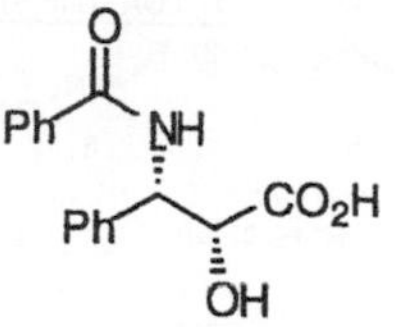

4

N-Benzoyl-(2*R*,3*S*)-3-
phenylisoserine

first semisynthesis of paclitaxel by Potier et al., which was based on the functionalization of 13-cinnamoylbaccatin III,[22] the semisynthesis of paclitaxel and docetaxel includes two elements: (i) asymmetric synthesis of the (2′*R*,3′*S*)-*N*-benzoyl-3-phenylisoserine moiety, and (ii) coupling of the phenylisoserine moiety with DAB (**3**). There have been numerous reports on the asymmetric syntheses of (2′*R*,3′*S*)-*N*-benzoyl-3-phenylisoserine and its congeners using a variety of synthetic methods.[23] On the contrary, there have been only a few methods for the coupling of the phenylisoserine moiety with DAB (**3**), i.e. Green–Potier esterification,[18] Holton oxazinone coupling,[23,24] Ojima-Holton β-lactam coupling,[23,25,26] Commerçon oxazolidinecarboxylic acid coupling,[27] and Kingston oxazolinecarboxylic acid coupling.[28] Among those coupling methods, the Ojima–Holton β-lactam coupling has been proved to be the most efficient and versatile method and thus most frequently used for the total synthesis of paclitaxel[11–15] and SAR study of taxoids. The Commerçon oxazolidinecarboxylic acid coupling has also been commonly used.

Since several excellent reviews on the semisyntheses of paclitaxel, docetaxel, and taxoids have been published,[6,23,29–32] we limit discussions here to our own contributions to the development of the efficient and practical semisyntheses of paclitaxel and its congeners.[25,26,32–46]

Our strategy was to apply the β-Lactam Synthon Method (β-LSM)[47–49] developed in these laboratories for (i) the asymmetric synthesis of the (2′*R*,3′*S*)-*N*-benzoyl-3-phenylisoserine moiety with excellent enantiomeric purity in high yield, and (ii) the ring-opening coupling of *N*-acyl-β-lactams with DAB (**3**).

After examining the feasibility of an asymmetric [2 + 2] ketene–imine cycloaddition route and an asymmetric ester enolate–imine cyclocondensation route, we chose the latter route for the efficient asymmetric synthesis of (3*R*,4*S*)-3-hydroxy-4-phenylazetidin-2-one and (2*R*,3*S*)-*N*-benzoyl-3-phenylisoserine. The cyclocondensation of the lithium enolate of (−)-(1*R*,2*S*)-2-phenyl-1-cyclohexyltriisopropylsiloxyacetate (**5a**: P = (*i*-Pr)$_3$Si (TIPS) and R* = (−)-(1*R*,2*S*)-2-phenyl-1-cyclohexyl) with *N*-trimethylsilylbenzaldimine (**6a**: R^1 = Ph) in THF at −78 °C gives (3*R*,4*S*)-3-triisopropylsiloxy-4-phenylazetidin-2-one (**7a**: P = TIPS and R^1 = Ph)

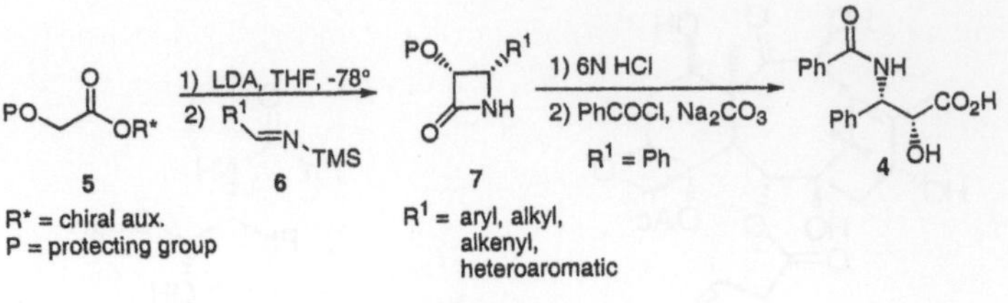

Scheme 1.

with 98% ee in more than 90% yield in one step under the optimized conditions (Scheme 1). The simple acidic hydrolysis of **7a**, followed by N-benzoylation by Schotten–Baumann procedure gives enantiopure $(2R,3S)$-N-benzoyl-3-phenylisoserine in excellent yield.[25,33,50] This asymmetric ester enolate–imine cyclocondensation method has proven to be very versatile and thus applicable to the efficient asymmetric syntheses of $(3R,4S)$- or $(3S,4R)$-3-hydroxyazetidin-2-ones bearing a variety of substituents at the C-4 position (Scheme 1).[49–53]

The coupling of N-benzoyl-2-O-EE-phenylisoserine (EE = ethoxyethyl) with 7-TES-baccatin III (TES = triethylsilyl) had been reported[18] by Green–Potier's group when we were investigating an efficient method for the introduction of $(2R,3S)$-N-benzoyl-3-phenylisoserine moiety at the sterically demanding hydroxyl group at the C-13 position of baccatin III. However, this coupling method suffered from low conversion in spite of using 5–6 equivalents of N-benzoyl-2-O-EE-phenylisoserine (**9**) to 7-TES-baccatin III (**8**) and considerable epimerization at the hydroxyl group at the C-2 position of the phenylisoserine moiety during the coupling (Scheme 2). The situation was essentially the same for the coupling of N-tert-butoxycarbonyl-2-O-EE-phenylisoserine with 7,10-di-Troc-DAB (**10**) (Troc = 2,2,2-trichloroethoxycarbonyl) for docetaxel synthesis.[18] Therefore, a new and efficient coupling method was necessary for the practical semisynthesis of paclitaxel and docetaxel.

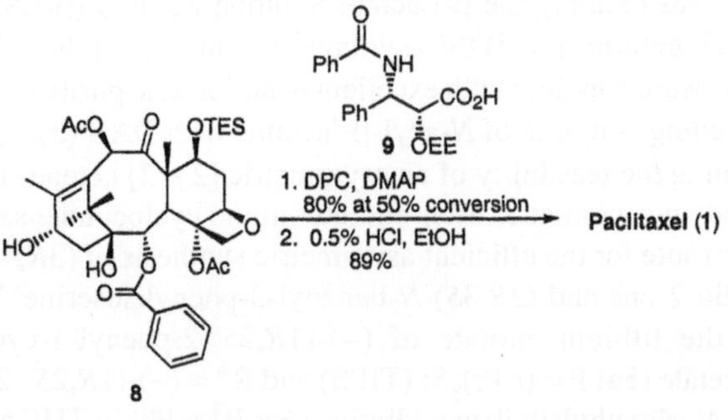

Scheme 2. Green–Potier coupling.

As mentioned above, we were planning to apply the β-LSM for the ring-opening coupling with DAB (3) at that time. Since monocyclic β-lactams are fairly stable against nucleophilic attack, we thought that it would be necessary to activate (3R,4S)-3-triisopropylsiloxy-4-phenylazetidin-2-one (7a) or its congener by acylating the N-1 position. Moreover, the retrosynthetic analysis indicated that 1-benzoyl-(3R,4S)-3-(protected hydroxy)-4-phenylazetidin-2-one (11) and 1-*tert*-butoxycarbonyl-(3R,4S)-3-(protected hydroxy)-4-phenylazetidin-2-one (12) would be ideal precursors of the *N*-acylphenylisoserine moieties of paclitaxel and docetaxel, respectively. Thus, we synthesized these *N*-acyl-β-lactams by acylating 7a with benzoyl chloride and di(*tert*-butyl)carbonate in excellent yields (Scheme 3).[25,26,34]

Holton claimed in a patent application that (3R,4S)-*N*-benzoyl-3-*O*-EE-β-lactam 11 (5 equiv), obtained through tedious classical optical resolution of racemic *cis*-3-hydroxy-4-phenylazetidin-2-one, could be directly coupled with 7-TES-baccatin III (8) in the presence of 4-dimethylaminopyridine (DMAP) and pyridine and the subsequent deprotection afforded paclitaxel in ca. 82% yield.[54] Although this procedure was proved to work by us and by others, the use of a large excess of β-lactam is obviously inefficient. Moreover, the Holton procedure did not work at all when *N*-*t*-Boc-β-lactam 12 was used for our attempted syntheses of docetaxel and its 10-acetyl analogue. This is due to the lack of reactivity of the *N*-*t*-Boc-β-lactam 12 toward the C-13 hydroxyl group of a protected baccatin III under the Holton conditions. The lack of reactivity is ascribed to the substantially weaker

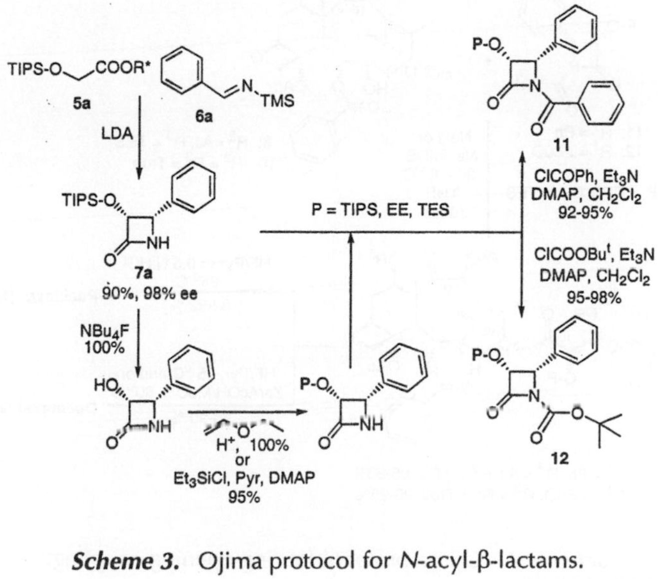

Scheme 3. Ojima protocol for *N*-acyl-β-lactams.

electron-withdrawing ability of the *tert*-butoxycarbonyl group than that of the benzoyl group.

In order to overcome this difficulty, we employed the 13-*O*-metalated derivatives of 7-TES-baccatin III (8) and 7,10-di-Troc-DAB (10). We examined NaH in THF and DME suspension,[25,26,34] BuLi, LDA, LiHMDS, NaHMDS, and KHMDS in THF solution, and found that NaHMDS is the best base for these ring-opening coupling of *N*-acyl-β-lactams with baccatins.[26,34] The ring-opening coupling proceeds smoothly at 0 °C for 8 or at –30 °C for 10 using only slight excess of an *N*-acyl-β-lactam (1.2 equiv) to a baccatin to give the coupling product 13a or 13b within 30 min in excellent yield, and the subsequent deprotection affords paclitaxel and docetaxel in high overall yields (Scheme 4).

Our protocol using NaH as the base[25] has been optimized by Georg's group and successfully applied to the syntheses of many paclitaxel analogues.[29,55] Our protocol using NaHMDS or LiHMDS as the base has been employed for the final step in paclitaxel total syntheses[13–15,17] as well as for the syntheses of numerous taxoids by us and other research groups worldwide. It should be noted that Holton and his collaborators independently developed a virtually identical ring-opening coupling process using 13-*O*-lithiated baccatin III with *N*-benzoyl-β-lactam 11 for paclitaxel synthesis, which was adopted by Bristol-Myers-Squibb for the commercial synthesis of paclitaxel.[23] Thus, the supply of paclitaxel for extensive clinical use was secured though the ring-opening coupling protocol developed by us and the Holton–BMS group. It is also worth mentioning that the Ojima–Holton protocol

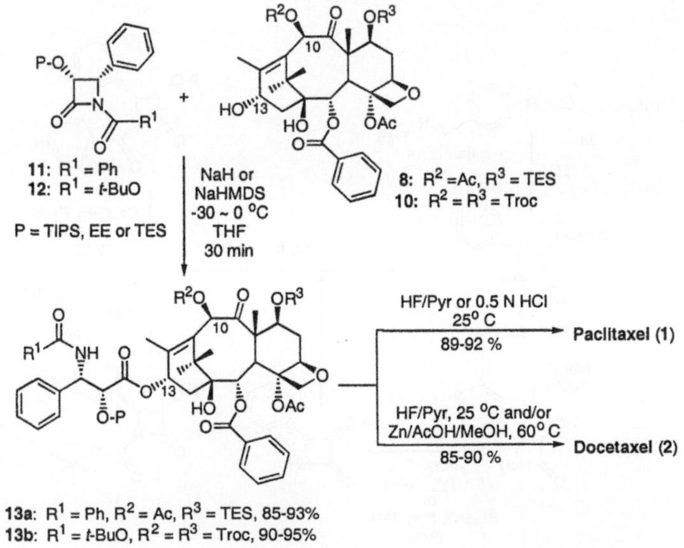

11: R^1 = Ph
12: R^1 = *t*-BuO

P = TIPS, EE or TES

NaH or
NaHMDS
-30 ~ 0 °C
THF
30 min

8: R^2 = Ac, R^3 = TES
10: R^2 = R^3 = Troc

HF/Pyr or 0.5 N HCl
25° C
89-92 % → Paclitaxel (1)

HF/Pyr, 25 °C and/or
Zn/AcOH/MeOH, 60° C
85-90 % → Docetaxel (2)

13a: R^1 = Ph, R^2 = Ac, R^3 = TES, 85-93%
13b: R^1 = *t*-BuO, R^2 = R^3 = Troc, 90-95%

Scheme 4. Ojima protocol for ring-opening coupling.

opened an extremely efficient and practical route to a diverse array of new taxoids, paving the way for extensive structure–activity relationship (SAR) studies of those taxoid anticancer agents.

Our extensive SAR study of paclitaxel and its congeners, i.e. taxoids, focused on the discovery of new taxoids exhibiting excellent cytotoxicity against different tumor types including those expressing the MDR phenotype, improved pharmacological properties, and reduced undesirable side effects. Our efforts toward the discovery and development of new taxoids possessing excellent anticancer activity against drug-resistant cancer cells and tumors have led to the development of the *second-generation taxoids*, which may be one of our most significant contributions to this field of research. The following sections will review our studies on the medicinal and biological chemistry of taxoid anticancer agents.

III. MODIFICATIONS AT THE 3′- AND 2-POSITIONS: FROM PACLITAXEL TO THE NONAROMATIC TAXOIDS

A. Cyclohexyl Analogues: Are Phenyl Groups Necessary for Anticancer Activity?

We began our SAR study by investigating the role of the phenyl rings at the C-2 and C-3′ positions of paclitaxel in its cytotoxicity and tubulin binding properties.[38] Accordingly, we examined the effect of the reduction of the aromatic ring to a cycloalkyl ring. 4-Phenyl-β-lactam **12** (P = EE) and 4-cyclohexyl-β-lactam **14** were coupled with 7,10-di-Troc-DAB (**10**) and 2-(hexahydro)-7,10-di-Troc-DAB (**15**) using the protocol discussed above to afford docetaxel analogues **16–18** bearing a cyclohexyl group in place of the C-3′ phenyl and/or a cyclohexanecarboxylate in place of the C-2 benzoate (Scheme 5).[38] The paclitaxel analogue **20**, bearing a cyclohexylcarboxylate group replacing the benzoate group at C-2, was obtained in a similar fashion by the coupling of the β-lactam **11** (P = TES) with 7-TES-2-(hexahydro)baccatin III (**19**) (Scheme 6). The β-lactam **14**, 7,10-di-Troc-baccatin **15** and 7-TES-baccatin **19** bearing cyclohexyl groups were prepared from β-lactam **7a** and

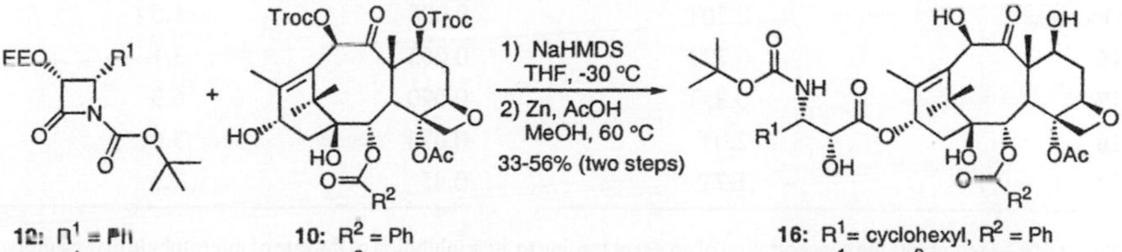

12: R^1 = Ph
14: R^1 = Cy

10: R^2 = Ph
15: R^2 = Cy

16: R^1= cyclohexyl, R^2 = Ph
17: R^1=Ph, R^2 = cyclohexyl
18: R^1= R^2 = cyclohexyl

Scheme 5.

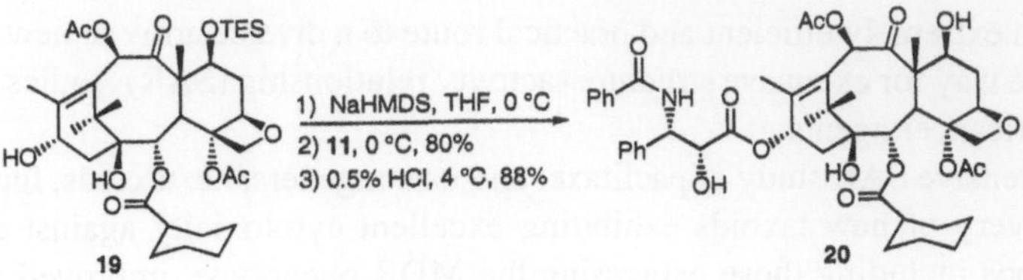

1) NaHMDS, THF, 0 °C
2) 11, 0 °C, 80%
3) 0.5% HCl, 4 °C, 88%

19   **20**

Scheme 6.

DAB (**3**) through efficient hydrogenation over rhodium on carbon.[38] Similar paclitaxel and docetaxel analogues containing cyclohexyl groups were independently reported by Georg and coworkers.[56]

Biological activities of the taxoids **16–18** and **20** were evaluated in three different assay systems, i.e. inhibition of microtubule disassembly, cytotoxicity against a murine P388 leukemia cell line, and a doxorubicin-resistant P388 cell line.[38] Results clearly indicate that a phenyl group at C-3′ or C-2 is not a requisite for strong binding to microtubules. As Table 1 shows, 3′-(cyclohexyl)docetaxel (**16**) (0.72T) and 2-(hexahydro)docetaxel (**17**) (0.85T) possess strong inhibitory activity for microtubule disassembly equivalent to that of docetaxel (0.7T), which is more potent than paclitaxel (1.0T). This finding has opened an avenue for development of new nonaromatic analogues of docetaxel and paclitaxel. 3′-Cyclohexyl-2-(hexahydro)docetaxel (**18**) turns out to have a substantially reduced activity. The cytotoxicities of **16–18** against P388 are, however, in the same range, i.e. 8–12 times weaker than docetaxel and 4–6 times weaker than paclitaxel. Thus, **18** shows

Table 1. Microtubule Disassembly Inhibitory Activity and Cytotoxicity of Cyclohexyl Analogues **16–18** and **20**

Taxoid	Microtubule Disassembly Inhibitory Activity[a] IC_{50}/IC_{50} (paclitaxel)	P388 Cell Line[b] ($\mu g/mL$)	P388/Dox[c] ($\mu g/mL$)
Docetaxel	0.70T	0.008	1.5
16	0.72T	0.063	3.1
17	0.85T	0.090	6.3
18	2.0T	0.076	3.6
20	1.7T	0.45	7.5

Notes: [a]IC_{50} represents the concentration of an agent leading to 50% inhibition of the rate of microtubule disassembly. IC_{50} (paclitaxel) is the IC_{50} value of paclitaxel in the same assay. In the same assay, the IC_{50} of paclitaxel is 0.015 μM.
[b]IC_{50} represents the concentration that inhibits 50% of cell proliferation.
[c]P388/Dox = doxorubicin resistant murine leukemia P388 cell line.

equivalent cytotoxicity to that of **16** or **17** in spite of much lower microtubule disassembly inhibitory activity. The cytotoxicities of these new taxoids against the P388/Dox cell line are only 2–2.5 times lower than that of docetaxel. The potency of 2-(hexahydro)paclitaxel (**20**) for these assays is much lower than that of the docetaxel counterpart **17**.[38]

B. Modifications at the 3'-Position: 3'-Alkyl and 3'-Alkenyl Taxoids

The promisingly high cytotoxicity of the cyclohexyl analogues **16–18** and **20** prompted us to further investigate the introduction of other nonaromatic substituents at the C-2 and C-3' positions. This section describes our findings on the effects of alkyl and alkenyl substituents at the C-3' position on cytotoxicity. A series of 3'-alkyl and 3'-alkenyl taxoids were synthesized in the same manner as described above using the β-LSM.[39] For the synthesis of 4-alkyl- and 4-alkenyl-β-lactams, the use of N-(4-methoxyphenyl)aldimines **21** instead of N-TMS-aldimines **6** was necessary in most cases to obtain high yields and enantioselectivity (Scheme 7). The 4-alkyl- and 4-alkenyl-β-lactams **22a–e** were subsequently coupled to baccat-

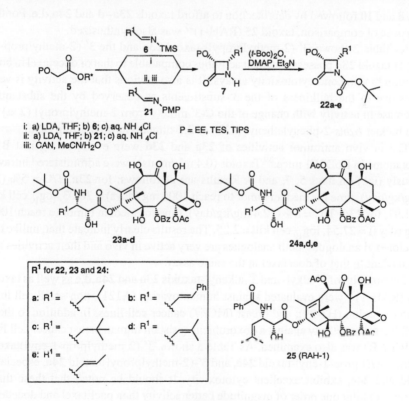

Scheme 7.

Table 2. Microtubule Disassembly Inhibitory Activity and Cytotoxicity of 3'-Alkyl and 3'-Alkenyl Taxoids

Taxoids	Microtubule Disassembly Inhibitory Activity[a] IC_{50}/IC_{50} (paclitaxel)	P388 Cell Line[b] IC_{50} (μM)	P388/Dox[b] IC_{50} (μM)
Docetaxel	0.70T	9.9	1.86
23a	0.78T	12.2	1.59
23b	1.45T	264	7.49
23c	1.45T	125	6.88
23d	0.64T	12.8	2.36

Notes: [a]IC_{50} represents the concentration of an agent leading to 50% inhibition of the rate of microtubule disassembly. IC_{50}(paclitaxel) is the IC_{50} value of paclitaxel in the same assay. In the same assay, the IC_{50} of paclitaxel is 0.015 μM.
[b]IC_{50} represents the concentration that inhibits 50% of cell proliferation.

ins **8** and **10** followed by deprotection to afford taxoids **23a–d** and **24a,d,e**. For the purpose of comparison, taxoid **25** (RAH-1)[57] was also synthesized.

As Table 2 shows, 3'-(2-methylpropyl)-taxoid **23a** and the 3'-(2-methylprop-1-enyl)-taxoid **23d** possess excellent activities comparable to that of docetaxel in both tubulin binding and cytotoxicity assays. It is also obvious that the activity is very sensitive to the bulkiness of the 3'-substituents as observed by the substantial decrease in activity with change of the C-3' moiety from 2-methylpropyl (**23a**) to the bulkier *trans*-2-phenylethenyl (**23b**) or 2,2-dimethylpropyl group (**23c**).

The in vivo antitumor activities of **23a** and **23d** were evaluated against B16 melanoma in B6D2F1 mice.[39] Taxoids (0.4 mL/mouse) were administered intravenously (i.v.) on days 5, 7, and 9. Results are as follows: for **23a**, T/C = 5% (20 mg/kg/day), time for median tumor to reach 1000 mg (days) = 26.03, $\log_{10}$ cell kill = 1.97; for **23d**, T/C = 8% (12.4 mg/kg/day), time for median tumor to reach 1000 mg (days) = 27.54, $\log_{10}$ cell kill = 2.25. The results clearly indicate that, unlike the cyclohexyl analogues, both analogues are very active in vivo and their activities are equivalent to that of docetaxel in the same assay.

Cytotoxicity of 3'-alkyl- and 3'-alkenyl-taxoids **23a** and **24a,d,e** as well as taxoid **25** (RAH-1)[57] were evaluated against human ovarian (A121), non-small cell lung (A549), colon (HT-29), and breast (MCF7) cancer cell lines. In addition to these cell lines, the activity against a doxorubicin-resistant human breast cancer cell line (MCF7-R) was also examined. As Table 3 shows, 3'-(2-methylprop-1-enyl)taxoid **24d**, 3'-(E)-prop-1-enyl-taxoid **24e**, and 3'-(2-methylpropyl)taxoid **24a**, especially **24d** and **24e**, exhibit excellent cytotoxicity. It should be noted that these three taxoids exhibit one order of magnitude better activity than paclitaxel and docetaxel against the doxorubicin-resistant MCF7-R. This finding provided significant infor-

Table 3. Cytotoxicity of 3′-Alkyl and 3′-Alkenyl Taxoids (IC$_{50}$, nM)[a]

Taxoid	A121[a] (ovarian)	A549[a] (NSCL)	HT-29[a] (colon)	MCF7[a] (breast)	MCF7-R[a,b] (breast)
Paclitaxel	6.1	3.6	3.2	1.7	300
Docetaxel	1.2	1.0	1.2	1.0	235
25 (RAH-1)	1.4	0.45	0.96	0.54	113
23a	1.9	0.70	0.50	0.80	107
24a	3.8	0.98	3.2	4.0	36
24d (SB-T-1212)	0.46	0.27	0.63	0.55	12
24e	0.90	0.54	0.76	0.51	14

Notes: [a]The concentration of compound which inhibits 50% (IC$_{50}$ nM) of the growth of human tumor cell line after 72 h drug exposure.
[b]MCF7-R mammary carcinoma cells, 180-fold resistant to doxorubicin.

mation for the development of newer antitumor agents effective against tumors expressing the MDR phenotype.

Next, we looked at the activity of **24d** (SB-T-1212) against doxorubicin-resistant (A2780-DX5), oxaliplatin-resistant (A2780-C25), and cisplatin-resistant (A2780-CP3) ovarian cancer cell lines (Table 4) in order to secure the relevance of using MCF7-R as the probe cell line to evaluate activity against drug-resistant cancer cell lines. As Table 4 shows, paclitaxel, docetaxel, and **24d** maintain their excellent activity against the oxaliplatin- and cisplatin-resistant cancer cells. However, paclitaxel and docetaxel suffer from a substantial (100–200 times) cross-resistance against the doxorubicin-resistant cancer cell line. In stark contrast to these two drugs, **24d** retains excellent activity (IC$_{50}$ = 5.7 nM). These results clearly indicate

Table 4. Cytotoxicity of Selected Antitumor Agents Against Ovarian and Drug-Resistant Ovarian Cancer Cell Lines (IC$_{50}$ nM)[a]

Taxoid	A2780-WT[b]	A2780-DX5[c]	A2780-C25[d]	A2780-CP3[e]
Cisplatin	450	280	1600	9000
Doxorubicin	5	357	63	56
Paclitaxel	2.7	547	3.4	4.1
Docetaxel	1.2	122	1.0	1.4
24d (SB-T-1212)	0.4	5.7	0.3	0.36

Notes: [a]See the footnote of Table 3.
[b]Human ovarian carcinoma.
[c]Doxorubicin-resistant ovarian carcinoma.
[d]Oxaliplatin-resistant ovarian carcinoma.
[e]Cisplatin-resistant ovarian carcinoma.

the validity of using the MCF7-R assay for the evaluation of the cytotoxicity of the new taxoids against drug-resistant cancer cells expressing the MDR phenotype.

The remarkable enhancement in activity against the drug-resistant human breast cancer cell line, MCF7-R, observed for taxoids 24 can be ascribed to a combination of three structural parameters, i.e. (i) an alkyl or alkenyl group with appropriate size at the C-3′ position, (ii) a *tert*-butoxycarbonyl group rather than a benzoyl group at the C3′-N position as evident from the comparison of 25 with 24d, and (iii) an acetyl group at C-10, which increases the activity by two- to threefold as observed in the comparison of 23a with 24a. This finding prompted us to examine the effects of C-10 modification of the 3′-alkyl and 3′-alkenyl taxoids on the cytotoxicity, and led us to discover a new series of exceptionally potent taxoids, *the second-generation taxoids*,[41] that possess two orders of magnitude higher potency than that of paclitaxel and docetaxel against MCF7-R cancer cells.

C. The Second-Generation Taxoids: Modifications at the C-10 Position of 3′-Alkyl and 3′-Alkenyl Taxoids

We synthesized a series of taxoids with various substituents at C-10 including ether, ester, carbamate, and carbonate functional groups to examine the effects of the C-10 modification on the activity against MCF7-R.[41] The C-10 modifications of 7-TES-DAB (8) were carried out with acyl, alkoxycarbonyl, N,N-dialkylcarbamoyl, and alkyl halides using LiHMDS as the base.[58,59] The reactions proceeded uneventfully to give the corresponding 10-modified 7-TES-DABs (26a–m) in good yields (Scheme 8). This procedure was not applicable when N-alkyl- and N-allylisocyanates were used as electrophiles.[60] Accordingly, we employed cuprous chloride as the activator[61] to obtain the desired 10-carbamoyl-7-TES-DABs (26n–s) in moderate to high yields (Scheme 8). Coupling of these 10-modified DABs with β-lactam 22d gave adducts 27a–s, and the subsequent deprotection afforded taxoids 28a–s. Taxoids 29a–e,s were obtained quantitatively by hydrogenation of the 2-methylprop-1-enyl moiety of taxoids 28a–c,s over palladium on carbon (Scheme 9).

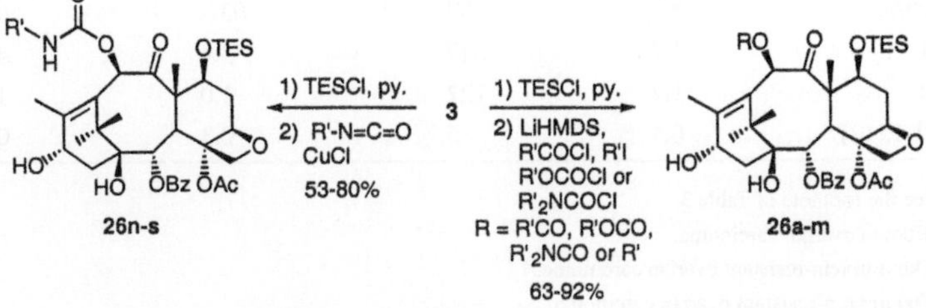

Scheme 8.

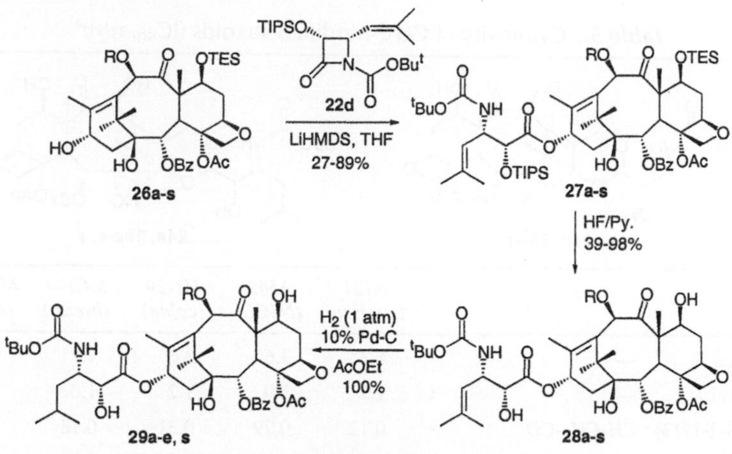

Scheme 9.

As Table 5 shows, taxoids **28** and **29** possess excellent cytotoxicity against human cancer cell lines, superior to those of paclitaxel and docetaxel except for taxoid **29s**. A number of these taxoids show one order of magnitude stronger activity than paclitaxel and docetaxel. The most cytotoxic taxoid against normal human cancer cell lines in this assay is 10-morpholine-*N*-carbonyl taxoid **28m** which exhibits an IC_{50} value of 0.09 nM against ovarian (A121) and breast (MCF7) cancer cell lines. The most significant result, however, is the remarkable activity ($IC_{50} = 2.1–9.1$ nM) against the drug-resistant cancer cell line, MCF7-R, exhibited by a number of the 10-modified taxoids. Among these, the three new taxoids, **28a** (SB-T-1213), **28b** (SB-T-1214), and **29a**, are exceptionally potent, possessing *two orders of magnitude better activity than paclitaxel and docetaxel.* Because of the observed exceptional activity against MCF7-R expressing the MDR phenotype, which is distinct from paclitaxel and docetaxel, we call these new taxoids "second-generation taxoids." It is noteworthy that the observed exceptional activity is clearly ascribed to the modification at C-10 by comparing the activities of these second-generation taxoids with the corresponding 10-acetyl taxoids **24d** (SB-T-1212, $IC_{50} = 12$ nM) and **24a** ($IC_{50} = 36$ nM). The observed substantial effects of C-10 modification on cytotoxicity forms a sharp contrast to other reports concluding that the C-10 hydroxyl deletion or acylation does not modulate the activity.[55,58,62]

The SARs of the taxoids **28** and **29** *against MCF7-R* show that the bulkiness of the C-10 substitution exhibits considerable effects on cytotoxicity. It appears that the propanoyl and cyclopropanecarbonyl groups are optimal as demonstrated by the exceptionally high potency of **28a** (SB-T-1213), **28b** (SB-T-1214), and **29a**. As the size of the C-10 modifier increases, the activity shows a fairly steady decline as exemplified by the comparison of 10-dimethylcarbamoyl taxoid **28c** ($IC_{50} = 4.9$

Table 5. Cytotoxity of C-10-Modified Taxoids (IC$_{50}$, nM)[a]

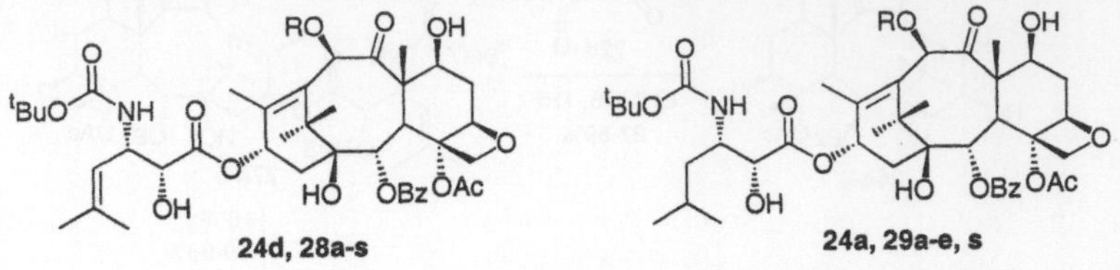

Taxiod	R	A121 (ovarian)	A549 (NSCL)	HT-29 (colon)	MCF-7 (breast)	MCF7-R (breast)
Paclitaxel	—	6.3	3.6	3.6	1.7	299
Docetaxel	—	1.2	1.0	1.2	1.0	235
28a (SB-T-1213)	CH$_3$CH$_2$-CO	0.12	0.29	0.31	0.18	2.2
28b (SB-T-1214)	cyclopropane-CO	0.26	0.57	0.36	0.20	2.1
28c	(CH$_3$)$_2$N-CO	0.30	0.60	0.5	0.13	4.9
28d	CH$_3$O-CO	0.23	0.28	0.30	0.14	5.3
28e	CH$_3$	0.70	0.90	0.90	0.37	123
28f	CH$_3$(CH$_2$)$_3$-CO	0.60	0.60	0.60	0.50	30
28g	CH$_3$(CH$_2$)$_4$-CO	0.50	0.57	0.90	0.40	17.3
28h	(CH$_3$)$_2$CHCH$_2$-CO	0.60	0.40	0.60	0.40	8.5
28i	(CH$_3$)$_3$CCH$_2$-CO	0.34	0.50	0.50	0.40	10.5
28j	cyclohexane-CO	0.50	0.75	1.05	0.46	22.3
28k	(E)-CH$_3$CH=CH-CO	0.45	1.70	0.60	0.26	3.4
28l	(CH$_3$CH$_2$)$_2$N-CO	0.17	0.20	0.40	0.20	9.1
28m	morpholine-4-CO	0.09	0.22	0.18	0.09	12.5
28n	CH$_3$NH-CO	1.2	0.7	1.2	0.4	162
28o	CH$_3$CH$_2$NH-CO	0.29	0.15	0.41	0.31	87
28p	CH$_3$CH$_2$CH$_2$NH-CO	0.34	0.22	0.55	0.36	65
28q	(CH$_3$)$_2$CHNH-CO	0.34	0.36	0.52	0.46	76
28r	CH$_2$=CHCH$_2$NH-CO	0.44	0.35	0.53	0.44	48
28s	cyclohexyl-NH-CO	0.28	0.20	0.46	0.33	20
29a	CH$_3$CH$_2$-CO	0.41	0.53	0.53	0.35	2.8
29b	cyclopropane-CO	0.51	1.1	0.78	0.51	4.3
29c	(CH$_3$)$_2$N-CO	0.40	0.50	0.60	0.36	5.8
29d	CH$_3$O-CO	0.18	0.35	0.44	0.28	6.4
29e	CH$_3$	0.70	0.60	0.60	0.33	214
29s	cyclohexyl-NH-CO	5.0	3.7	5.0	1.6	33

Note: [a]See the footnote of Table 3.

nM) with 10-diethylcarbamoyl taxoid **28l** (IC_{50} = 9.1 nM) as well as 10-hexanoyl taxoid **28g** (IC_{50} = 17.3 nM) with 10-cyclohexanecarbonyl taxoid **28j** (IC_{50} = 22.3 nM). However, it is worthy of note that the electronic and conformational factors also exert substantial influence on the activity as observed in the comparison of 10-cyclohexanecarbonyl taxoid (**28j**, IC_{50} = 22.3 nM) with 10-morpholine-*N*-carbonyl taxoid (**28m**, IC_{50} = 12.5 nM). It should also be pointed out that **28m** is not among the most potent analogues active against MCF7-R, although its activity against normal cancer cell lines is the highest in the series of 10-modified taxoids. 10-Methyl taxoids **28e** (IC_{50} = 123 nM) and **29e** (IC_{50} = 214 nM) resulted in two orders of magnitude decrease in activity. This result implies the importance of the carbonyl functionality at this position for exceptionally high activity. Although 10-*N,N*-dialkylcarbamoyl taxoids **28c** and **28l** show nanomolar-level IC_{50} values, taxoids bearing *N-mono*alkylcarbamoyl and *N-mono*allylcarbamoyl groups at C-10, **28n–s** and **29s**, show, uniformly, a substantial decrease in activity (20–162 nM). It should be noted that cyclohexylcarbamoyl, a bulky carbamoyl group, brings about better activity than small size carbamoyl groups as exemplified by the comparison of **28s** (IC_{50} = 20 nM) or **29s** (IC_{50} = 33 nM) with **28n** (R = CH_3NHCO, IC_{50} = 162 nM).

It is worthy of note that the activity against MCF7-R expressing MDR phenotype is very sensitive to the structure of the C-10 modifier, while this structural variation generally has little effect on the activity against the normal cancer cell lines. The fact clearly indicates that there is no apparent correlation between the activities of the taxoids against the normal cancer cell lines and the drug-resistant cell line.

Multidrug resistance (MDR), a common and serious problem in cancer chemotherapy, is a property by which tumor cells become resistant to various hydrophobic cytotoxic agents through the action of P-glycoprotein, a membrane bound protein that is overexpressed in MDR cancer cells. P-Glycoprotein acts as an efflux pump that binds and eliminates cytotoxins from the cell membrane and cytosol, thereby keeping the intracellular drug concentration at innocuous levels.[63–65] It is reasonable to assume that the observed SAR of the 10-modified taxoids that is unique to the drug-resistant cell line MCF7-R is related, at least in part, to the binding ability of these taxoids to P-glycoprotein. All the taxoids **28** and **29**, except **29s**, possess subnanomolar level IC_{50} values against the normal cancer cell lines (as an average value for four cell lines). Clearly most of the C-10 modifications studied are well tolerated for cytotoxicity. On the contrary, the binding of these taxoids to P-glycoprotein is strongly affected by the structure of the C-10 modifier, i.e. it appears that the C-10 position is crucial for P-glycoprotein to recognize and bind taxoid anticancer agents. To elucidate the role of P-glycoprotein in MDR, we have developed efficient photoaffinity labels, i.e. tritiated photoreactive paclitaxel analogues (see Section VI).[66–68] We are also actively developing novel and effective taxane-based MDR reversal agents that are capable of interfering with the action of P-glycoprotein and restore the concentrations of the anticancer agents in the cytosol of drug-resistant cancer cells.[69]

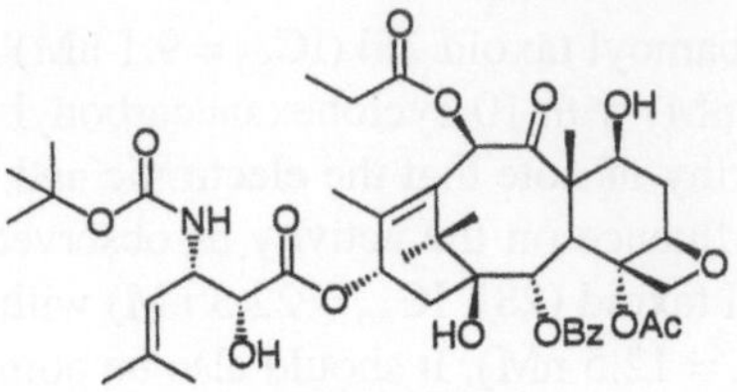

28a (SB-T-1213)

The in vivo antitumor activity of **28a** (SB-T-1213) was evaluated against B16 melanoma implanted subcutaneously in B6D2F$_1$ mice.[41,70] The results indicated that **28a** (SB-T-1213) is extremely active in vivo against this tumor. Although docetaxel showed a higher log cell kill value at optimal dosage, **28a** is more potent than docetaxel on a mg/kg basis, reflecting the exceptional activity of this agent in the in vitro cell line assay (*vide supra*). Based on this encouraging result, several of the second-generation taxoids are currently under evaluation for their in vivo activity against a series of drug-resistant tumors (human cancer xenografts) in nude mice.

D. Further SAR Study on 3′-Modified Taxoids: Novel Taxoids Bearing Methano- and Oxanorstatine Residues

In order to study further the steric and electronic effects at C-3′ on cytotoxicity, we synthesized analogues of taxoid **23d** in which the alkene moiety was converted to the corresponding cyclopropane or epoxide. First, we developed highly efficient synthetic routes to enantiopure β-lactams bearing 2,2-dimethylcyclopropyl (**30**) and 2,2-dimethyl-1,2-oxapropyl (**31**) groups at C-4.[71] The reaction of N-PMP-3-hydroxy-β-lactam **33** with Et$_2$Zn and CH$_2$I$_2$ (10 equiv) in 1,2-dichloroethane afforded 4-((S)-2,2-dimethylcyclopropyl)-β-lactam **34** as the single isomer in 93% yield (Scheme 10). The subsequent protection of the C-3 hydroxyl group as TIPS ether, oxidative cleavage of the PMP group, and t-Boc protection gave β-lactam **30** in high yield. The epoxidation of β-lactam **22d** with m-chloroperoxybenzoic acid (m-CPBA) afforded a 1:1 mixture of **31-S** and **31-R** which were readily separated by flash chromatography on silica gel (Scheme 10). In sharp contrast with this, the reaction of N-t-Boc-3-hydroxy-β-lactam **36** with m-CPBA gave 4-((R)-2-methyl-1,2-epoxypropyl)-β-lactam **32** as the sole product in 92% yield (Scheme 10).

The stereochemistry of the cyclopropyl as well as oxirane moiety in these β-lactams was determined based on the single crystal X-ray analysis of β-lactam **34** and the 4-nitrobenzoyl derivative (**37**) of 3-hydroxy-β-lactam **32**.[71] The extremely high diastereoselectivity observed in these cyclopropanation and epoxidation reactions can be explained by taking into account the highly organized transition state structures illustrated in Figure 1. The directing effect of the C-3

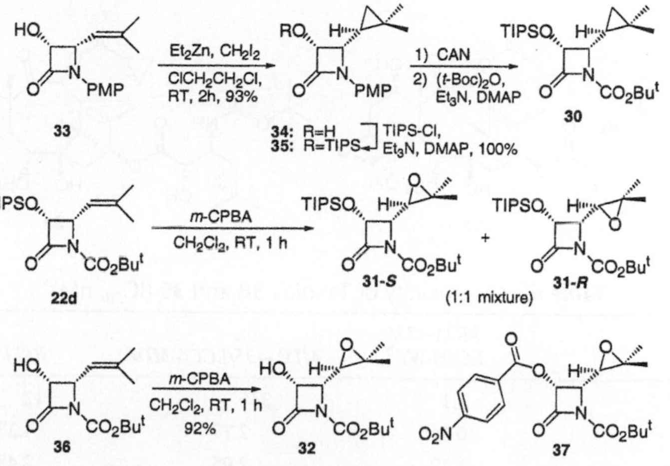

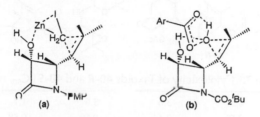

Scheme 10.

hydroxyl group, by virtue of the constrained β-lactam skeleton, is maximized to achieve the exclusive formation of β-lactams **34** and **32**.

Novel taxoids bearing the methanonorstatine or oxanorstatine residue at the C-13 position of baccatin were synthesized through coupling of β-lactam **30** or **31** with baccatins **8** and **26b** using our standard coupling conditions (*vide supra*). The cytotoxicity of new taxoids **38** and **39** bearing the methanonorstatine residue were evaluated against a human tumor breast carcinoma cell line (MDA-435/LCC6-WT) and an MDR1 transduced cell line (MDR-435/LCC6-MDR).[71] These taxoids exhibited extremely high potency, particularly against the drug-resistant cell line (Table 6). It is worthy of note that the ratio of the activities against drug-resistant cells and drug-sensitive cells of 2.48 (resistant/sensitive) for **39** is the best ratio ever reported to date for any taxoid. The cytotoxicity of these taxoids against a number of other cancer cell lines is currently underway.

Figure 1. Proposed transition state structure for the formation of β-lactams **34** (a) and **32** (b).

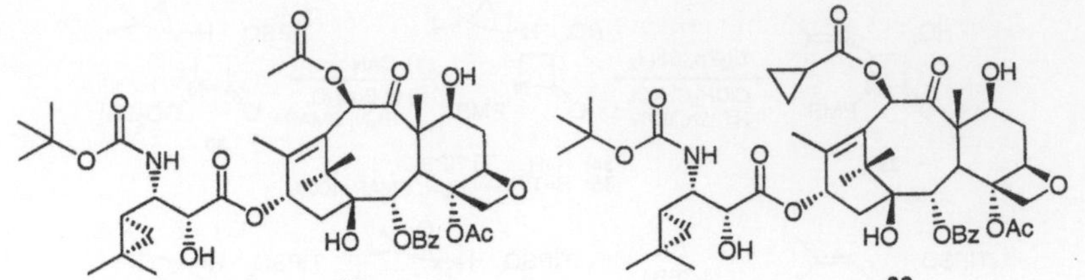

38   39

Table 6. Cytotoxicity of Taxoids **38** and **39** (IC$_{50}$, nM)[a]

Taxoid	MDA-435/ LCC6-WT[b]	MDA-435/LCC6-MDR[c]	R/S Ratio[d]
Paclitaxel	3.1	346	112
38	0.64	2.77	4.33
39	1.19	2.95	2.48

Notes: [a]See the footnote of Table 3.
     [b]Human breast carcinoma cell line.
     [c]Doxorubicin-resistant human breast carcinoma cell line.
     [d]Ratio of cytotoxicity (drug-resistant cell line)/(drug-sensitive cell line).

The cytotoxicity of taxoids **40-R** and **40-S** bearing the oxanorstatine residue were evaluated in our standard five human cancer cell line assay.[71] As Table 7 shows, these two stereoisomers were found to possess considerably different cytotoxicity, which is rather unexpected. Taxoid **40-R** showed a 3- to 10-fold increase in activity as compared to paclitaxel, while the other isomer, taxoid **40-S**, exhibited a 2- to 4-fold decrease in activity, i.e. one to two orders of magnitude difference in activity is observed just by changing the stereochemistry of the epoxide moiety at C-3'. This

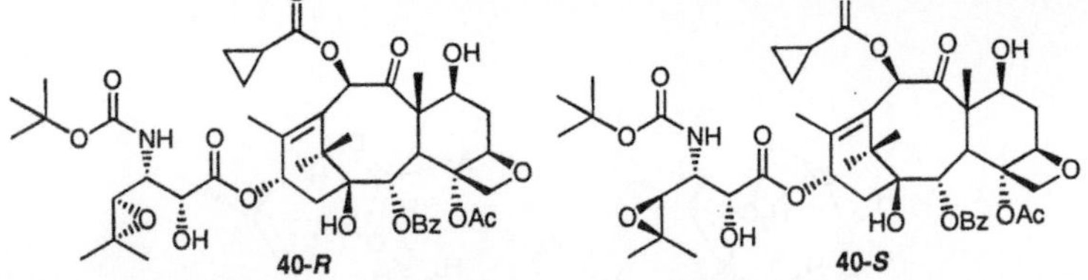

40-R   40-S

Table 7. Cytotoxicity of Taxoids **40-R** and **40-S** (IC$_{50}$, nM)[a]

Paclitaxel	6.3	3.6	3.6	1.7	299
40-R	0.63	0.44	0.59	0.58	72
40-S	21.7	14.5	8.37	6.50	1066

Note: [a]See the footnote of Table 3.

observation is very intriguing and confirms that the cytotoxicity of taxoids is highly sensitive to the structure of substituents at the C-3' position.

E. Modifications at the C-2 Benzoate Moiety: Nonaromatic Taxoids

The improved activity of 3'-alkyl and 3'-alkenyl taxoids described above clearly establishes the dispensability of aromatic character at the C-3' position. Next, we investigated the effects of C-2 modification on cytotoxicity by replacing the C-2 benzoate moiety with nonaromatic ester groups. Replacement of the 2-benzoate with simple alkyl and alkenyl esters in conjunction with modification at C-3' provides a series of novel taxoids devoid of all the aromatic groups of paclitaxel and docetaxel.[44]

With regard to the modification of the C-2 benzoate, several methods were reported for the introduction of other aryl or hetero-aromatic groups. However, those methods turned out to be not compatible to our purpose, which required modifications at C-2 as well as C-3'. The successful routes to 2-modified baccatins **44** and **45** using 13-oxo-7-TES-DAB (**41**)[72] are illustrated in Scheme 11.[44] We also used 7-TES-(hexahydro)baccatin III (**19**) for the synthesis of 3'-modified 2-cyclohexanecarbonyl taxoids.

The ring-opening coupling of β-lactams **22a,d,e** and **12** (P = TIPS) with baccatins **44, 45**, and **19** under the standard conditions (*vide supra*), followed by deprotection, afforded taxoids **46–52** in 51–93% yields.[44] Taxoid **53** was obtained in quantitative

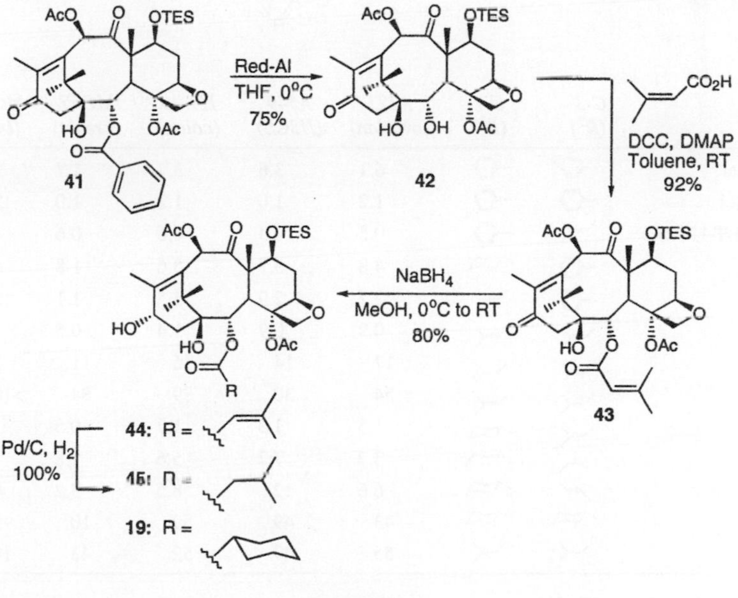

Scheme 11.

yield via hydrogenation of the 3′-phenyl ring of **46** over Pt/C under ambient pressure of hydrogen in EtOAc.[44] Taxoids **54** and **55** were prepared in a similar manner from taxoids **50** and **51**, respectively, through hydrogenation over Pd/C.[44]

The cytotoxicity of taxoids **46–55** were evaluated against our standard five human cancer cell lines. Results are summarized in Table 8 with the values for paclitaxel, docetaxel, and taxoid **24d** (SB-T-1212) for comparison.

As Table 8 shows, several of the new taxoids exhibit similar or enhanced activity as compared to paclitaxel and docetaxel. Taxoid **48** (SB-T-104221: nonataxel) with 2-methylprop-1-enyl groups at both C-3′ and C-2 is more active than both paclitaxel and docetaxel, clearly establishing that a benzoate group at C-2 is not necessary for strong cytotoxicity in lieu of the appropriate alkyl, alkenyl, or cyclohexyl group. Taxoids **49–51** bearing a combination of 2-methylprop-1-enyl and 2-methylpropyl groups at C-3′ and C-2 were, however, all less potent than **48**. The results clearly indicate the considerable sensitivity for the modification at these positions. Based

Table 8. Cytotoxicity of Non-Aromatic Taxoids (IC$_{50}$, nM)[a]

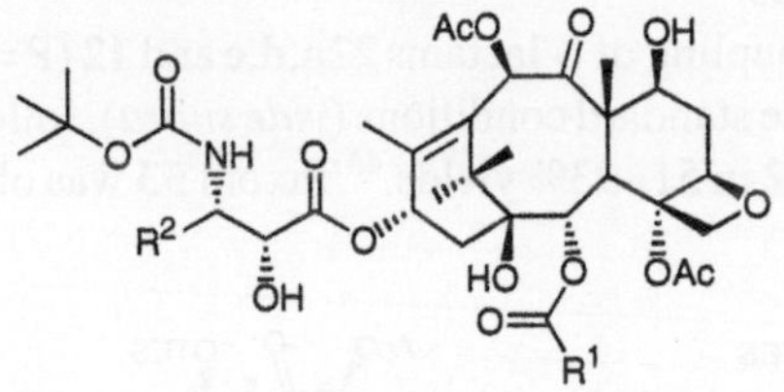

Taxoid	C-3′ (R²)	C-2 (R¹)	A121 (ovarian)	A549 (NSCL)	HT-29 (colon)	MCF7 (breast)	MCF7-R (breast)
Paclitaxel			6.1	3.6	3.2	1.7	299
Docetaxel		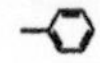	1.2	1.0	1.2	1.0	235
24d (SB-T-1212)		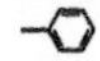	0.5	0.3	0.6	0.6	12
46			4.8	5.7	5.6	1.8	412
47			2.7	2.9	3.5	1.1	140
48			0.9	0.9	1.4	0.5	38
49			17	14	16	11	589
50			54	30	89	84	>1000
51			1.5	1.3	1.7	0.5	26
55		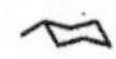	3.2	7.7	5.6	1.8	57
52			6.6	12	8.5	3.2	446
53	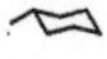		43	49	59	10	591
54		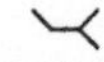	55	47	52	43	1081

Note: [a] See the footnote of Table 3.

on the comparison of **51–53**, **55**, and **46**, it is clear that the substituent at the C-3′ position plays a predominant role in determining the activity in the case of 2-cyclohexanecarbonyl taxoids. For example, taxoid **51** shows better activity than docetaxel, especially against MCF7 and MCF7-R.

Although the size of the group at C-3′ is found to exert marked effects on the cytotoxicity, the C-2 substituent has prominent effects on the activity as well. While our earlier studies showed that a cyclohexanecarbonyl group at C-2 diminishes the activity (see Section III.A),[38,56] the placement of the proper alkyl or alkenyl group at the C-3′ position overrides this negative effect and brings about an increase in activity comparable to that of docetaxel. This is clearly seen from the results for taxoids **51**, **52**, and **55**. Thus, a rigid moiety with an appropriate size, such as a ring or an olefin, at the C-3′ and C-2 positions appears to be a requisite for strong activity. In fact, the introduction of alkyl groups that can freely rotate such as 2-methylpropyl to these positions (taxoids **49**, **50**, and **54**), discernibly reduces the activity.

The requirement for a certain level of rigidity of the C-2 and C-3′ substituents led us to investigate the solution conformation of taxoid **48** (nonataxel) in aqueous media by 2-D NMR and molecular modeling techniques. We found that the two 2-methylprop-1-enyl moieties of this molecule orient themselves with respect to each other quite specifically to form a strong hydrophobic clustering (see Section IV.C).[73]

IV. FLUORINE-CONTAINING TAXOIDS: SAR STUDY AND FLUORINE PROBE APPROACH TO THE IDENTIFICATION OF BIOACTIVE CONFORMATIONS

In the course of our study on the rational design, synthesis and SAR of new antitumor taxoids, we became interested in incorporating fluorine(s) into paclitaxel and taxoids in order to investigate the effects of fluorine on cytotoxicity, blocking of known metabolic pathways, and its use as a probe for identifying bioactive conformation(s) of taxane antitumor agents.

Studies on the metabolism of paclitaxel have shown that the *para* position of the C-3′ phenyl, the *meta* position of the C-2 benzoate, the C-6 methylene, and the C-19 methyl groups are the primary sites of hydroxylation by the P450 family of enzymes.[74,75] The predominant one among these is the hydroxylation of the C-3′ phenyl at the *para* position, presumably by the cytochrome 3A family.[74] Substitution of a C–H bond with a C–F bond has been shown to substantially slow down the enzymatic oxidation in general.[76] Accordingly, we synthesized a series of fluorine-containing taxoids with the expectation of obtaining analogues with improved metabolic stability and cytotoxicity.[40,45] Also discussed in this section is the successful use of fluorine-containing taxoids as probes for the study of solution and solid-state (microtubule-bound) conformations of paclitaxel. The conformational analyses of fluorine-containing taxoids based on VT-^{1}H NMR, VT-^{19}F NMR, and molecular modeling have provided extremely valuable information towards the identification of the recognition and binding conformations of paclitaxel at its binding site on the microtubules.[77,78]

A. Fluorine-Containing Taxoids: Synthesis and Biological Activity

4-(4-Fluorophenyl)-β-lactams **56a** and **56b** were synthesized in excellent yields with high enantiopurity through the asymmetric ester enolate–imine cyclocondensation discussed in Section II, which uses (–)-*trans*-2-phenylcyclohexanol[79] as the chiral auxiliary. However, this chiral auxiliary did not work efficiently for the asymmetric synthesis of *N-t*-Boc-trifluoroalkyl-β-lactam **57**, giving the intermediate β-lactam **58** with only 23% ee. Fortunately, the use of (–)-10-dicyclohexylsulfamoyl-D-isoborneol[80] as the chiral auxiliary solved the problem, giving **58** with 93% ee (Scheme 12), which was readily converted to **57** in good overall yield.[40]

4-Trifluoromethyl-β-lactam **63** was prepared in racemic form via a ketene–imine [2 + 2] cycloaddition, following previously published methods[81] with modifications.[46] The subsequent oxidative cleavage of PMP, acylation of NH with $(t\text{-Boc})_2O$, hydrogenolysis over Pd/C, and protection as 1-ethoxyethyl ether gave *N-t*-Boc-β-lactam **63** in good overall yield (Scheme 13).

Fluorine-containing taxoids **64–69** were synthesized in good yields through the ring-opening coupling of β-lactams **56** and **57** with baccatins **8**, **10**, and 7,10-di-TES-DAB (**70**) under the standard conditions (*vide supra*). Cytotoxicity of taxoids **64–69** was assayed against our standard five human cancer cell lines (Table 9). As Table 9 shows, F-docetaxel analogues, **66** and **67**, possess excellent activity against

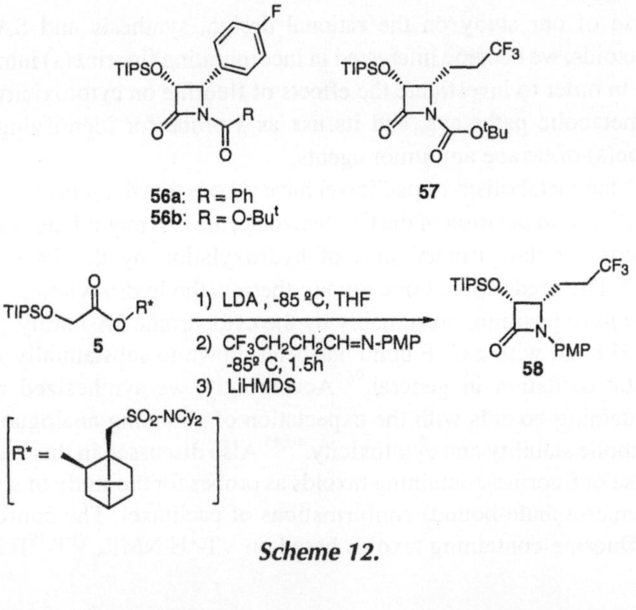
F
TIPSO
O
N
R
O
CF3
TIPSO
O
N
O-tBu
O
56a: R = Ph
56b: R = O-Bu^t
57
TIPSO
O
R*
5
1) LDA, -85 ºC, THF
2) CF3CH2CH2CH=N-PMP
-85º C, 1.5h
3) LiHMDS
CF3
TIPSO
O
N
PMP
58
SO2-NCy2
R* =

Scheme 12.

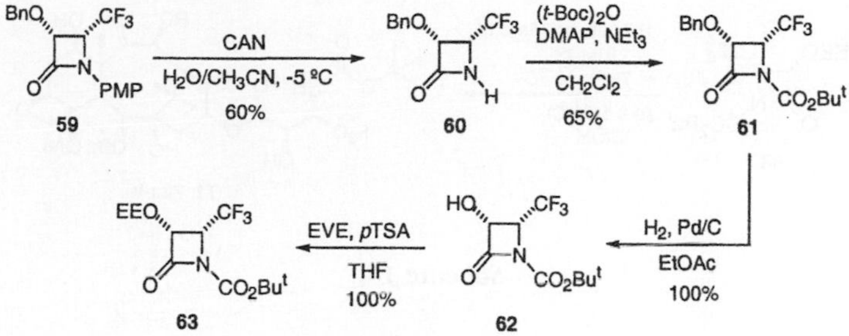
BnO
CF3
N
PMP
O
59
CAN
H2O/CH3CN, -5 °C
60%
N
H
60
(t-Boc)2O
DMAP, NEt3
CH2Cl2
65%
CO2But
61
H2, Pd/C
EtOAc
100%
HO
62
EVE, pTSA
THF
100%
EEO
63

Scheme 13.

all cell lines tested, especially against A549 and MCF7.[40] F-paclitaxel analogue **64** has an activity comparable to paclitaxel except against MCF7-R, but F_2-paclitaxel **65** shows reduced activity. The 3′-(3,3,3-trifluoropropyl) taxoid **68** shows reduced activity, but its 10-acetyl analogue **69** recovers considerable activity.

For the synthesis of a series of 3′-CF_3-taxoids **71** and **72a–h**, the ring-opening coupling of racemic *N*-*t*-Boc-β-lactam **63** (excess) with 7-TES-baccatins **70**, **8**, and

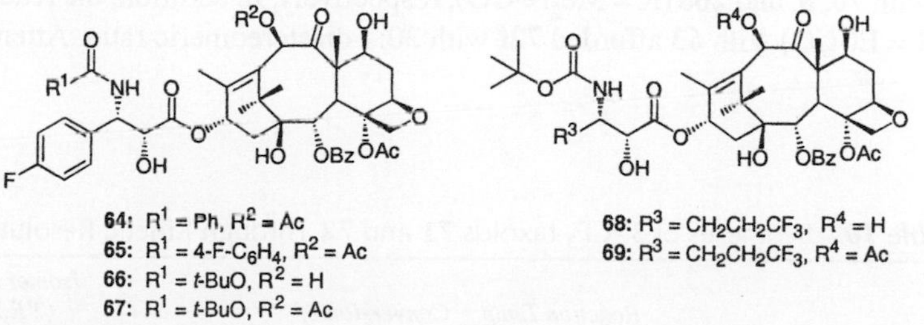
R1
NH
O
R2O
OH
F
OH
HO
OBz
OAc
O
R4O
R3
64: R1 = Ph, R2 = Ac
65: R1 = 4-F-C6H4, R2 = Ac
66: R1 = t-BuO, R2 = H
67: R1 = t-BuO, R2 = Ac
68: R3 = CH2CH2CF3, R4 = H
69: R3 = CH2CH2CF3, R4 = Ac

Table 9. Cytotoxicity of Fluorine-Containing Taxoids (IC_{50}, nM)[a]

F-Taxoid	*A121 (ovarian)*	*A549 (NSCL)*	*HT-29 (colon)*	*MCF7 (breast)*	*MCF7-R (breast)*
Paclitaxel	6.3	3.6	3.6	1.7	300
Docetaxel	1.2	1.0	1.2	1.0	235
64	6.3	4.2	14.5	5.1	>1000
65	76	35	51	45	>1000
66	1.3	0.49	3.9	0.40	477
67	1.2	0.47	3.5	0.42	315
68	78	36	44	36	>1000
69	10	14	7.1	9.3	219

Note: [a]See the footnote in Table 3.

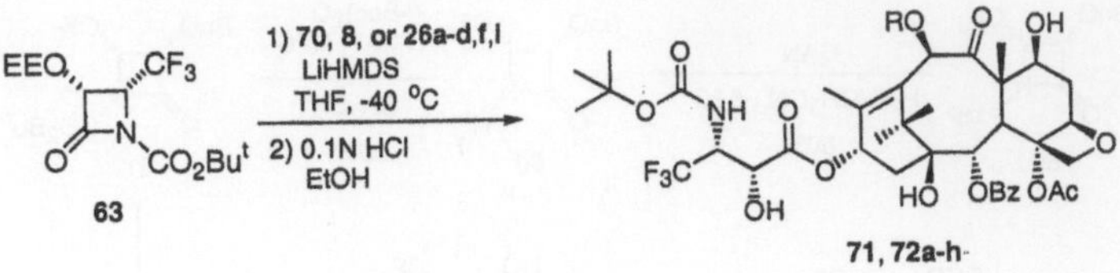

Scheme 14.

26a–d,f,i was examined with expectation of the occurrence of high-level kinetic resolution at low temperatures (Scheme 14). This type of kinetic resolution in the coupling of racemic β-lactam intermediates and metalated baccatin III was previously noted in a patent literature by Holton and Biediger.[23,82] However, the examples shown were all for 4-aryl-β-lactams and thus it was not predictable what level kinetic resolution might take place for this particular 4-CF$_3$-β-lactam **63**. The coupling reactions proceeded smoothly to give the desired 3′-CF$_3$-taxoids **71** and **72a–h** with diastereomeric ratios of 9:1 to >30:1 (by ^{19}F NMR analysis) in fairly good overall yields after deprotection.[45,46] As Table 10 illustrates, a high level of kinetic resolution of the racemic β-lactam **63** was observed in all cases. In particular, (2′R, 3′R) taxoids **71**, **72a**, and **72b** were obtained exclusively through the reactions of **63** with **70**, **8**, and **26c** (R = Me$_2$N-CO), respectively. In addition, the reaction of **26a** (R = Et-CO) with **63** afforded **72f** with 30:1 diastereomeric ratio. Attempts to

Table 10. Syntheses of 3′-CF$_3$-taxoids **71** and **72** Through Kinetic Resolution

Taxoid	R	Reaction Temp. (°C)	Conversion of Coupling (%)^a	Yield (%)^b	Isomer Ratio^c (2′R,3′S): (2′S,3′R)
71	H	−40 to −20	80	54 (67)	single isomer
72a	Ac	−40 to −10	72	41 (57)	single isomer
72b	Me$_2$N–CO	−40 to 0	100	63	single isomer
72c	cyclopropane-CO	−40 to 0	100	64	10:1
72d	MeO-CO	−40 to −20	93	54 (58)	24:1
72e	morpholine-4-CO	−40 to −20	100	60	23:1
72f	Et-CO	−40 to −15	100	74	>30:1
72g	CH$_3$(CH$_2$)$_3$–CO	−40 to −10	100	56	9:1
72h	(CH$_3$)$_3$CCH$_2$–CO	−40 to −20	91	59 (65)	22:1

Notes: ^{a}Based on consumed baccatin.

^{b}Two-step yield. Values in parentheses are conversion yields.

^{c}Determined by ^{19}F NMR analysis.

separate out the minor diastereomers by column chromatography have not been successful.

Cytotoxicity of 3'-CF$_3$-taxoids **71** and **72a–h** thus obtained were evaluated against human cancer cell lines and the results are summarized in Table 11.[45] As Table 11 shows, all these taxoids possess excellent activities, and are substantially more potent than either paclitaxel or docetaxel in virtually every case. The most remarkable results are, however, one order of magnitude better activities of the 10-acylated taxoids **72a–h** as compared to paclitaxel and docetaxel against the drug-resistant breast cancer cell line, MCF7-R. The marked difference in cytotoxicity observed between 3'-(2-CF$_3$-ethyl) taxoid, **68** or **69**, and 3'-CF$_3$-taxoids **72** reconfirms high sensitivity of the C-3' position to the size of substituent for the biological activity.

The comparison of 10-OH taxoid **71** with 10-acyl taxoids **72a–h** clearly indicates that the observed remarkable activity enhancement against MCF7-R can be attributed to modifications at the C-10 position, while this modification has little effect on the activity against the normal cancer cell lines, i.e. there is no apparent relationship between the activity against the normal cancer cell lines and that against the drug-resistant cell line. As described earlier, similar effects of 10-modification have been observed in the 3'-alkyl and 3'-alkenyl series of taxoids, e.g. **24**, **28**, and **29**, in which the activity against MCF7-R is very sensitive to the bulkiness of the 10-modifier (Section III.C).[41] In these 3'-CF$_3$-taxoids, however, the activity is not so sensitive to the steric bulk of the 10-modifier, e.g. **72g** (R = 3,3-dimethyl-butanoyl) and **72f** (R = Et-CO) exhibit virtually the same level of activity.

Metabolism studies on these fluorine-containing taxoids to obtain evidence for their ability to block metabolic pathways, e.g. suppression of P-450 oxidation (*vide supra*), have just started. Results will be reported elsewhere in the future.

Table 11. Cytotoxicity of 3'-CF$_3$-taxoids (IC$_{50}$, nM)[a]

Taxoid	R	A121 (ovarian)	A549 (NSCL)	HT-29 (colon)	MCF7 (breast)	MCF7-R (breast)
Paclitaxel	—	6.3	3.6	3.6	1.7	299
Docetaxel	—	1.2	1.0	1.2	1.0	235
71	H	1.15	0.44	0.65	0.44	156
72a	Ac	0.3	0.2	0.4	0.6	17
72b	Me$_2$N-CO	0.3	0.2	0.4	0.3	21
72c	cyclopropane-CO	0.4	0.4	0.5	0.5	16
72d	MeO-CO	0.3	0.2	0.4	0.4	21
72e	morpholine-4-CO	0.5	0.4	0.4	0.4	48
72f	Et-CO	0.3	0.3	0.4	0.3	14
72g	CH$_3$(CH$_2$)$_3$-CO	0.7	0.8	1.4	0.7	26
72h	(CH$_3$)$_3$CCH$_2$-CO	0.5	0.5	0.6	0.5	12

Note: [a]See footnote of Table 3.

B. The Fluorine Probe Approach: Solution-Phase Structure and Dynamics of Taxoids

The rational design of new generation taxoid anticancer agents would be greatly facilitated by the development of reasonable models for the biologically relevant conformations of paclitaxel. In this regard, we recognized early on that the design and synthesis of fluorinated taxoids would have an extremely useful offshoot of providing us with the capability of studying bioactive conformations of taxoids using a combination of ^{19}F/^{1}H NMR techniques and molecular modeling.[77]

Previous studies in the conformational analysis of paclitaxel and docetaxel have largely identified two major conformations, with minor variations between studies.[83] Structure **A**, characterized by a *gauche* conformation with a H2′–C2′–C3′–H3′ dihedral angle of ca. 60°, is based on the X-ray crystal structure of docetaxel,[84] and is believed to be commonly observed in aprotic solvents (Figure 2).[83] Structure **B**, characterized by the *anti* conformation with a H2′–C2′–C3′–H3′ dihedral angle of ca. 180°, has been observed in theoretical conformational analysis[85,86] as well as 2-D NMR analyses,[87] and has also been found in the X-ray structure of the crystal obtained from a dioxane/H$_2$O/xylene solution (Figure 2).[88] Despite extensive structural studies, no systematic study on the dynamics of these two and other possible bioactive conformations of paclitaxel had been reported when we started our study on this problem. The relevance of the "fluorine probe" approach to study dynamic properties prompted us to conduct a detailed investigation into the solution dynamics of fluorine-containing paclitaxel and docetaxel analogues.

The use of ^{19}F NMR for a variable temperature (VT) NMR study of fluorinated taxoids is obviously advantageous over the use of ^{1}H NMR because of the wide dispersion of the ^{19}F chemical shifts that allows fast dynamic processes to be frozen out. Accordingly, F$_2$-paclitaxel **65** and F-docetaxel **66** were selected as probes for the study of the solution structures and dynamic behavior of paclitaxel and docetaxel, respectively, in protic and aprotic solvent systems.[77] The inactive 2′,10-diacetyldocetaxel (**73**) was also prepared to investigate the role of the 2′-hydroxyl moiety in the conformational dynamics.[89] While molecular modeling and NMR analyses (at room temperature) of **73** indicate that there is no significant conformational changes as compared to paclitaxel, the ^{19}F NMR VT study clearly indicates that this modification exerts marked effects on the dynamic behavior of the molecule.[77]

Analysis of the low temperature VT NMR (^{19}F and ^{1}H) and ^{19}F-^{1}H heteronuclear NOE spectra of **65** and **66** in conjunction with molecular modeling has revealed the presence of an equilibrium between two conformers in protic solvent systems. Interpretation of the temperature dependence of the coupling constants between H2′ and H3′ for **65** shows that one of these conformers (conformer C) has an unusual near-eclipsed arrangement around the H2′–C2′–C3′–H3′ dihedral angle ($J_{H2'-H3'} = 5.2$ Hz, corresponding to the H2′–C2′–C3′–H3′ torsion angle of 124° based on the MM2 calculation), and is found to be more prevalent at ambient temperatures. The

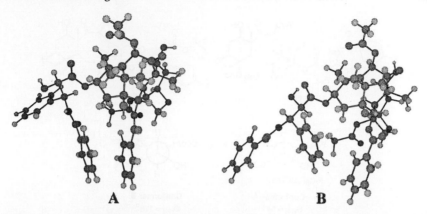

Figure 2. Conformation of paclitaxel based on the X-ray structure of docetaxel and proposed for nonpolar aprotic organic solvents (structure **A**) and the conformation based on the X-ray structure of paclitaxel (structure **B**) and proposed for aqueous solvents.

other one corresponds to the *anti* conformer (conformer B, $J_{\text{H2}'-\text{H3}'} = 10.1$ Hz, corresponding to the H2'–C2'–C3'–H3' torsion angle of 178° based on the MM2 calculation) and is quite closely related to the structure **B** in Figure 2. These conformers are different from the one observed in aprotic solvents (conformer A, H2'–C2'–C3'–H3' torsion angle of 54°) that is related to the X-ray crystal structure of docetaxel represented by the structure **A** in Figure 2.[84] Figure 3 shows the Newman projections for these three conformers; representative low-energy structures for the three conformers are shown in Figure 4.

Restrained molecular dynamics (RMD) studies presented evidence for the hydrophobic clustering of the 3'-phenyl and 2-benzoate (Ph) for both conformers B and C. Although the conformer C possesses the rather unusual semieclipsed arrangement around the C2'–C3' bond, the unfavorable interaction associated with such a conformation is apparently offset by significant solvation stabilization, observed in the comparative RMD study in a simulated aqueous environment for the three conformers. The solvation stabilization term for the conformer C was estimated to be about 10 kcal/mol greater than those for the conformers A and B. Accordingly, the "fluorine probe" approach has succeeded in finding a new conformer that has never been predicted by the previous NMR and molecular modeling studies.[83]

Strong support for the conformer C can be found in its close resemblance to a proposed solution structure of a water-soluble paclitaxel analogue, paclitaxel-7-MPA (MPA = *N*-methylpyridinium acetate),[90] in which the H2'–C2'–C3'–H3' torsion angle of the *N*-phenylisoserine moiety is 127°, which is only a few degrees different from the value for the conformer C.

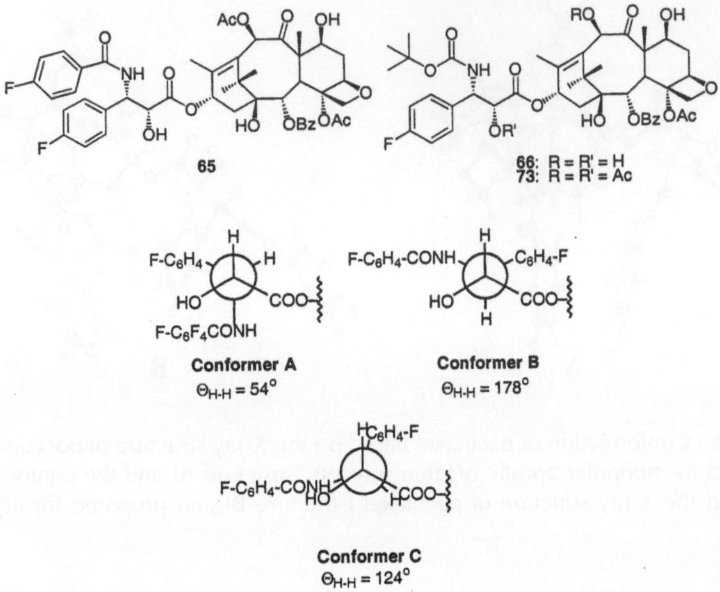

Structures 65, 66, 73

Conformer A
$\Theta_{H-H} = 54°$

Conformer B
$\Theta_{H-H} = 178°$

Conformer C
$\Theta_{H-H} = 124°$

Figure 3. Newman projections of the three conformers for F_2-paclitaxel **65**.

Thus, the "fluorine probe" approach has proved useful for the conformational analysis of paclitaxel and taxoids in connection with the determination of possible bioactive conformations.[77] The previously unrecognized conformer C might be the molecular structure first recognized by the β-tubulin binding site on microtubules.

The fluorine probe approach[77] confirms hydrophobic clustering to be one of the main driving forces for the conformational stabilization of fluorine-containing paclitaxel and docetaxel analogues. It is obvious that the size of the hydrophobic moieties must play a very important role in determining the extent of clustering, and hence the conformational equilibrium. In this regard, we became interested in determining the conformational preference for the highly potent 3'-CF_3-taxoid **72a**. The ^1H and ^{19}F VT NMR study has revealed that **72a** does not exhibit different conformations in DMSO–D_2O, MeOD, and CD_2Cl_2.[45] In fact, the $J_{H2'-H3'}$ value remains ca. 2 Hz in all the protic and aprotic solvents examined. This value corresponds to a H2'–C2'–C3'–H3' torsion angle of ca. 60° with no detectable temperature dependence. These results imply that this molecule does not form "hydrophobic cluster" even in DMSO–D_2O. Systematic conformational search (Sybyl 6.04) for **72a** identifies the gauche conformers with H2'–C2'–C3'–H3' torsion angle of 60° and -60° to be more stable than the *anti* conformer (180°) by 2–5 kcal/mol, while the exact opposite is true for paclitaxel for which the *anti* conformer is more stable than the *gauche* conformers.[91] Chem 3D representation of the most likely solution conformation of **72a** based on the H2'–C2'–C3'–H3'

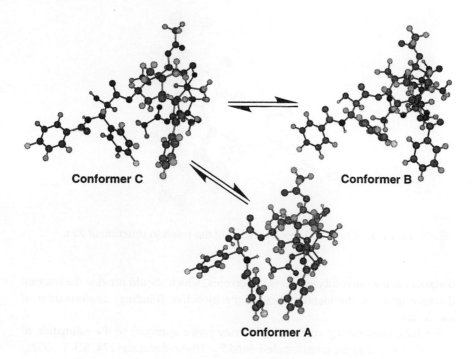

Conformer C

Conformer B

Conformer A

Figure 4. Conformers of **65** observed in protic and aprotic media.

torsion angle of 60° is shown in Figure 5. This conformation is similar to the X-ray structure of docetaxel,[84] although the 3′-phenyl is replaced by a sterically and electronically very different CF_3 group. The fact that **72a** and related CF_3–taxoids are extremely active warrants further investigation on the interaction of this molecule with microtubules using the CF_3 group as a probe.

C. Determination of the Binding Conformation of Taxoids to Microtubules Using Fluorine Probes

The knowledge of the solution structures and dynamics of paclitaxel and its analogues is necessary for a good understanding of the recognition and binding processes between paclitaxel and its binding site on the microtubules, which also provides crucial information for the design of future generation anticancer agents.[73] However, the elucidation of the microtubule-bound conformation of paclitaxel is critical for the rational design of efficient inhibitors of microtubule disassembly. The lack of information about the three-dimensional tubulin binding site has prompted us to apply our fluorine probe approach to the determination of the F–F

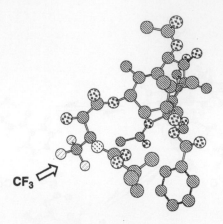

Figure 5. Chem 3-D representation of the solution structure of **72a**.

distances in the microtubule-bound F_2–taxoids, which should provide the relevant distance map for the identification of the bioactive (binding) conformation of paclitaxel.

We have successfully applied the fluorine probe approach to the estimation of the F–F distance in the microtubule-bound F_2–10-Ac-docetaxel (**74**, SB-T-30021) using the solid-state magic angle spinning (SS MAS) ^{19}F NMR coupled with the radio frequency driven dipolar recoupling (RFDR) technique in our preliminary study.[78]

F_2–10-Ac-docetaxel (**74**) was first studied in a polycrystalline form by the RFDR technique. Based on the standard simulation curves derived from molecules with known F–F distances (distance markers), the F–F distance of two fluorine atoms in **74** was estimated to be 5.0 ± 0.5 Å (Figure 6). This value corresponds quite closely to the estimated F–F distances for the conformers B and C (F–F distance is ca. 4.5 Å for both conformers) based on our RMD studies for F_2–paclitaxel (**65**) (*vide supra*). This means that the microcrystalline structure of **74** is consistent with the hydrophobic clustering conformer B or C, but not with the conformer A in which the F–F distance is ca. 9.0 Å.

74 (SB-T-30021)

The microtubule-bound complex of **74** revealed the F–F distance to be 6.5 ± 0.5 Å (Figure 6), which is larger than that observed in the polycrystalline form by ca. 1 Å. It is very likely that the microtubule-bound conformation of **74** is achieved by a small distortion of the solution conformation (the *recognition conformation*), the latter being described by either conformer B or C (Figure 4).

Restrained high-temperature molecular dynamics in vacuum were conducted for **74** while maintaining a distance restraint of 6.5 Å between the two fluorine atoms in the minimization step for each sampled conformer. This study revealed that the distance of 6.5 Å between the two fluorine atoms could be maintained by energetically similar conformers with H2'–C2'–C3'–H3' torsions of 180°, 60°, and –60° (Figure 7). Our investigation into the identification of the common pharmacophore of paclitaxel and epothilones has recently revealed a highly plausible common pharmacophore structure.[73,92] When we screened the four low-energy conformers shown in Figure 7, conformer I was singled out as the most likely microtubule-bound conformation of paclitaxel.

We are currently evaluating several taxoid analogues containing fluorines at different positions by SS MAS ^{19}F NMR. These studies are geared towards generating a detailed distance map that will help to finally pinpoint the microtubule-bound conformation of paclitaxel. The above account has demonstrated the power of the fluorine probe approach, that is clearly evident from its ability to supply extremely valuable and precise information about both bound and dynamic conformations of biologically active molecules, especially useful in the absence of knowledge about the three-dimensional structure of their binding site.

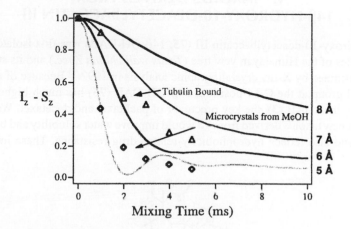

Figure 6. Magnetization Exchange Experiment for microtubule-bound **74** (solid lines represent curves for distance markers).

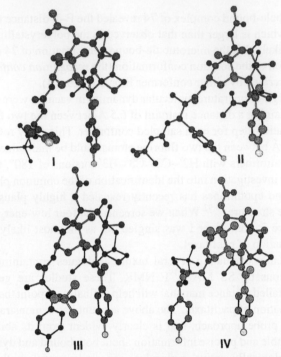

Figure 7. Four low-energy conformers obtained from the distance-restrained MD of **74** (H2'-C2'-C3'-H3' torsions in parentheses): **I** (180°), **II** (180°), **III** (60°), and **IV** (–60°).

V. TAXOIDS DERIVED FROM
14β-HYDROXY-10-DEACETYLBACCATIN III

14β-Hydroxy-10-deacetylbaccatin III (**75**, 14β-OH-DAB) was first isolated from the needles of the Himalayan yew tree (*Taxus wallichiana Zucc.*) and its structure was determined by X-ray crystallographic analysis in 1992.[93] Because of an extra hydroxyl group at the C-14 position, 14β-OH-DAB (**75**) has much higher water solubility than DAB (**3**), the key precursor of paclitaxel and docetaxel. We envisaged that new taxoids derived from **75** would improve water solubility and bioavailability, and also reduce hydrophobicity-related drug resistance. These improved

14β-Hydroxy-10-deacetylbaccatin III (**75**)

pharmacological properties may well be related to the modification of undesirable toxicity and activity spectra against different cancer types.

A. Initial Studies: 14β-OH-Taxoids Bearing 1,14-Carbonate and Pseudo-Taxoids

Our initial study on taxoids derived from 14β-OH-DAB (**75**) dealt with the syntheses of new taxoids bearing an *N*-acylphenylisoserine residue attached to either the C-13 or the C-14 position (*pseudo*-taxoids).[43] We have found that the reactivity of the five hydroxyl group of **75** decreases in the order C-7 > C-10 > C-14 > C-13. Accordingly, the coupling of the baccatin at the 13-position with an *N*-acylphenylisoserine precursor requires an appropriate protection of the hydroxyl groups at the C-7, C-10, and C-14 positions. 7,10-di-Troc-14-OH-DAB (**76**), in which the 7- and 10-hydroxyl groups are selectively protected, was prepared by reacting **75** with Troc-Cl in pyridine. Reaction of **75** with a larger excess of Troc-Cl afforded the 7,10-di-Troc-14-OH-DAB-1,14-carbonate (**77**) (Scheme 15).

The coupling of baccatins **76** and **77** with *N*-benzoyl-β-lactam **11** and the *N-t*-Boc-β-lactam **12** followed by deprotection using Zn in AcOH afforded taxoids **78–81** in fair to good yields. These taxoids possess strong to modest cytotoxicity against our standard five human cancer cell lines (Table 12). The microtubule disassembly inhibitory activity of these new taxoids was also evaluated. Taxoid **78** (SB-T-1011) exhibits activity same to or better than that of paclitaxel. Results are listed in Table 12. The attachment of the *N*-acylphenylisoserine residue at the C-14 position (taxoids **80** and **81**) instead of the original C-13 position results in ca. 10-fold decrease in cytotoxicity, but **80** still retains 10 nM level (IC$_{50}$) cytotoxicity.

The in vivo antitumor activity of **78** was evaluated against A151 human ovarian carcinoma xenograft in nude athymic mice, and was found to be equivalent or slightly better than that of paclitaxel, causing total regression of the tumor.[94]

A molecular modeling study shows that the conformation of pseudo-docetaxel **80** overlaps well with that of docetaxel based on the X-ray crystal structure,[84] except that the two hydrophobic groups, phenyl at C-3′ and *tert*-butoxycarbonyl group at C3′–N, have exchanged their positions almost perfectly. The ability of pseudo-

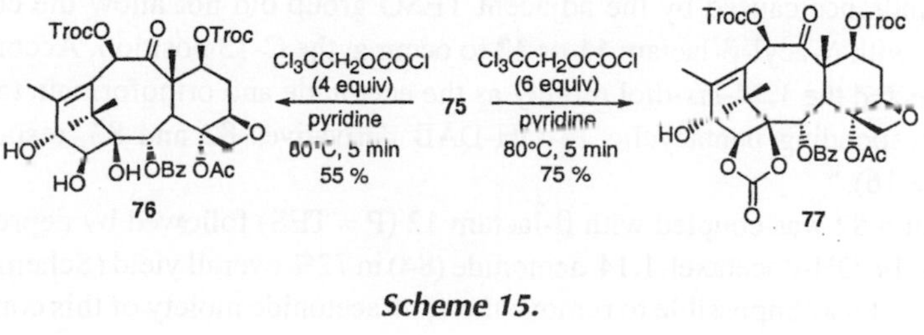

Scheme 15.

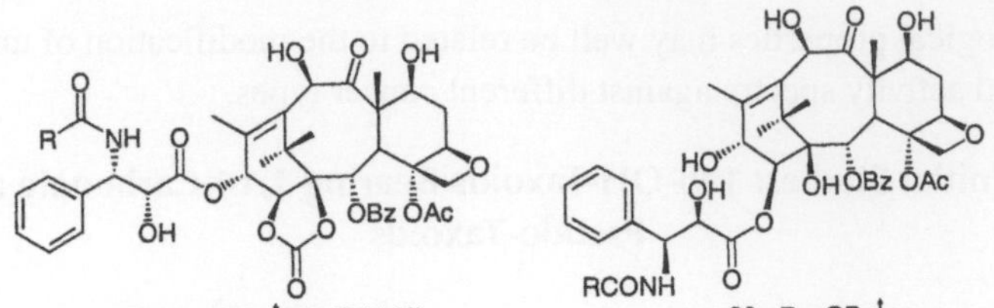

78: R = OBu^t (SB-T-1011)
79: R = Ph

80: R = OBu^t
81: R = Ph

Table 12. Cytotoxicity of Taxoids and *Pseudo*-Taxoids Derived from
14β-OH-DAB (IC$_{50}$, nM)[a]

Taxoid	A121 (ovarian)	A549 (NSCL)	HT-29 (colon)	MCF7 (breast)	MCF7-R (breast)	Microtubule Disassembly Inhibition
Paclitaxel	6.1	3.6	3.2	1.7	299	1.0T
78 (SB-T-1011)	6.2	2.1	1.8	1.8	543	0.9T
79	105	33	24	11	>1000	3T
80	80	31	50	26	>1000	>100T
81	1044	512	248	269	>1000	>100T

Note: [a]See footnote of Table 3.

taxoid **80** to mimic the conformation of docetaxel provides a rationale for retaining 10 nM level cytotoxicity.

B. 14β-Hydroxydocetaxel

A docetaxel analogue that has an additional hydroxyl group in the molecule is an attractive target to synthesize because of its apparent advantage in water solubility and thus formulation of the agent as compared to paclitaxel and docetaxel. However, the synthesis of the 14-OH-free taxoid was not straightforward due to the inherent difficulty in selectively protecting the hydroxyl group at C-14. Attempted protection of the 14-hydroxyl group with excess Troc-Cl/pyridine led to the formation of the DAB-1,14-carbonate **77** (*vide supra*). Although selective protection of the C-14 hydroxyl group as the triethylsilyl (TES) ether was possible, the steric hindrance caused by the adjacent TESO group did not allow the coupling reaction with *N*-acyl-β-lactam **11** or **12** to occur at the C-13 position. Accordingly, we protected the 1,14-*cis*-diol moiety as the acetonide and orthoformate to obtain the corresponding pentacyclic 14-OH-DAB derivatives **82** and **83**, respectively (Scheme 16).[36]

Baccatin **82** was coupled with β-lactam **12** (P = TES) followed by deprotection to afford 14-OH-docetaxel-1,14-acetonide (**84**) in 72% overall yield (Scheme 17).[36] However, it was impossible to remove the 1,14-acetonide moiety of this compound

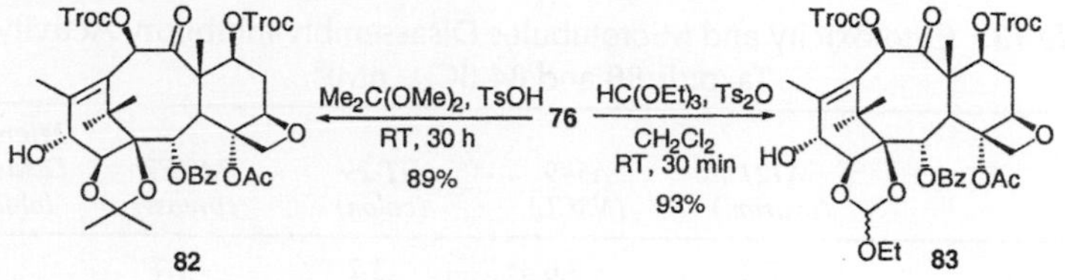

Scheme 16.

without opening the D-ring. Thus, we coupled baccatin **83** with β-lactam **12** under the standard conditions (*vide supra*) to obtain **85** in 78% yield. Deprotection of the 2'-TES group with 0.5% HCl, followed by reaction with formic acid afforded 14-formyl-taxoid **86**. Taxoid **86** was then deprotected *in situ* by treatment with 1% aqueous NaHCO$_3$ to yield **87** (73% for two steps). Finally, the removal of the Troc groups at C-7 and C-10 using Zn/HCl gave 14β-hydroxydocetaxel (**88**, SB-T-1001) in 73% yield (Scheme 17), completing the first synthesis of this compound.[36,95]

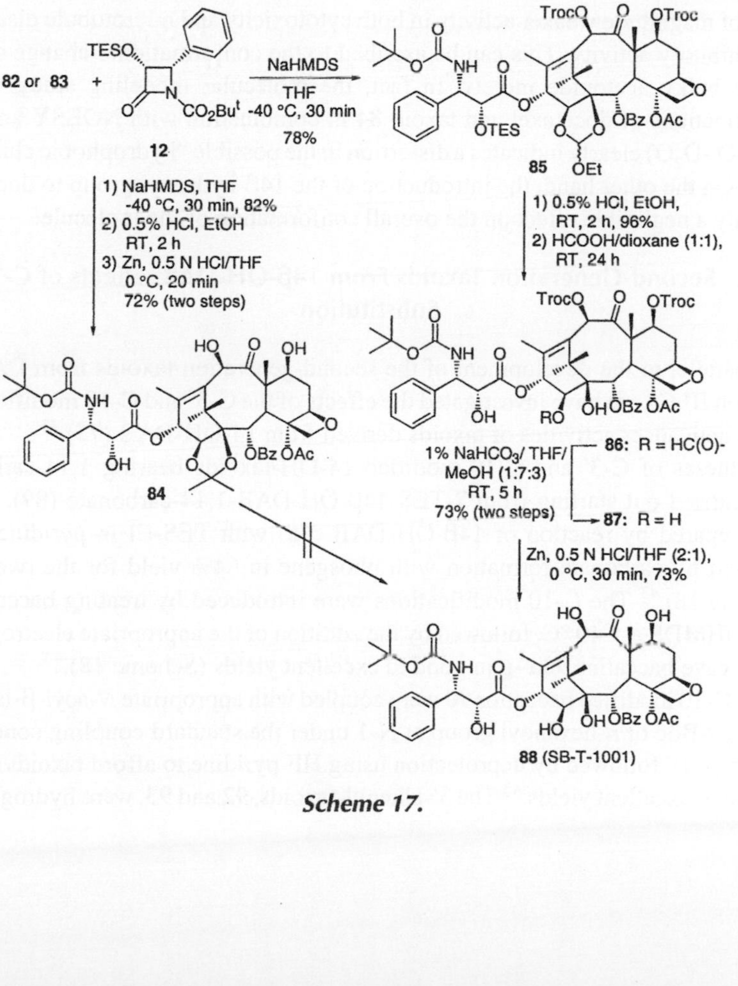

Scheme 17.

Table 13. Cytotoxicity and Microtubules Disassembly Inhibitory Activity of
Taxoids **88** and **84** (IC$_{50}$, nM)[a]

Taxoid	A121 (ovarian)	A549 (NSCL)	HT-29 (colon)	MCF7 (breast)	Microtubule Disassembly Inhibition[b]
Paclitaxel	6.1	3.6	3.2	1.7	1.0T
Docetaxel	1.2	1.0	1.2	1.0	0.7T
88 (SB-T-1001)	3.3	0.8	2.1	1.9	0.8T
84	46	18	21	34	3T

Notes: [a]See the footnote in Table 3.
[b]See footnote a in Table 1.

14β-Hydroxydocetaxel (**88**) possesses strong cytotoxicity in between that of docetaxel and paclitaxel, except for the activity against the A549 human non-small cell lung cancer cell line (IC$_{50}$ = 0.8 nM) which is slightly better than that of docetaxel (Table 13). Taxoid **84** bearing a 1,14-acetonide, however, showed one order of magnitude weaker activity in both cytotoxicity and microtubule disassembly inhibitory activity. This can be ascribed to the conformational change caused by the bulky acetonide moiety. In fact, the molecular modeling study on the conformations of docetaxel and taxoid **84** in combination with NOESY analyses (DMSO–D$_2$O) clearly indicates a distortion in the possible "hydrophobic cluster"[87] of **84**. On the other hand, the introduction of the 14β-hydroxyl group to docetaxel has only a negligible effect on the overall conformation of the molecule.

C. Second-Generation Taxoids From 14β-OH-DAB: Effects of C-10 Substitution

In parallel to the development of the second-generation taxoids from DAB (**3**) (Section III.C), we have investigated the effects of the C-3' and C-10 modifications on the anticancer activities of taxoids derived from 14-OH-DAB (**75**).[43]

Syntheses of C-3' and C-10 modified 14-OH-taxoids bearing 1,14-carbonate were carried out starting from 7-TES-14β-OH-DAB-1,14-carbonate (**89**), which was prepared by reaction of 14β-OH-DAB (**75**) with TES-Cl in pyridine/DMF followed by carbonate formation with phosgene in 64% yield for the two steps (Scheme 18).[43] The C-10 modifications were introduced by treating baccatin **89** with LiHMDS at –40 °C, followed by the addition of the appropriate electrophiles, which gave baccatins **90a–g** in good to excellent yields (Scheme 18).[43]

The C-10 modified baccatins **90** were coupled with appropriate N-acyl-β-lactams bearing t-Boc or n-hexanoyl group at N-1 under the standard coupling conditions (*vide supra*), followed by deprotection using HF-pyridine to afford taxoids **91–96** in good to excellent yields.[43] The 3'-alkenyl-taxoids, **92** and **93**, were hydrogenated

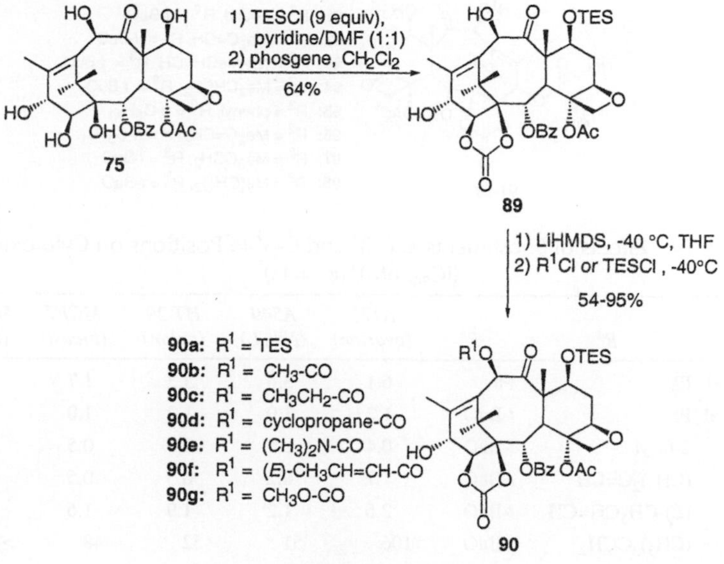

Scheme 18.

over Pd/C to give the corresponding 3'-alkyl-taxoids, **97** and **98**, in quantitative yields.[43] The anticancer activities of these 1,14-carbonate taxoids were evaluated against our standard five human cancer cell lines.

Table 14 summarizes the effects of the substituents at the C-3' and C3'-N positions for analogues in which the C-10 substituent is maintained as the free hydroxyl.[43] When R^2 is phenyl (**78, 79, 88,** and **95a**), *t*-Boc is the best substituent as R^3CO among *t*-Boc, benzoyl, and hexanoyl groups, i.e. the cytotoxicity decreases in the order *t*-Boc > PhCO > $C_5H_{11}CO$. It is obvious that the C-3' substituents including the 2-furyl (**91a**), 2-methyl-1-propenyl (**92a**), (*E*)-1-propenyl (**93a**), 2-methylpropyl (**97a**), and propyl (**98a**) substantially increase cytotoxicity. It is also clear that the 2,2-dimethylpropyl group at the C-3' position markedly decreases cytotoxicity (**94a**). Among these 1,14-carbonate taxoids listed in Table 14, **91a** and **92a** exhibit excellent overall cytotoxicity, considerably better than paclitaxel and docetaxel against different cancer cell lines.

Table 15 summarizes the effects of the modifications at C-10 on cytotoxicity. The substituent at C-10 exerts a remarkable effect on the activity against the drug-resistant human breast cancer cell line, MCF7-R, i.e. the activity increases, dramatically in some cases (**91a** vs. **91b**; **93a** vs. **93b**; **97a** vs. **97e**), by replacing the hydrogen of the C-10 hydroxyl with an acyl group or *N,N*-dimethylcarbamoyl group. The activities against normal cancer cell lines are also somewhat influenced by the substituent at C-10, but to a much lesser extent. Three taxoids, **91b, 92c,** and **92g**, exhibit sub-nanomolar IC_{50} values against all normal cancer cell lines examined as

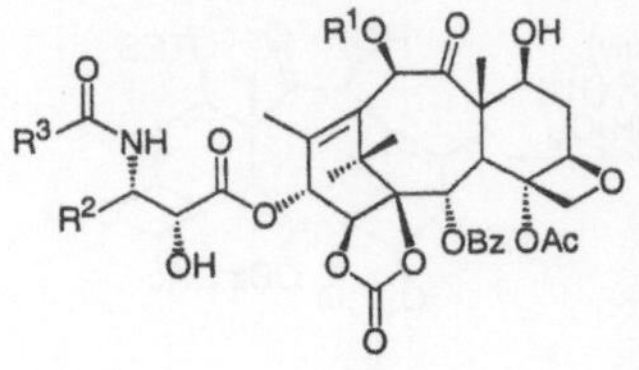

91: R^2 = 2-furyl, R^3 = t-BuO
92: R^2 = Me$_2$C=CH, R^3 = t-BuO
93: R^2 = (E)-MeCH=CH, R^3 = t-BuO
94: R^2 = Me$_2$CHCH$_2$, R^3 = t-BuO
95: R^2 = phenyl, R^3 = n-C$_5$H$_{11}$
96: R^2 = Me$_2$C=CH, R^3 = n-C$_5$H$_{11}$
97: R^2 = Me$_3$CCH$_2$, R^3 = t-BuO
98: R^2 = Me(CH$_2$)$_2$, R^3 = t-BuO

91-98

Table 14. Effects of Substituents at C-3′ and C-3′-N Positions on Cytotoxicity $(IC_{50}, nM)^a (R^1 = H)$

Taxoid	R^2	R^3	A121 (ovarian)	A549 (NSCL)	HT-29 (colon)	MCF7 (breast)	MCF7-R (breast)
Paclitaxel	Ph	Ph	6.1	3.6	3.2	1.7	299
Docetaxel	Ph	t-BuO	1.2	1.0	1.2	1.0	235
91a	2-furyl	t-BuO	0.4	0.5	0.6	0.5	135
92a	(CH$_3$)$_2$C=CH	t-BuO	1.7	0.2	0.5	0.5	54
93a	(E)-CH$_3$CH=CH	t-BuO	2.6	1.2	1.9	1.6	762
94a	(CH$_3$)$_3$CCH$_2$	t-BuO	106	51	52	48	>1000
95a	Ph	n-C$_5$H$_{11}$	330	389	157	13	>1000
96a	(CH$_3$)$_2$C=CH	n-C$_5$H$_{11}$	12	18	16	3.2	321
97a	(CH$_3$)$_2$CHCH$_2$	t-BuO	1.0	1.8	3.4	2.7	385
98a	CH$_3$(CH$_2$)$_2$	t-BuO	2.1	4.4	3.8	4.2	265

Note: [a]See the footnote in Table 3.

well as 10 nanomolar level IC_{50} values against MCF7-R. The taxoid **92f** shows the highest activity against MCF7-R (IC_{50} = 17 nM).

This finding is significant since neither paclitaxel nor docetaxel possesses strong activity against MCF7-R which expresses MDR phenotype. The relative activity index, i.e. IC_{50}MCF7-R/IC_{50}MCF7, of paclitaxel and docetaxel are 176 and 235, respectively. In this regard, **92b** has the best relative activity index (IC_{50}MCF7-R/IC_{50}MCF7 = 11) while keeping the cytotoxicity against normal cancer cell lines at the lower nanomolar level IC_{50} values.

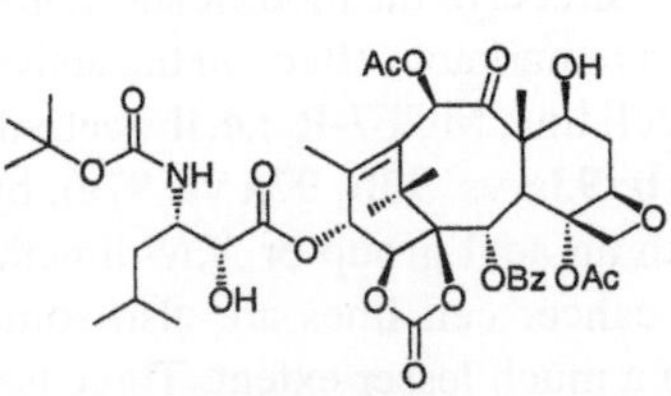

97b (SB-T-101131)

As discussed previously for the second-generation taxoids derived from DAB (**3**) (Section III.C), an appropriate combined modification of C-3′ and C-10 substituents has dramatic effects on the activity against the drug-resistant cancer cells. Our current hypothesis for this phenomenon, that involves the binding ability of taxoids to P-glycoprotein, will be discussed more in detail in Section VI. These results, together with the strong in vivo antitumor activity of 1,14-carbonate taxoid **78** (SB-T-1011, Section V.A), make it highly attractive to consider taxoids derived from 14-OH-DAB as clinical candidates for cancer chemotherapy. As Table 15 shows, several 1,14-carbonate taxoids in this series possess one order of magnitude better cytotoxicity than **78**, and a couple of these taxoids possess excellent in vivo antitumor activity against human ovarian A121 tumor xenografts in nude mice. For example, taxoid **97b** shows a clear dose response and has a maximum tolerated dose (MTD) equivalent to that of paclitaxel while keeping much higher potency than paclitaxel, i.e. this taxoid has a wide therapeutic window. Also, taxoid **97b**

Table 15. Effects of the Substituents at C-10 on Cytotoxicity (IC$_{50}$, nM)[a]

Taxoid	R^1	A121 (ovarian)	A549 (NSCL)	HT-29 (colon)	MCF7 (breast)	MCF7-R (breast)
Paclitaxel	CH$_3$-CO	6.1	3.6	3.2	1.7	299
Docetaxel	H	1.2	1.0	1.2	1.0	235
91a	H	0.4	0.5	0.6	0.5	135
91b	CH$_3$-CO	0.4	0.5	0.6	0.5	49
92a	H	1.7	0.2	0.5	0.5	72
92b	CH$_3$–CO	1.5	1.4	2.4	3.3	36
92c	CH$_3$CH$_2$–CO	0.7	0.5	0.6	0.2	26
92d	cyclopropane-CO	0.5	0.5	1.0	0.4	28
92e	(CH$_3$)$_2$N–CO	0.7	0.6	1.2	0.4	33
92f	(*E*)-CH$_3$CH–CHCO	0.8	0.5	1.5	0.4	17
92g	CH$_3$O-CO	0.6	0.5	0.7	0.3	38
93a	H	2.6	1.2	1.9	1.6	762
93b	CH$_3$-CO	2.4	0.4	3.0	1.6	72
96d	cyclopropane-CO	12	18	16	3.2	321
96e	(CH$_3$)$_2$N–CO	4.8	5.5	5.1	1.7	295
96f	(*E*)-CH$_3$CH–CHCO	17	25	28	2.1	137
97a	H	2.5	1.4	4.2	3.0	189
97b	CH$_3$–CO	1.2	0.7	1.5	1.1	36
97d	cyclopropane-CO	1.1	1.2	3.3	0.7	22
97c	(CH$_3$)$_2$N–CO	0.6	0.6	1.3	0.6	22
98a	H	2.1	4.4	3.8	4.2	385
98b	CH$_3$–CO	2.3	4.7	4.8	4.6	201

Note: [a]See the footnote in Table 3.

exhibits 50–80 times higher potency than paclitaxel in the apoptosis assay against drug-resistant cancer cells, MCF7-R (breast), and CEM VBL-R (leukemia). These data are extremely encouraging and thus taxoid **97b** (SB-T-101131) is currently our leading candidate to enter clinical trials.

VI. NOR-SECO TAXOIDS: PROBING THE MINIMUM STRUCTURAL REQUIREMENTS FOR ANTICANCER ACTIVITY

It is very important to clarify the minimum structural requirements for paclitaxel and taxoids to exhibit anticancer activity by looking at simplified structure analogues. Along this line, we have investigated the role of the A ring by synthesizing novel nor-seco analogues of paclitaxel and docetaxel.[37,96]

Novel nor-seco baccatin **99** was synthesized in 92% yield through oxidative cleavage of the A ring of 14-OH-DAB (**75**) with periodic acid via the hydroxy ketone intermediate **100** (Scheme 19). Protection of the 7-hydroxyl of **99** as TES ether followed by reduction of the aldehyde with sodium borohydride yielded nor-seco baccatin alcohol **101** in 80% yield. Nor-seco 13-amino-baccatins **102** and **103** were synthesized from **101** and **99a** via the Mitsunobu reaction and reductive amination, respectively, in high yields (Scheme 20).

7-TES-Nor-seco baccatin alcohol **101** was coupled to β-lactams **11** and **12** (P = EE) under the standard conditions (*vide supra*), followed by deprotection to give novel nor-seco paclitaxel **104a** and nor-seco docetaxel **104b** in fairly good to excellent yields.[37] Nor-seco taxoids **105a–d** bearing a C-13 amide linkage were synthesized in high yields through the ring-opening coupling of β-lactams **11**, **22a** and **22d** (P = TIPS) with nor-seco baccatin amine **102** under neutral conditions at 40 °C.[96] In a similar manner, nor-seco taxoid **106** with an *N*-Me amide linkage at C-13 was obtained in excellent yield through the coupling of the β-lactam **11** (P =

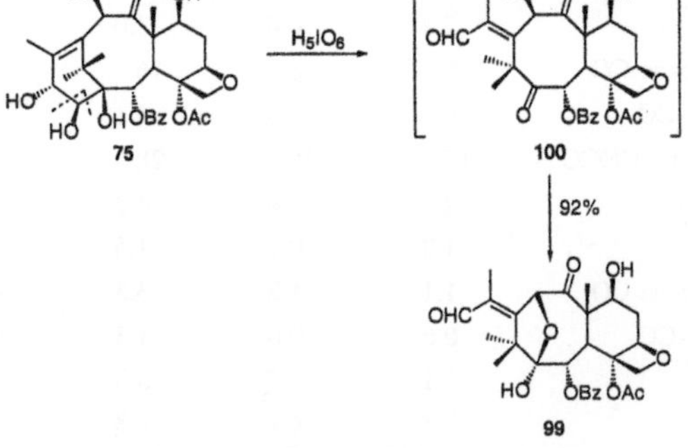

Scheme 19.

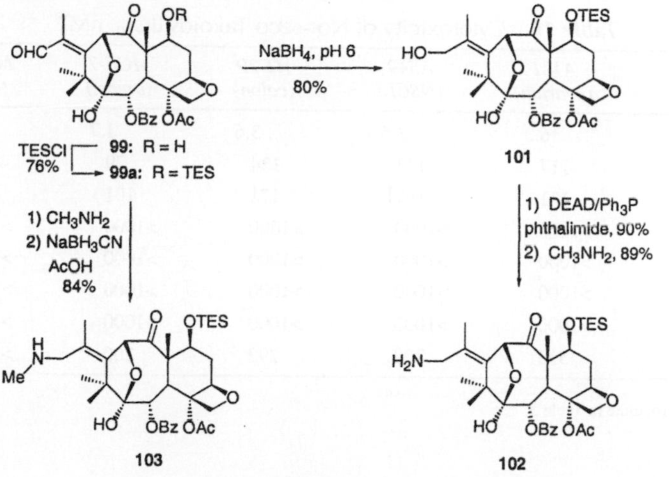

Scheme 20.

H) with the nor-seco baccatin *N*-methylamine **103** under neutral conditions at room temperature.

As Table 16 shows, nor-seco taxoids with a C-13 ester linkage, **104a** and **104b**, possess 80–130 nM and 100–170 nM level IC_{50} values, respectively. These taxoids are 20–40 times less active than paclitaxel, but they clearly retain a certain level of cytotoxicity. It is worth mentioning that **104a** and **104b** exhibit IC_{50} values of 471 nM and 360 nM, respectively, against the doxorubicin-resistant breast cancer cell line (MCF7-R), i.e. **104a** and **104b** are comparable to paclitaxel. The nor-seco taxoid **106** bearing an *N*-Me amide linkage possesses 200–700 nM level IC_{50} values, retaining a certain level of cytotoxicity. In contrast to nor-seco taxoids **104a**, **104b**, and **106**, the nor-seco taxoids with an *N*-H amide linkage, **105a–d**, are all virtually inactive (IC_{50} >1000 nM).

The results obtained clearly indicate the importance of the A ring for the strong cytotoxicity of taxoids.[19] However, the fact that the reduced-structure analogues, **104a** and **104b**, retain a certain level of cytotoxicity and their potency against

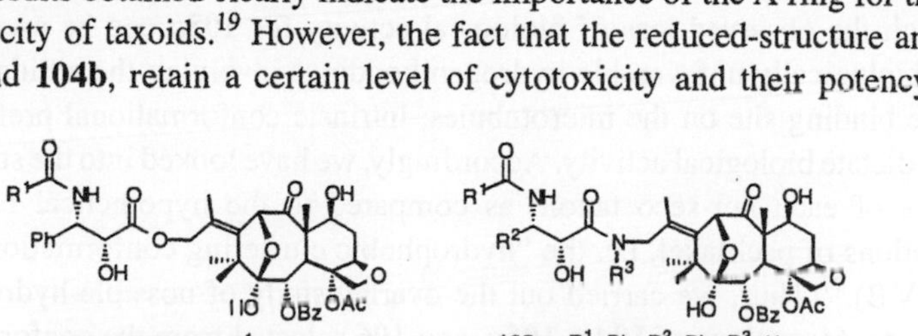

104a: SB-T-2002: R^1=Ph
104b: SB-T-2001: R^1=*t*-BuO

105a: R^1=Ph, R^2=Ph, R^3=H
105b: R^1=*t*-BuO, R^2=Ph, R^3=H
105c: R^1=*t*-BuO, R^2=(CH$_3$)$_2$C=CH, R^3=H
105d: R^1=*t*-BuO, R^2=(CH$_3$)$_2$CHCH$_2$, R^3=H
106: R^1=Ph, R^2=Ph, R^3=CH$_3$

Table 16. Cytotoxicity of Nor-seco Taxoids (IC$_{50}$, nM)^a

Taxoid	A121 (ovarian)	A549 (NSCL)	HT-29 (colon)	MCF-7 (breast)	MCF7-R (breast)
Paclitaxel	6.3	3.6	3.6	1.7	299
104a	117	133	134	79	471
104b	131	169	171	101	360
105a	>1000	>1000	>1000	>1000	>1000
105b	>1000	>1000	>1000	>1000	>1000
105c	>1000	>1000	>1000	>1000	>1000
105d	>1000	>1000	>1000	>1000	>1000
106	535	708	292	200	>1000

Note: ^{a}See footnote in Table 3.

MCF7-R is comparable to that of paclitaxel is rather surprising. Replacement of the C-13 ester linkage with an *N*-H amide linkage is deleterious to cytotoxicity: similar findings have been reported by Chen et al.[97] for paclitaxel analogues bearing an *N*-H amide linkage. It is quite unexpected and intriguing that the introduction of *N*-Me group to the amide linkage, i.e. nor-seco taxoid **106**, recovers cytotoxicity substantially.

In order to explain the intriguing structure–activity relationships observed, we carried out molecular modeling studies on the representative analogues **104a** (ester linkage), **105a** (*N*-H amide linkage) and **106** (*N*-Me amide linkage). The RMD study on these three analogues confirmed the greater dynamism of the phenylisoserine moiety at the C-13 position of the nor-seco taxoids than that of paclitaxel[77] as expected. The cleavage of the taxane A ring allows for free rotation around the C12–C13 bond, resulting in different conformations for the phenylisoserine moiety mostly not related to potentially bioactive conformations of paclitaxel. We believe this conformational dynamism is responsible for the lack of or low activity of the nor-seco taxoids.

Although the observed loss of biological activity for **105a** can be ascribed to possible biologically unfavorable hydrogen bondings involving the amide hydrogen at the binding site on the microtubules, intrinsic conformational preferences may well dictate biological activity. Accordingly, we have looked into the structural preference of each nor-seco taxoid as compared to the hypothetical bioactive conformations of paclitaxel, i.e. the "hydrophobic clustering conformations" (see Section IV.B).[77] Thus, we carried out the overlay study of possible hydrophobic clustering conformations of **104a**, **105a**, and **106** selected from the conformations obtained from the RMD study mentioned above, albeit in low dynamic population with that of paclitaxel which is predominant in water.[77] These studies have revealed that (i) the reduced yet significant activity of nor-seco taxoids **104a** (C-13 ester linkage) and **106** (C-13 *N*-Me amide linkage) is explained by their ability to adopt the hydrophobic clustering conformation of paclitaxel, which is proposed to be responsible for its strong cytotoxicity, and (ii) the low-energy hydrophobic clustering conformation of nor-seco taxoid **105a** (C-13 *N*-H amide linkage) overlaps poorly with that of paclitaxel, that is reflected in the loss of activity.

VI. PHOTOAFFINITY LABELING STUDIES: PROBING THE MICROTUBULE AND P-GLYCOPROTEIN BINDING DOMAINS

Careful analysis of the structural requirements for strong cytotoxicity should provide the basis for the rational design of the next generation taxoids with improved activity, bioavailability, and negligible drug resistance. However, the lack of direct structural information about the three-dimensional tubulin binding site on the microtubules for paclitaxel makes rational drug design difficult. Nevertheless, photoaffinity labeling and protein sequencing can provide direct information about the structure of the drug binding site,[66,98–100] although X-ray analysis or electron crystallography may eventually provide such information.[101] Recently, photoaffinity analogues of paclitaxel, 3'-(4-azidobenzamido)paclitaxel[98–100] and 2-(3-azidobenzoyl)paclitaxel[102] were found to interact with the N-terminal amino acid residues (1–31) and the 15 amino acid residues (217–231), respectively, of β-tubulin representing two different domains of the binding site for paclitaxel in β-tubulin. In this section, we describe our own contribution toward the characterization of the paclitaxel binding site of tubulin as well as that of P-glycoprotein using photoaffinity analogues of paclitaxel.[66]

We have developed a new photoreactive analogue of paclitaxel, 3'-*N*-BzDC-3'-*N*-debenzoylpaclitaxel (**109**) and its ditritiated derivative ([³H]-**109**) has been evaluated for its ability to photolabel tubulin and P-glycoprotein.[66] Radiolabeled photoreactive analogue [³H]-**109** was synthesized by *N*-acylation of 3'-*N*-debenzoyl-2',7-bis(*O*-TES)paclitaxel (**108**) with *N*-(2,3-ditritio-3-(4-benzoylphenyl)propanoyloxy)succinimide ([³H]-**107**), followed by purification on a reversed phase semipreparative HPLC using a C-18 column (Scheme 21).[66] Photoaffinity label [³H]-**109** was assessed to possess >99.9% radiochemical purity and a high specific radioactivity (34 Ci/mmol).

Although **109** was not as potent as paclitaxel, it was found to induce tubulin polymerization and stabilize the resulting microtubules in the same manner as paclitaxel, which allows for the use of [³H]-**109** as an excellent photoaffinity probe. Photoaffinity labeling of microtubules by [³H]-**109** showed the major radioactive band on the fluorograph between the α- and β-tubulin bands. It is very likely that [³H]-**109** is labeling β-tubulin but that the photoderivatized β-subunit has a somewhat altered mobility on the gel. The specificity of the binding of [³H]-**109** to tubulin was examined by a competition experiment using a 500-fold molar excess

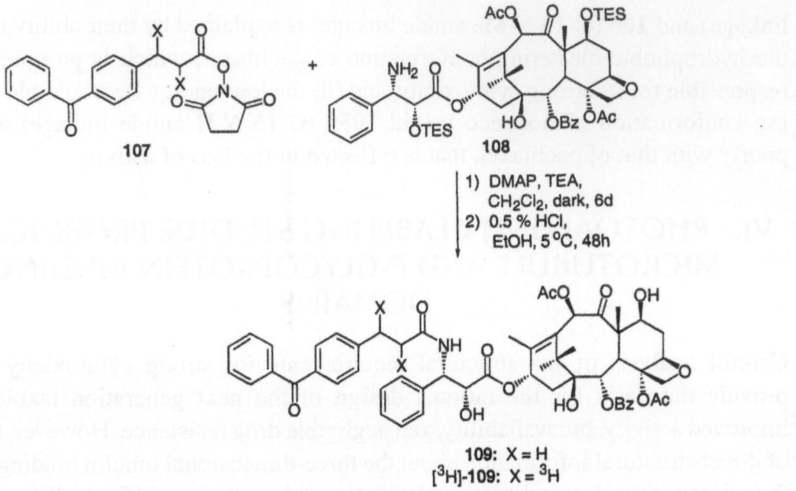

1) DMAP, TEA,
   CH$_2$Cl$_2$, dark, 6d
2) 0.5 % HCl,
   EtOH, 5 °C, 48h

107

108

109: X = H
[^{3}H]-109: X = ^{3}H

Scheme 21.

of paclitaxel. Results clearly indicate that they are indeed binding to the same or an overlapping binding site.

We encountered multidrug resistance (MDR) in the context of the development of the second-generation taxoids (Sections III.C, IV.A, and V.C) that are characterized by their remarkable potency against cancer cell lines expressing the MDR phenotype. In mammalian cancer cells, overexpressed P-glycoprotein, a transmembrane protein, is believed to be primarily responsible for MDR. P-Glycoprotein functions as a drug efflux pump for hydrophobic anticancer agents such as paclitaxel and doxorubicin, preventing their accumulation in the cytosol at cytotoxic level (Figure 8). Previous photoaffinity labeling studies of P-glycoprotein were not based on the photoaffinity labels of anticancer drugs, but on the photolabeled calcium-channel blockers which allowed for the mapping of two drug-binding domains of P-glycoprotein.[103–105] Thus, [^{3}H]-109 was the first photoaffinity analogue of an anticancer drug (paclitaxel) which was used for the photolabeling of P-glycoprotein.[66]

The photoreactive analogue [^{3}H]-109 specifically photolabeled different isoforms of murine P-glycoprotein present in drug-resistant cell lines, T1 (paclitaxel-resistant), V1, and V3 (vinblastine-resistant). Western blot analysis using a specific antibody to P-glycoprotein confirmed that the radiolabeled bands were indeed P-glycoprotein. The J7 parental drug-sensitive cells do not express P-glycoprotein in appreciable amounts, and thus no membrane associated protein was photolabled by [^{3}H]-109 in these cancer cells. A 25-fold molar excess of paclitaxel effectively displaced the [^{3}H]-109 from the P-glycoprotein binding site. To the best of our

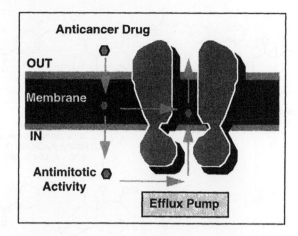

Figure 8. Mechanism of action of P-glycoprotein in causing multidrug resistance.

knowledge, this is the first successful photoaffinity labeling of P-glycoprotein with a photoreactive analogue of an anticancer agent. [66]

We synthesized the second ditritiated photoaffinity analogue [³H]-**114**, with the benzophenone photoreactive group at the C-7 position of paclitaxel (Scheme 22). The photolabeling of P-glycoprotein with [³H]-**114** was successfully carried out in the same manner as the case of [³H]-**109**. The subsequent protein sequencing of the labeled P-glycoproteins using [³H]-**109** and [³H]-**114** identified two separate binding domains. It was found that [³H]-**109** was incorporated into the peptide sequence 985–1088, while [³H]-**114** into the peptide sequence 683–760. The results strongly suggest that there is a specific binding site for paclitaxel in P-glycoprotein. Although preliminary, this information should eventually lead to identification of the binding pocket(s) in P-glycoprotein for paclitaxel, and may reveal the molecular architecture of the transporter domain.

The results of the photolabeling experiments on P-glycoprotein strongly indicate that the excellent cytotoxicity of the second-generation taxoids e.g. **28a** (SB-T-

28a (SB-T-1213)

97b (SB-T-101131)

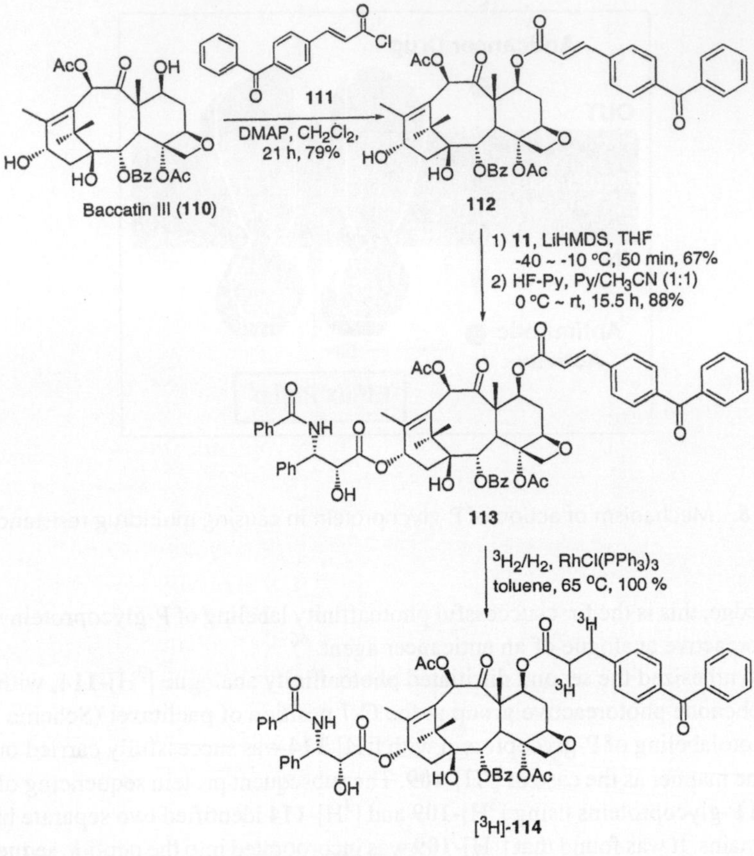

Scheme 22.

1213) and **97b** (SB-T-101131), against the multidrug-resistant cells is very likely to be related to the substantially muted binding ability of the second-generation taxoids to P-glycoprotein. Consequently, it is strongly suggested that the remarkable effects of the C-3′ and C-10 modifications, observed in these taxoids, on the activity against drug-resistant cancer cells expressing the MDR phenotype may well be ascribed to the effective inhibition of the P-glycoprotein binding by these modifications, especially the acylation at C-10 as well as the introduction of nonaromatic substituent at C-3′ that decreases lipophilicity, leading to the circumvention of the efflux mechanism.

Further studies are actively underway into this exciting area of research that promises to further enhance our understanding of the complex mechanisms of drug resistance and thereby allowing the design of even more promising next generation taxoid anticancer agents.

VII. FURTHER STUDIES ON THE BIOLOGICAL ACTIVITY OF TAXOIDS: ALTERNATIVE MECHANISMS OF ACTION

It has been shown that paclitaxel, in addition to its known unique mechanism of action on microtubules, mimics certain effects of bacterial lipopolysaccharide (LPS) on murine macrophages (Mϕ).[106] Similar to LPS, paclitaxel also stimulates macrophages to produce active mediators such as nitric oxide (NO) and tumor necrosis factor (TNF), causing endotoxic shock.[107,108] Thus paclitaxel is an LPS-receptor agonist. We investigated the structure–activity relationships for the taxoid-induced Mϕ activation to gain new insight into what might be an entirely new mechanism of action, albeit minor or auxiliary, for these compounds.[109]

A series of taxoids modified at the C-3′ and the C3′–N position (Figure 9) was synthesized and assayed for their ability to induce NO or TNF production by murine C3H/HeN (LPS-responsive) and C3H/HeJ (LPS-hyporesponsive) Mϕ and for inhibition of the growth of Mϕ-like cell lines, as well as that of LPS-responsive J774.1 and its LPS-hyporesponsive mutant J7.DEF3 cell lines.[109] The SAR study revealed the structural requirements to be entirely different from those for strong cytotoxicity.

We have found that the substituent at the C3′–N position (R²) of the taxoid plays the most important role for the activation of C3H/HeN macrophages to induce NO/TNF production. Thus, benzoyl, 4-fluorobenzoyl, 4-methylbenzoyl, and naphthalenecarbonyl groups at the C3′–N position exhibit strong activity, while others show a substantial decrease in activity. There is good correlation between NO and TNF production, with the exception of the *tert*-butoxycarbonyl group or the thiophene-2-carbonyl group that favors TNF over NO production. As described in the previous sections, the SAR study of taxoids has shown that *N-tert*-butoxy-carbonyl at the C3′-N position is a substantially better acyl group than *N*-benzoyl group for cytotoxicity, whereas the aromatic acyl group at this position appears to be critical for effective macrophage activation.

The substituent at the C-3′ position (R³) is found to have little effect on the production of TNF or NO. This finding forms a sharp contrast to the SAR of the C-3′ substituent for cytotoxicity. The acetyl group at the C-10 position appears to be important for better activity for macrophage activation.

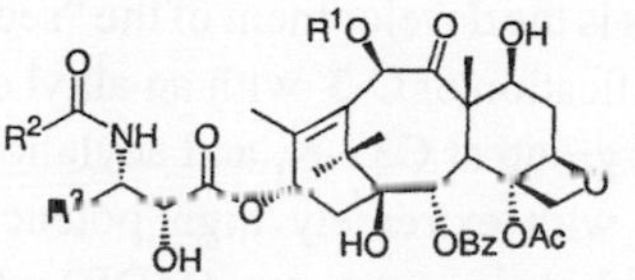

R¹ = H, Ac
R² = Ph, t-BuO, 4-F-Ph, 2-thienyl, 2-furyl, biphenyl, styryl, cyclohexyl, 1-naphtyl, 2-naphthyl
R³ = Ph, 2-methylprop-1-enyl, 4-F-Ph

Figure 9. C-3′ and C-3′-N modified taxoids for the SAR study of TNF/NO induction by murine macrophages.

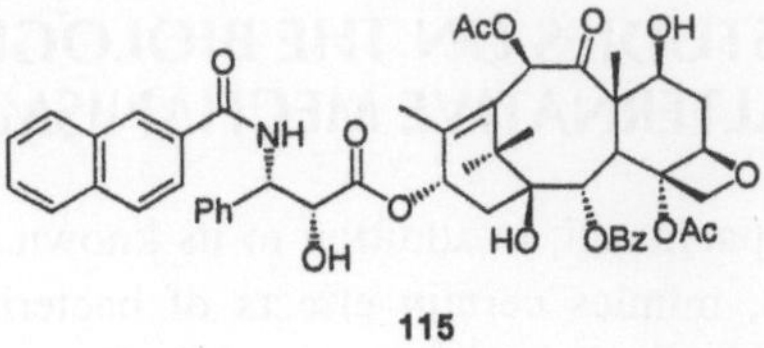

115

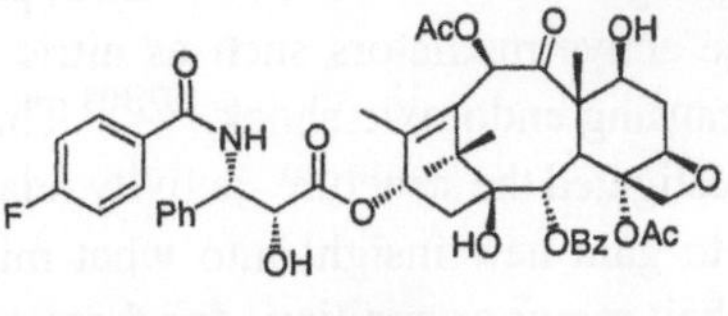

116

Figure 10. Potent inducers of TNF/NO production by murine macrophages.

As expected, when the taxoids were assayed for their activities against the LPS-hyporesponsive C3H/HeJ macrophages, there was no production of either TNF or NO. Figure 10 shows two potent inducers of TNF/NO production by murine macrophages found in this SAR study.

The growth inhibitory activity, i.e. antiproliferative activity, assay of this series of taxoids against LPS-responsive J774.1 cell lines has disclosed the absence of correlation and in some cases the presence of inverse correlation between the structural requirements for TNF/NO inducibility and growth inhibitory activity. Certain taxoids also show differences in their ability of inducing TNF and NO production, indicating the presence of two independent mechanisms for activation of macrophage to induce the production of TNF and NO. These preliminary results are very encouraging and intriguing; they warrant further investigation of this possible alternative mechanism of action for paclitaxel and taxoids based on the immune system.

VIII. SUMMARY AND FUTURE PROSPECTS

The SAR studies of taxoids in our laboratory has clearly highlighted the importance of several functionalities in the taxoids for optimum cytotoxicity. One of the most significant contributions out of these studies is the development of the "second-generation taxoids" in which appropriate modification of C-3' with an alkyl or alkenyl group, placement of a *tert*-butoxycarbonyl group at C3'–N, and acylation at C-10 with different modifiers provide taxoids with extremely high potency against drug-resistant cancer cells expressing multidrug-resistance (MDR) phenotype. These second-generation taxoids are highly promising candidates for clinical trials because of their excellent in vivo activities. In addition, the modification at the C-2 benzoate with alkenyl and alkyl groups has resulted in highly active, entirely nonaromatic second-generation taxoids.

Our studies on fluorine-containing taxoids have produced several highly active and exciting lead compounds, in particular 3'-trifluoromethyl taxoids. Studies on the role of the fluorine atom in inhibiting metabolic pathways as well as the SAR for their ability to induce apoptosis is actively underway. Moreover, we have introduced a novel fluorine probe approach toward conformational analysis. Fluorine atoms are placed at key positions of paclitaxel and its congeners that have negligible effects on the activity or conformational disposition of the parent drug, and are used in a variety of NMR methods to analyze solution and solid-state conformations. The fluorine probe approach has proven highly effective for the identification of bioactive conformations either at the first recognition by the tubulin binding site or in the microtubule-bound status.

Development of the 14β-hydroxy taxoid series is a highly successful venture as well. We have developed new chemistry and obtained some advanced candidates for clinical trials which may indeed have improved bioavailability and unique tumor specificity profiles.

The β-lactam synthon method (β-LSM) including our highly efficient coupling protocol with baccatins has enabled us to have a rapid access to a variety of new taxoids. This protocol has also been applied to the synthesis of excellent photoaffinity labels for tubulin and P-glycoprotein. P-Glycoprotein photoaffinity labeling with two radiolabeled photoreactive paclitaxel analogues and the subsequent protein sequencing have identified two specific domains for paclitaxel binding. Further studies on these are actively underway which may lead to the better understanding of the MDR mechanism as well as to the development of the anticancer drugs and/or MDR reversal agents of the future.

Our SAR study of paclitaxel analogues on their ability to activate macrophage, inducing the production of NO and TNF, has revealed stark differences in the structural requirements for cytotoxicity vs. macrophage activation. The results warrant a great deal of further study on the possible alternative or auxiliary mechanism of action for paclitaxel and taxoids, which might lead to the discovery of a new series of taxoid anticancer agents with unique mechanism of action.

It is obvious that there are more exciting new chemistry and biology to be explored by further investigations on taxoid anticancer agents. Extensive research along this line is actively underway in these laboratories, and will be reported in future accounts.

ACKNOWLEDGMENTS

This work has been supported by a grant from the National Institutes of Health (NIGMS). Generous support from Indena SpA, Rhône-Poulenc Rorer, and the Center for Biotechnology at Stony Brook which is sponsored by the New York State Science and Technology Foundation is acknowledged. The authors would like to extend their sincere thanks to Dr.

Ralph J. Bernacki, Paula Pera, and Jean M. Veith of the Grace Cancer Drug Center, Roswell Park Memorial Research Institute, Dr. Ezio Bombardelli of Indena, SpA, Drs. François Lavelle and Marie C. Bissery of Rhône-Poulenc Rorer, Professors Susan B. Horwitz and George Orr of the Albert Einstein College of Medicine, Professor Ann McDermott and Dr. Lane Gilchrist of Columbia University, and Dr. Teruo Kirikae of the Jichi Medical School for their productive collaborations. The authors thank Drs. John J. Walsh, Pierre-Yves Bounaud, and Michael L. Miller for their assistance in writing this chapter.

REFERENCES

1. Georg, G. I.; Chen, T. T.; Ojima, I.; Vyas, D. M. *Taxane Anticancer Agents: Basic Science and Current Status*; American Chemical Society: Washington DC, 1995.
2. Suffness, M. *Taxol: Science and Applications*; CRC Press: New York, 1995.
3. Rowinsky, E. K.; Onetto, N.; Canetta, R. M.; Arbuck, S. G. *Seminars in Oncology* 1992, *19*, 646–662.
4. Rowinsky, E. K. *Ann. Rev. Med.* 1997, *48*, 353–374.
5. Wani, M. C.; Taylor, H. L.; Wall, M. E.; Coggon, P.; McPhail, A. T. *J. Am. Chem. Soc.* 1971, *93*, 2325–2327.
6. Guénard, D.; Guéritte-Vogelein, F.; Potier, P. *Acc. Chem. Res.* 1993, *26*, 160–167.
7. Shiff, P. B.; Fant, J.; Horwitz, S. B. *Nature* 1979, *277*, 665–667.
8. Shiff, P. B.; Horwitz, S. B. *Proc. Natl. Acad. Sci. USA* 1980, *77*, 1561–1565.
9. Vallee, R. B. In *Taxol®: Science and Applications*; M. Suffness, Ed.; CRC Press: New York, 1995, pp. 259–274.
10. Ringel, I.; Horwitz, S. B. *J. Natl. Cancer Inst.* 1991, *83*, 288–291.
11. Holton, R. A.; Somoza, C.; Kim, H.-B.; Liang, F.; Biediger, R. J.; Boatman, P. D.; Shindo, M.; Smith, C. C.; Kim, S.; Nadizadeh, H.; Suzuki, Y.; Tao, C.; Vu, P.; Tang, S.; Zhang, P.; Murthi, K. K.; Gentile, L. N.; Liu, J. H., *J. Am. Chem. Soc.* 1994, *116*, 1597–1598.
12. Holton, R. A.; Kim, H.-B.; Somoza, C.; Liang, F.; Biediger, R. J.; Boatman, P. D.; Shindo, M.; Smith, C. C.; Kim, S.; Nadizadeh, H.; Suzuki, Y.; Tao, C.; Vu, P.; Tang, S.; Zhang, P.; Murthi, K. K.; Gentile, L. N.; Liu, J. H. *J. Am. Chem. Soc.* 1994, *116*, 1599–1600.
13. Nicolaou, K. C.; Yang, Z.; Liu, J. J.; Ueno, H.; Nantermet, P. G.; Guy, R. K.; Claiborne, C. F.; Renaud, J.; Couladouros, E. A.; Paulvannan, K.; Sorensen, E. J. *Nature* 1994, *367*, 630–634.
14. Danishefsky, S.; Masters, J.; Young, W.; Link, J.; Snyder, L.; Magee, T.; Jung, D.; Isaacs, R.; Bornmann, W.; Alaimo, C.; Coburn, C.; Di Grandi, M. *J. Am. Chem. Soc.* 1996, *118*, 2843–2859.
15. Wender, P. A.; Badham, N. F.; Conway, S. P.; Floreancig, P. E.; Glass, T. E.; Houze, J. B.; Krauss, N. E.; Lee, D.; Marquess, D. G.; McGrane, P. L.; Meng, W.; Natchus, M. G.; Shuker, A. J.; Sutton, J. C.; Taylor, R. E. *J. Am. Chem. Soc.* 1997, *119*, 2757–2758.
16. Mukaiyama, T.; Shiina, I.; Iwadare, H.; Sakoh, H.; Tani, Y.; Hasegawa, M.; Saitoh, K. *Proc. Japan Acad.* 1997, *73, Ser. B*, 95–100.
17. Kawahara, S.; Nishimori, T.; Kusama, H.; Kuwajima, I. *7th International Kyoto Conference on Organic Chemistry* 1997, *Abstracts*, OP-44.
18. Denis, J.-N.; Greene, A. E.; Guénard, D.; Guéritte-Voegelein, F.; Mangatal, L.; Potier, P. *J. Am. Chem. Soc.* 1988, *110*, 5917–5919.
19. Kingston, D. G. I. *Pharmacol. Ther.* 1991, *52*, 1–34 and references cited therein.
20. Suffness, M. In *Annual Reports in Medicinal Chemistry*; Bristol, J. A., Ed.; Academic Press: San Diego, 1993, Vol. 28; Chap. 32, pp. 305–314.
21. Holms, F. A.; Kudelka, A. P.; Kavanagh, J. J.; Huber, M. H.; Ajani, J. A.; Valero, V. In *Taxane Anticancer Agents: Basic Science and Current Status*; *ACS Symp. Ser. 583*; Georg, G. I.; Chen, T. T.; Ojima, I., Vyas, D. M., Ed.; American Chemical Society: Washington, DC, 1995, pp 31–57.

22. Mangatal, L.; Adeline, M. T.; Guénard, D.; Guéritte-Voegelein, F.; Potier, P. *Tetrahedron* **1989**, *45*, 4177–4190.

23. Holton, R. A.; Biediger, R. J.; Boatman, P. D. In *Taxol®: Science and Applications*; Suffness, M, Ed.; CRC Press: New York, 1995, pp 97–121.

24. Holton, R. A. "Method for Preparation of Taxol Using Oxazinone," U.S. Patent, 1991, 5,175,315.

25. Ojima, I.; Habus, I.; Zhao, M.; Zucco, M.; Park, Y. H.; Sun, C.-M.; Brigaud, T. *Tetrahedron* **1992**, *48*, 6985–7012.

26. Ojima, I.; Sun, C. M.; Zucco, M.; Park, Y. H.; Duclos, O.; Kuduk, S. D. *Tetrahedron Lett.* **1993**, *34*, 4149–4152.

27. Commerçon, A.; Bourzat, J. D.; Didier, E.; Lavelle, F., In *Taxane Anticancer Agents: Basic Science and Current Status*; Georg, G. I.; Chan, T. T; Ojima, I.; Vyas, D. M., Ed.; American Chemical Society: Washington, DC, 1995, pp 233–246.

28. Kingston, D. G. I.; Chaudhary, A. G.; Gunatilaka, A. A. L.; Middleton, M. L. *Tetrahedron Lett.* **1994**, *35*, 4486–4489.

29. Georg, G. I.; Harriman, G. C. B.; Vander Velde, D. G.; Boge, T. C.; Cheruvallath, Z. S.; Datta, A.; Hepperle, M.; Park, H.; Himes, R. H.; Jayasinghe, L. In *Taxane Anticancer Agents: Basic Science and Current Status*; Georg, G. I.; Chen, T. T.; Ojima, I.; Vyas, D. M., Ed.; *ACS Symp. Series 583*; American Chemical Society: Washington DC, 1995, pp 217–232.

30. Kingston, D. G. I. In *Taxane Anticancer Agents: Basic Science and Current Status; ACS Symp. Ser. 583*; Georg, G. I.; Chen, T. T.; Ojima, I.; Vyas, D. M., Ed.; American Chemical Society: Washington, DC, 1995, pp 203–216.

31. Klein, L. L.; Li, L.; Yeung, C. M.; Maring, C. J.; Thomas, S. A.; Grampovnik, D. J.; Plattner, J. J. In *Taxane Anticancer Agents: Basic Science and Current Status; ACS Symp. Ser. 583*; Georg, G. I.; Chen, T. T; Ojima, I.; Vyas, D. M., Ed.; American Chemical Society: Washington, DC, 1995, pp 276–287.

32. Ojima, I.; Park, Y. H.; Fenoglio, I.; Duclos, O.; Sun, C.-M.; Kuduk, S. D.; Zucco, M.; Appendino, G.; Pera, P.; Veith, J. M.; Bernacki, R. J.; Bissery, M.-C.; Combeau, C.; Vrignaud, P.; Riou, J. F.; Lavelle, F. In *Taxane Anticancer Agents: Basic Science and Current Status; ACS Symp. Ser. 583*; Georg, G. I.; Chen, T. T.; Ojima, I.; Vyas, D. M, Ed.; American Chemical Society: Washington, D. C., 1995, pp 262–275.

33. Ojima, I.; Habus, I.; Zhao, M.; Georg, G. I.; Jayasinghe, R. *J. Org. Chem.* **1991**, *56*, 1681–1684.

34. Ojima, I.; Zucco, M.; Duclos, O.; Kuduk, S. D.; Sun, C.-M.; Park, Y. H. *Bioorg. Med. Chem. Lett.* **1993**, *3*, 2479–2482.

35. Ojima, I.; Park, Y. H.; Sun, C.-M.; Fenoglio, I.; Appendino, G.; Pera, P.; Bernacki, R. J. *J. Med. Chem.* **1994**, *37*, 1408–1410.

36. Ojima, I.; Fenoglio, I.; Park, Y. H.; Pera, P.; Bernacki, R. J. *Bioorg. Med. Chem. Lett.* **1994**, *4*, 1571–1576.

37. Ojima, I.; Fenoglio, I.; Park, Y. H.; Sun, C.-M.; Appendino, G.; Pera, P.; Bernacki, R. J. *J. Org. Chem.* **1994**, *59*, 515–517.

38. Ojima, I.; Duclos, O.; Zucco, M.; Bissery, M.-C.; Combeau, C.; Vrignaud, P.; Riou, J. F.; Lavelle, F. *J. Med. Chem.* **1994**, *37*, 2602–2608.

39. Ojima, I.; Duclos, O.; Kuduk, S. D.; Sun, C.-M.; Slater, J. C.; Lavelle, F.; Veith, J. M.; Bernacki, R. J. *Bioorg. Med. Chem. Lett.* **1994**, *4*, 2631–2634.

40. Ojima, I.; Kuduk, S. D.; Slater, J. C.; Gimi, R. H.; Sun, C. M. *Tetrahedron* **1996**, *52*, 209–224.

41. Ojima, I.; Slater, J. C.; Michaud, E.; Kuduk, S. D.; Bounaud, P.-Y.; Vrignaud, P.; Bissery, M.-C.; Veith, J.; Pera, P.; Bernacki, R. J. *J. Med. Chem.* **1996**, *39*, 3889–3896.

42. Ojima, I.; Kuduk, S. D.; Slater, J. C.; Gimi, R. H.; Sun, C. M.; Chakravarty, S.; Ourevich, M.; Abouabdellah, A.; Bonnet-Delpon, D.; Bégué, J.-P.; Veith, J. M.; Pera, P.; Bernacki, R.J. In *Biomedical Frontiers of Fluorine Chemistry, ACS Symp. Ser. 639*; Ojima, I.; McCarthy, J. R.; Welch, J. T., Ed.; American Chemical Society: Washington, DC, 1996, pp 228–243.

43. Ojima, I.; Slater, J. S.; Kuduk, S. D.; Takeuchi, C. S.; Gimi, R. H.; Sun, C.-M.; Park, Y. H.; Pera, P.; Veith, J. M.; Bernacki, R. J. *J. Med. Chem.* **1997**, *40*, 267–278.

44. Ojima, I.; Kuduk, S. D.; Pera, P.; Veith, J. M.; Bernacki, R. J. *J. Med. Chem.* **1997**, *40*, 279–285.

45. Ojima, I.; Slater, J. C.; Pera, P.; Veith, J. M.; Abouabdellah, A.; Bégué, J.-P.; Bernacki, R. J., "Synthesis and Biological Activity of Novel 3'-Trifluoromethyl Taxoids", *Bioorg. Med. Chem. Lett.* **1997**, *7*, 209–214.

46. Ojima, I.; Slater, J. C. *Chirality* **1997**, *9*, 487–494.

47. Ojima, I. In *The Organic Chemistry of β-Lactam Antibiotics*; Georg, G. I., Ed.; VCH Publishers: New York, 1992, pp 197–255.

48. Ojima, I.; Sun, C. M.; Park, Y. H. *J. Org. Chem.* **1994**, *59*, 1249–1250.

49. Ojima, I. *Acc. Chem. Res.* **1995**, *28*, 383–389 and references cited therein.

50. Ojima, I., "Process for the Production of Chiral Hydroxy-β-Lactams and Hydroxyamino Acids Derived Therefrom," U.S. Patent, 1994, 5,294,737.

51. Ojima, I.; Park, Y. H.; Sun, C. M.; Brigaud, T.; Zhao, M. *Tetrahedron Lett.* **1992**, *33*, 5737–5740.

52. Ojima, I. In *Advances in Asymmetric Synthesis*; Hassner, A., Ed.; JAI Press: Greenwich, 1995, Vol. 1, pp 95–146.

53. Ojima, I.; Delaloge, F. *Chem. Soc. Rev.* **1997**, *26*, 377–386.

54. Holton, R. A., "Method for Preparation of Taxol," Eur. Pat. Appl. 1990, U.S. Patent, 1992, 5,175,315; EP 400,971, 1990: *Chem. Abstr.* **1990**, *114*, 164568q.

55. Georg, G. I.; Boge, T. C.; Cheruvallath, Z. S.; Clowers, J. S.; Harriman, G. C. B.; Hepperle, M.; Park, H. In *Taxol®: Science and Applications*; M. Suffness, Ed.; CRC Press: New York, 1995, pp 317–375.

56. Boge, T. C.; Himes, R. H.; Vander Velde, D. G.; Georg, G. I. *J. Med. Chem.* **1994**, *37*, 3337–3343.

57. Holton, R. A.; Chai, K. B.; Nadizadeh, H. "Preparation of Butenyl-substituted Taxanes as Anticancer Drugs," U.S. Patent, 1994, 5,284,864: *Chem. Abstr.* **1994**, *121*, 301103v.

58. Kant, J.; O'Keeffe, W. S.; Chen, S.-H.; Farina, V.; Fairchild, C.; Johnston, K.; Kadow, J. F.; Long, B. H.; Vyas, D. *Tetrahedron Lett.* **1994**, *35*, 5543–5546.

59. Rao, K. V.; Bhakuni, R. S.; Oruganti, R. S. *J. Med. Chem.* **1995**, *38*, 3411–3414.

60. Chen, S. H.; Kant, J.; Mamber, S. W.; Roth, G. P.; Wei, J.; Marshall, D.; Vyas, D.; Farina, V. *BioMed. Chem. Lett.* **1994**, *4*, 2223–2228.

61. Duggan, M. E.; Imagire, J. S. *Synthesis* **1989**, 131.

62. Chen, S.-H.; Fairchild, C.; Mamber, S. W.; Farina, V. *J. Org. Chem.* **1993**, *58*, 2927–2928.

63. Kirschner, L. S.; Greenberger, L. M.; Hsu, S. I.-H.; Yang, C.-P. H.; Cohen, D.; Piekarz, R. L.; Castillo, G.; Han, E. K.-H. H.; Yu, L.; Horwitz, S. B. *Biochem. Pharm.* **1992**, *43*, 77–78.

64. Gottesman, M. M.; Pastan, I. *Ann. Rev. Biochem.* **1993**, *62*, 385–427.

65. Sikic, B. I. *J. Clin. Oncol.* **1993**, *11*, 1629–1635.

66. Ojima, I.; Duclos, D.; Dormán, G.; Simonot, B.; Prestwich, G. D.; Rao, S.; Lerro, K. A.; Horwitz, S. B. *J. Med. Chem.* **1995**, *38*, 3891–3894.

67. Bounaud, P.-Y.; Ojima, I. *212th American Chemical Society National Meeting, Orlando, FL, August 25–29, 1996, Abstracts*, ORGN 0073.

68. Wu, Q.; Bounaud, P.-Y.; Kuduk, S. D.; Yang, C.-P. H.; Ojima, I.; Horwitz, S. B.; Orr, G. A. *Biochemistry* **1998**, *37*, 11272–11279.

69. Ojima, I.; Bounaud, P.; Takeuchi, C. S.; Pera, P.; Bernacki, R. J. *Bioorg. Med. Chem. Lett.* **1998**, *8*. In press.

70. Geran, R. I.; Greenberg, N. H.; MacDonald, M. M.; Schumacher, A. M.; Abbott, B. J. *Cancer Chemother. Rep. Part 3* **1972**, *3*, 1–103.

71. Ojima, I.; Lin, S. *J. Org. Chem.* **1998**, *63*, 224–225.

72. Nicolaou, K. C.; Renaud, J.; Nantermet, P. G.; Couladouros, E. A.; Guy, R. K.; Wrasidlo, W. *J. Am. Chem. Soc.* **1995**, *117*, 2409–2420.

73. Chakravarty, S.; Ojima, I. *The 214th American Chemical Society National Meeting, Las Vegas, September 8–12, 1997, Abstracts*, MEDI 075.

74. Vuilhorgne, M.; Gaillard, C.; Sanderlink, G. J.; Royer, I.; Monsarrat, B.; Dubois, J.; Wright, M. In *Taxane Anticancer Agents: Basic Science and Current Status; ACS Symp. Series 583*; Georg, G. I.; Chen, T. T.; Ojima, I.; Vyas, D. M., Ed.; American Chemical Society: Washington DC, 1995, pp 98–110.

75. Monsarrat, B.; Mariel, E.; Cros, S.; Garès, M.; Guénard, D.; Guéritte-Voegelein, F.; Wright, M. *Drug Metab. Dispos.* **1990**, *18*, 895–901.

76. Filler, R.; Kobayashi, Y.; Yagupolskii, L. M. *Organofluorine Compounds in Medicinal Chemistry and Biomedical Applications*; Elsevier: Amsterdam, 1993, Vol. 48.

77. Ojima, I.; Kuduk, S. D.; Chakravarty, S.; Ourevitch, M.; Bégué, J.-P. *J. Am. Chem. Soc.* **1997**, *119*, 5519–5527.

78. Gilchrist, L.; McDermott, A.; Ojima, I.; Kuduk, S. D.; Walsh, J. J.; Chakravarty, S.; Orr, G.; Horwitz, S. B., 1997. Unpublished results.

79. Whitesell, J. K.; Lawrence, R. M. *Chimia* **1986**, *40*, 318–321.

80. Oppolzer, W.; Chapuis, C.; Bernardinelli, G. *Tetrahedron Lett.* **1984**, *25*, 5885–5888.

81. Abouabdellah, A.; Bégué, J.-P.; Bonnet-Delpon, D. *Synlett* **1996**, 399–400.

82. Holton, R. A.; Biediger, R. J. "Certain Alkoxy Substituted Taxanes and Pharmaceutical Compositions Containing Them," U.S. Patent 1993, 5,243,045.

83. Georg, G. I. For a comprehensive review on the conformational studies on paclitaxel and docetaxel, see ref. 29.

84. Guéritte-Voegelein, F.; Mangatal, L.; Guénard, D.; Potier, P.; Guilhem, J.; Cesario, M.; Pascard, C. *Acta Crstallogr.* **1990**, *C46*, 781–784.

85. Williams, H. J.; Scott, A. I.; Dieden, R. A.; Swindell, C. S.; Chirlian, L. E.; Francl, M. M.; Heerding, J. M.; Krauss, N. E. *Can. J. Chem.* **1994**, 252–260.

86. Williams, H. J.; Scott, A. I.; Dieden, R. A.; Swindell, C. S.; Chirlian, L. E.; Francl, M. M.; Heerding, J. M.; Krauss, N. E. *Tetrahedron* **1993**, *49*, 6545–6560.

87. Vander Velde, D. G.; Georg, G. I.; Grunewald, G. L.; Gunn, C. W.; Mitscher, L. A. *J. Am. Chem. Soc.* **1993**, *115*, 11650–11651.

88. Mastropaolo, D.; Camerman, A.; Luo, Y.; Brayer, G. D.; Camerman, N. *Proc. Natl. Acad. Sci. USA* **1995**, *92*, 6920–6924.

89. Williams, H. J.; Moyna, G.; Scott, A. I.; Swindell, C. S.; Chirlian, L. E.; Heerding, J. M.; Williams, D. K. *J. Med. Chem.* **1996**, *39*, 1555–1559.

90. Paloma, L. G.; Guy, R. K.; Wrasidlo, W.; Nicolaou, K. C. *Chem. Biol.* **1994**, *2*, 107–112.

91. Chakravarty, S.; Ojima, I., **1997**. Unpublished results.

92. Ojima, I. *7th International Kyoto Conference on Organic Chemistry, Kyoto, Japan, November 10–14*, 1997, *Abstracts*, IL-19.

93. Appendino, G.; Gariboldi, P.; Gabetta, B.; Pace, R.; Bombardelli, E.; Viterbo, D. *J. Chem. Soc., Perkin Trans. 1* **1992**, 2925–2929.

94. Sharma, A.; Straubinger, R. M.; Ojima, I.; Bernacki, R. J. *J. Pharm. Sci.* **1995**, *84*, 1400–1404.

95. Kant, J.; Farina, V.; Fairchild, C.; Kadow, J. F.; Langley, D. R.; Long, B. H.; Rose, W. C.; Vyas, D. M. *Bioorg. Med. Chem. Lett.* **1994**, *4*, 1565–1570.

96. Ojima, I.; Lin, S.; Chakravarty, S.; Fenoglio, I.; Park, Y. H.; Sun, C. M.; Appendino, G.; Pera, P.; Veith, J. M.; Bernacki, R. J. *J. Org. Chem.* **1998**, *63*, 1637–1645.

97. Chen, S.; Farina, V.; Vyas, D.M.; Doyle, T. W.; Long, B. H.; Fairchild, C. *J. Org. Chem.* **1996**, *61*, 2065–2070.

98. Rao, S.; Horwitz, S. B.; Ringel, I. *J. Natl. Cancer Inst.* **1992**, *84*, 785–788.

99. Rao, S.; Krauss, N. E.; Heerding, J. M.; Swindell, C. S.; Ringel, I.; Orr, G. A.; Horwitz, S. B. *J. Biol. Chem.* **1994**, *269*, 3132–3134.

100. Dasgupta, D.; Park, H.; Harriman, G. C. B.; Georg, G. I.; Himes, R. H. *J. Med. Chem.* **1994**, *37*, 2976–2980.

101. Nogales, E.; Wolf, S. G.; Khan, I. A.; Ludeña, R. F.; Downing, K. H. *Nature* **1995**, *375*, 424–427.

102. Rao, S.; Orr, G. A.; Chaudhary, A. G.; Kingston, D. G. I.; Horwitz, S. B. *J. Biol. Chem.* **1995**, *270*, 20235–20238.
103. Yoshimura, A.; Kuwazuru, Y.; Sumizawa, T.; Ichikawa, M.; Ikeda, S.-I.; Uda, T.; Akiyama, S.-I. *J. Biol. Chem.* **1989**, *264*, 16282–16291.
104. Greenberger, L. M.; Lisanti, C. J.; Silva, J. T.; Horwitz, S. B. *J. Biol. Chem.* **1991**, *266*, 20744–20751.
105. Bruggemann, E. P.; Germann, U. A.; Gottesman, M. M.; Pastan, I. *J. Biol. Chem.* **1989**, *264*, 15483–15488.
106. Ding, A. H.; Porteu, F.; Sanchez, E.; Nathan, C. F. *Science* **1990**, *248*, 370–372.
107. Morrison, D. C.; Ryan, J. L. *Annu. Rev. Med.* **1987**, *38*, 417–432.
108. Rietschel, E. T.; Kirikae, T.; Schade, F. U.; Mamat, U.; Schmidt, G.; Loppnow, H.; Ulmer, A. J.; Zahrigner, U.; Seydel, U.; di Padova, F.; Schreier, M.; Brade, H. *FASEB J.* **1994**, *8*, 217–225.
109. Kirikae, T.; Ojima, I.; Kirikae, F.; Ma, A.; Kuduk, S. D.; Slater, J. C.; Takeuchi, C. S.; Bounaud, P.-Y.; Nakano, M. *Biochem. Biophys. Res. Commun.* **1996**, *227*, 227–235.

SYNTHESIS AND STRUCTURE–ACTIVITY RELATIONSHIPS OF PEROXIDIC ANTIMALARIALS BASED ON ARTEMISININ

Mitchell A. Avery, Maria Alvim-Gaston, and John R. Woolfrey

Advances in Medicinal Chemistry
Volume 4, pages 125–217.
Copyright © 1999 by JAI Press Inc.
All rights of reproduction in any form reserved.
ISBN: 0-7623-0064-7

I. INTRODUCTION

Malaria is a leading cause of infant mortality worldwide.[1] The prevalent malaria parasite *Plasmodium falciparum* is increasingly resistant to traditional antimalarials and continues to spread.[2–5] New effective antimalarial agents are urgently needed and one of the few currently under development[6] was found to be the primary antimalarial constituent of the Chinese medicinal herb Qinghao,[7] derived from the plant *Artemisia annua L*. Isolation and structural characterization showed the novel natural product (+)-artemisinin (**1**; qinghaosu, QHS) to be a remarkably stable peroxide.[8–10] Unfortunately artemisinin is not readily available from the plant relative to its modest antimalarial activity, is insoluble in aqueous media, is not devoid of toxicity, and, perhaps most problematically, has only a short plasma half-life. Nevertheless, these shortcomings are mitigated by the ability of the artemisinin class of antimalarials to rapidly clear parasitemia, including severe cerebral cases, as well as having good relative potency against resistant strains of *P. falciparum*. The need to develop derivatives of the natural product which have superior pharmacological properties continues to motivate contributions to synthetic approaches,[11–14] total syntheses,[15–17] and analogue-based structure–activity relationship (SAR) studies. This review describes the growth of our broad program based on artemisinin and derivatives thereof, and does not intentionally omit the praiseworthy efforts of our colleagues (e.g. Acton, Brossi, Bunnelle, Bustos, Casteel, Haynes, Jefford, Jung, Kepler, Klayman, Lansbury, Lee, Li, Lin, Little, McChesney, O'Neill, Posner, Thebtaranonth, Roth, Vennerstrom, Venugopalan, Xu, Ye, Zhou, and Ziffer).

Initially, we sought a practical total synthesis of the natural enantiomer (+)-artemisinin (**1**) to support clinical therapeutic studies with artemisinin and derived prodrug congeners, such as artemether and artesunate, and by providing a radiolabeled version of artemisinin for metabolism and mode of action studies (Eq. 1). Associated model studies from our total synthesis resulted in numerous additional analogues for our fledgling SAR study and the conception of other structural classes. As we completed our practical total synthesis of (+)-artemisinin, the versatility and novelty of our approach was proven by placing substituents on the tetracyclic skeleton.

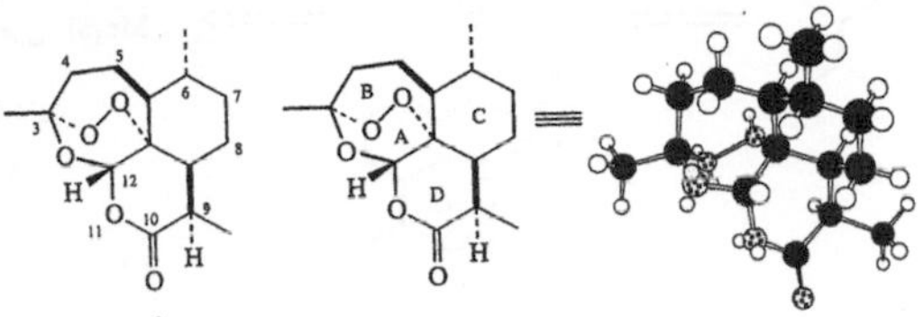

(1)

From the beginning of our interest in artemisinin, we suspected a site of action with structural requirements, or putative "receptor." Early studies showed that simple peroxides did not display comparable antimalarial activity.[18] The first analogues were derived from precious artemisinin itself, and some are currently under worldwide clinical development: notable dihydroartemisinin derivatives include artemether, arteether, artesunate, and artelinic acid.[19–22] Recently, various groups obtained (+)-artemisinin from (+)-artemisinic acid.[23–26] Synthetic methodology has significantly added to the numerous tetracyclic analogues of artemisinin: racemic 6,9-desmethylartemisinin,[27–29] 10-deoxo analogues,[30–33] and 10-substituted-10-deoxoartemisinin derivatives.[34,35] Since the mechanism of action of artemisinin is still under debate, a pharmacophore was sought via systematic dissection and elaboration of the artemisinin structure. Many groups including ours sought simplified analogues based on substructures as a potential practical alternative to lengthy total synthesis. Easily prepared substructures have been extensively featured in numerous communications and have included simple 1,2,4-trioxanes,[36–42] 1,2,4,43-tetraoxanes,[18,43] and fragments of the artemisinin ring system: a C/D portion,[44] A/B/C portion,[45–48] and the A/C/D portion.[49–51] We also describe herein our attempts at a broad QSAR model which reconcile the biological results of our numerous classes of analogues with the reported SAR from other laboratories.

II. TOTAL SYNTHESIS OF (+)-ARTEMISININ

From a retrosynthetic standpoint, we felt that the most obvious intermediate in the production of **1** would be the "unraveled" α-hydroperoxyaldehyde **2** because in a ketalization-like process, simple cyclodehydration of **2** should readily furnish the tetracyclic natural product **1**. The inherent synthetic challenge for **1** thus lies in the preparation of the unstable aldehyde **2** (Eq. 2), and in commendable fashion others have employed an enol ether photo-oxygenation as entry to that functional arrangement.[15,16,52] Other more biomimetic approaches[14,53] have been explored involving photo-oxygenation of artemisinic acid[23,24] or arteannuin B,[25] both natural products co-occurring with artemisinin in *A. annua*.

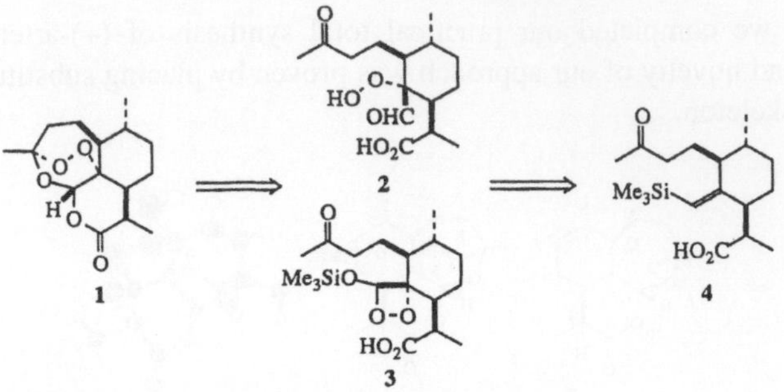

(2)

In contrast, we took advantage of the addition of ozone to a vinyl silane to produce the desired α-hydroperoxycarbonyl moiety as described by Büchi and Wüest.[54] Thus, the next retron in our analysis was the 2β, 6β-disubstituted cyclohexenyl silane 4, which, according to Büchi, would afford 2 or the synthetically equivalent dioxetane 3 upon exposure to ozone.

Initially we chose to test elements of this approach to artemisinin in an abbreviated version of 4 that lacked the 2β-(3-oxobutyl) and 3α-methyl groups. Hence, on low-temperature ozonolysis of the vinyl silane 5 in methanol, transient and stereo-exclusive formation of dioxetane 6 was observed upon immediate analysis by NMR (Eq. 3). On standing, the dioxetane 6 underwent rearrangement and cyclization to furnish hydroperoxy-lactone 7 in 54% isolated yield on a scale sufficient for X-ray structural study.[44]

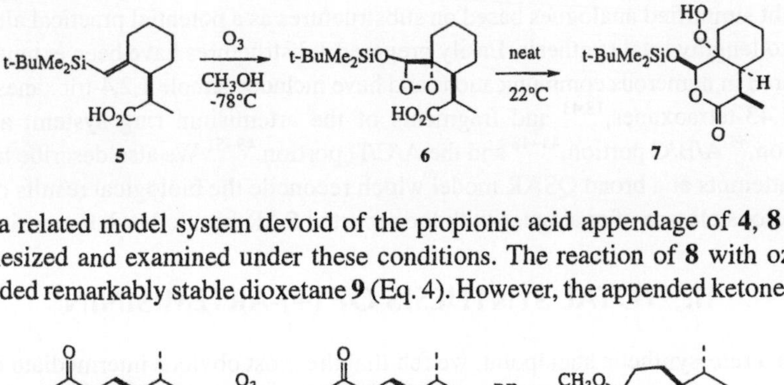

(3)

In a related model system devoid of the propionic acid appendage of 4, 8 was synthesized and examined under these conditions. The reaction of 8 with ozone provided remarkably stable dioxetane 9 (Eq. 4). However, the appended ketone was

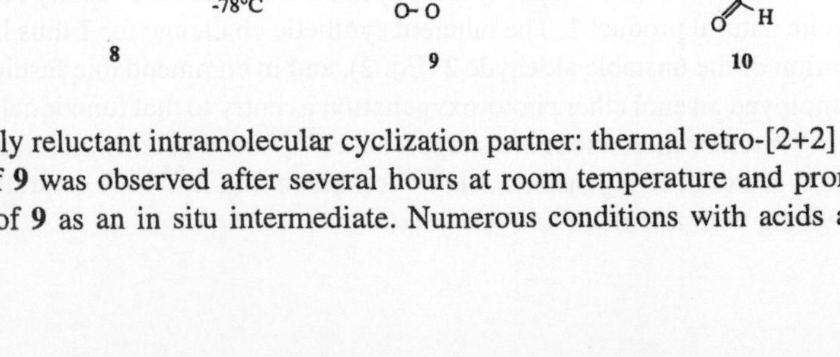

(4)

an initially reluctant intramolecular cyclization partner: thermal retro-[2+2] cyclization of 9 was observed after several hours at room temperature and prompted capture of 9 as an in situ intermediate. Numerous conditions with acids at low temperature in aprotic solvents were examined until dioxetane **9** was successfully intercepted in a methanolic solution containing boron trifluoride etherate to afford remarkably stable crystalline aldehyde **10** in 69% isolated yield.[45]

The favorable outcome of these model studies implied that keto-acid **4** would behave as desired on ozonolysis, initially giving forth a dioxetane, which would then undergo transformation to the natural product **1** on acidification.

A number of approaches to the requisite 2β, 6β-disubstituted cyclohexylidenyl-silane arrangement in **4** were considered but ultimately discarded. For example, an appropriately protected 2,6-disubstituted cyclohexanone could in principle provide vinylsilane **4** by a Wittig or Peterson olefination. Unfortunately, bis(trimethylsilyl) methyllithium only gives vinylsilanes from non-enolizable ketones.[55] Furthermore, even if the bulkier Wittig counterpart, trimethylsilylmethylidene triphenylphos-phorane, were to undergo addition, β-elimination is favored for silicon over phosphorus.[56] Although trimethylsilyl(dimethylmethoxysilyl)methyl lithium[57] is available, we have found that this reagent converts hindered 2-substituted cyclo-hexanones,[45] but not 2,6-disubstituted cyclohexanones.[49]

An alternative approach that we examined involved the conceptual joining of the terminal ends of both the 2 and 6 substituents of requisite cyclohexanone **12** (Eq. 5).[27] For example, known bicyclo[3.2.1]nonenone **13**[58] is incapable of deprotona-tion by the olefination reagent and indeed bis(trimethylsilyl)methyllithium[55] pro-vided pivotal vinyl silane **14** (Eq. 6).

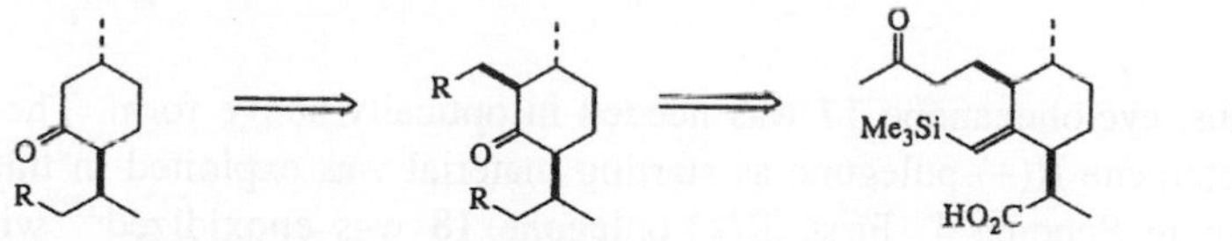

(5)

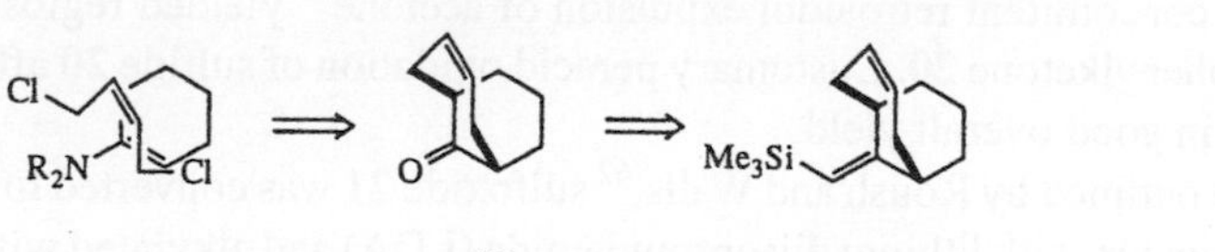

11　　　　**12**　　　　**4**

(6)

13　　　　**14**

Selective scission and subsequent processing of **14** afforded access to compounds structurally similar to **4**, but with one important drawback—incorporation of the 3α methyl group is not possible because the starting enamine reaction fails with β-alkyl substituents on the cyclohexyl ring.[58] Furthermore, the introduction of a cyclohexyl methyl group would introduce regiochemical problems. The use of bicyclic vinyl silanes such as **14** were therefore restricted to the preparation of a few racemic artemisinin analogues.

Consequently, it was necessary to employ a lengthier approach to the construction of the key vinylsilane **4**. We recognized that the termini of the carboxyl and vinyl silane moieties were situated in a 1,5 arrangement and that this system might be available by a Claisen ester–enolate rearrangement of **15** as shown in Eq. 7.

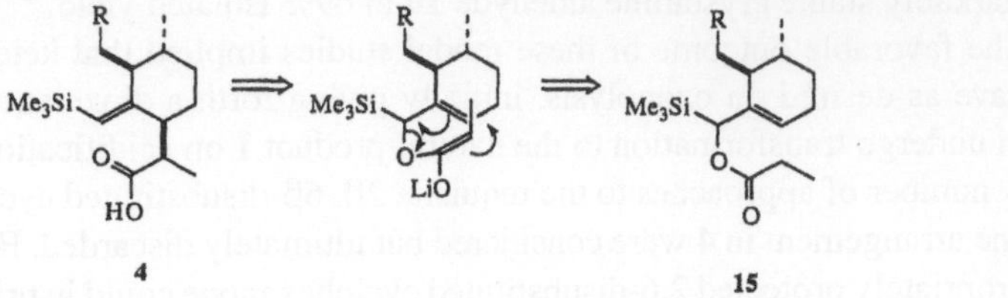

(7)

Previously, Ireland–Claisen ester–enolate rearrangement of the corresponding α-propionyloxy-allylsilane led to model system **5**.[44] Therefore, elaboration to **4** via rearrangement of **15** was pursued. To complete our retrosynthetic analysis, a plausible route to **15** was devised, involving straightforward homologation of 2β,3α-disubstituted cyclohexanone **17** to cyclohexene-carboxaldehyde **16**, which in turn undergoes silylanion addition and subsequent acylation (Eq. 8).

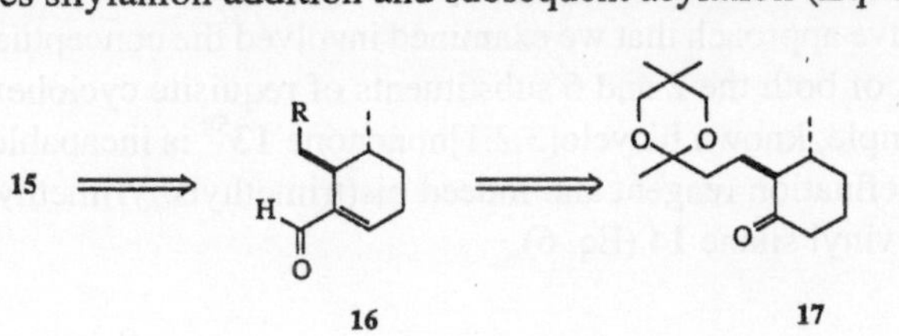

(8)

Thus, cyclohexanone **17** was needed in optically active form. The use of the monoterpene R(+)-pulegone as starting material was exploited in this regard as shown in Scheme 1. First, R(+)-pulegone **18** was epoxidized[59] with alkaline hydrogen peroxide, providing pulegone epoxide **19**. Thiophenoxide opening of **19** with concomitant retroaldol expulsion of acetone[60] yielded regiosomerically pure thiophenylketone **20**. Customary peracid oxidation of sulfide **20** afforded sulfoxide **21**[61] in good overall yield.

As outlined by Roush and Walts,[62] sulfoxide **21** was converted to the corresponding dianion with lithium diisopropylamide (LDA) and alkylated with n-butyl iodide to provide a diastereomerically complex mixture that was then converted directly to 2-butyl-3R-methylcyclohex-2-en-1-one (50% yield, 6:1 (β:α) mixture at C-2) upon thermolysis. We found that sulfoxide **21** could be alkylated with 2-(2-bromoethyl)-2,5,5-trimethyl-1,3-dioxane[63] and that the resultant complex mixture could be desulfurized with aluminum amalgam to furnish desired ketone **17** in 40–50% yield as 9:1 (β:α) mixture at C-2. Surprisingly, alkylation of this sulfoxide dianion was not improved using 2-(2-iodoethyl)-2,5,5-trimethyl-1,3-dioxane[17] in place of the bromide.

The alternate use of the dianion of β-ketoester **22** did not improve the alkylation yield because subsequent saponification of alkylation product **23** epimerized C-2

(6:4 mixture) (Eq. 9). Although the approach to **17** via sulfoxide **21** was used, it is clear that this functional arrangement could be made more efficiently.

(9)

With ketone **17** in hand, its homologation to unsaturated aldehyde **16** was pursued. We expected that a regioisomerically pure vinyl anion would be accessible from the corresponding hydrazone, and that this anion could be intercepted with dimethylformamide to provide **16**. Upon exposure of ketone **17** to *p*-toluenesulfonyl hydrazide in tetrahydrofuran (THF), solvolysis of the ketal group and subsequent hydrazone formation was observed. Under base catalysis with pyridine in THF, epimerization occurred at C-2 prior to hydrazone formation. Fortunately, if THF and pyridine were simply stripped away and the neat mixture was placed under vacuum, clean hydrazone formed in nearly quantitative yield. Subsequent treatment of hydrazone **24** in *N,N,N′N′*-tetramethylenediamine (TMEDA) with four equivalents of *n*-butyl lithium afforded a red solution of vinyl anion, which was quenched with dimethylformamide to afford the regiochemically pure $\Delta^{1,6}$-unsaturated aldehyde **16** in 70% yield (Scheme 1). At this stage the accompanying 2α-diastereomer was conveniently removed by chromatography.

Initial efforts to transform aldehyde **16** to Claisen precursor **15** involved silyl-Wittig/Brook rearrangement of the corresponding allylic silyl ether **25**. Aldehyde **16** underwent smooth 1,2-reduction with diisobutylaluminum hydride in ether and subsequent silylation provided Brook rearrangement precursor **25**. Standard conditions for effecting the deprotonation of an allylic silyl ether were employed:[64] treatment of **25** with *sec*-butyllithium in the presence of TMEDA furnished rearranged α-silyl alcohol **26**, but in modest yield, with the recovery of the balance of starting material **25**. In an attempt to improve the conversion, the use of *tert*-butyllithium promoted rearrangement in a somewhat better, but maximal, yield of 30%. The apparent disparity between the result for **25** and its 2,3-unsubstituted congener (Brook rearrangement yield of 74%)[44] is presumably caused by increased congestion from the large 2-butyl side chain in the transition state for the rearrangement of **25**.

Nevertheless, diastereomeric product **26** was used to test the validity of the Claisen approach. Thus, esterification of **26** with propionic anhydride offered the requisite ester **15** along with the inseparable diastereomer **15a** (1:1 ratio). Kinetic deprotonation of the mixture of **15/15a** was effected with lithium *N*-cyclohexyl-*N*-isopropylamide (LICA) and, upon warming to room temperature, Ireland-Claisen

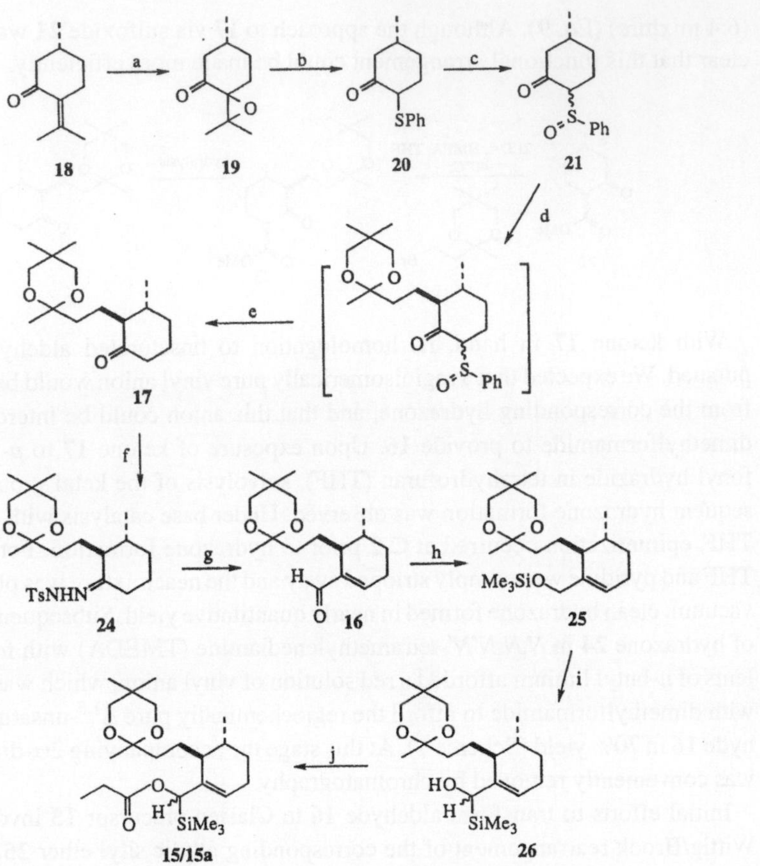

a → b → c →

18   **19**   **20**   **21**

d

e ←   [...]

17   O=S-Ph

f

g →   h →

24   **16**   **25**

i

j ←

15/15a   **26**

Scheme 1. Key: a) alkaline HOOH, THF; b) NaSPh, THF; c) m-CPBA, CH₂Cl₂, –78 °C; d) 2 LDA, HMPA or DMTP, THF, –35 °C; then 2-(2-bromoethyl)-2,5,5-trimethyl-1,3-dioxane; e) Al(Hg) amalgam, wet THF; f) p-CH₃PhSO₂NHNH₂, neat, 1 mm Hg; g) 4 BuLi, TMEDA, 0 °C; then DMF; h) DIBAH, Et₂O, –78 °C; then TMSCl, pyridine, CH₂Cl₂; i) t-BuLi, THF, –30 °C; then HOAc, –78 °C; j) propionic anhydride, DMAP, pyridine, CH₂Cl₂.

ester enolate rearrangement[65] provided a diastereomeric mixture of carboxylic acids **27** in moderate yield (Eq. 10).

Obviously, it was hoped that the acid **4** would constitute part of the mixture derived from the subsequent deketalization of **27** and that ozonolysis followed by acid treatment would afford artemisinin. However, after the aforemaitioned processing, no tetracylic products were observed. This suggested that the rearrangement had occurred through an "α" (si) face transition state leading to C-6α diastereomers

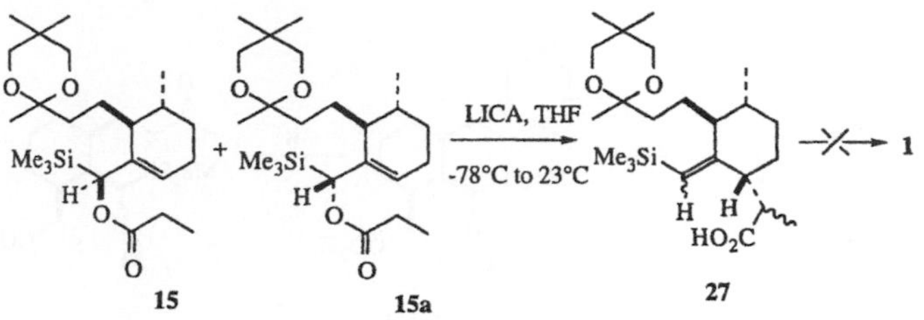

$$(10)$$

(i.e. **27**) that are not capable of ultimate cyclization to tetracyclic products. This result was not surprising in light of molecular mechanics calculations, which revealed a substantial interaction of the (Z)-enolate methyl group with the axially disposed C-2 butyl side chain in the "β" face transition state. This interaction was relieved in the "α" face transition state, thus explaining the exclusive formation of undesired "α"-oriented products (i.e. **27**).[66,67]

We examined two alternate approaches that would circumvent these interactions. First, generation of the alternate-(E)-enolate would remove the methyl–butyl diaxial-like interaction, making β products possible, but the product(s) would have threo geometry in the acid side chain. One result of this might be the production of 9-epiartemisinin (**29**), which we originally thought would epimerize to artemisinin under the cyclization conditions. Second, a methyl group could be removed from the transition state by using the acetate rather than the propionate ester, and the resultant product(s) would require a C-methylation to arrive at the requisite substrate. The first approach entailed treatment of the **15/15a** mixture with a different amide base, lithium hexamethyldisilazane (LHMDS), which is known to provide enolate geometry ratios in the opposite sense to LDA (Eq. 11).[68] The resulting

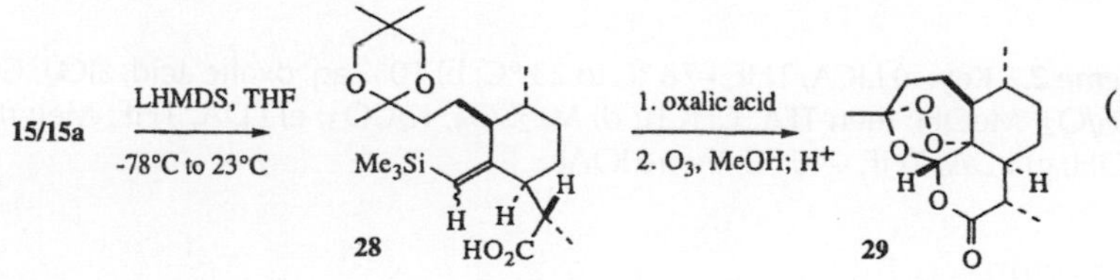

$$(11)$$

mixture of carboxylic acids **28**, produced in modest yield, was deketalized and ozonized to afford an unknown, but chromatographically related, tetracyclic product, which we assumed was 9-epiartemisinin **29**.

However, this product was not epimerizable to artemisinin and thus we moved on to the next approach. It was later proven that (+)-9-epiartemisinin had indeed been prepared in this sequence, albeit in low overall yield.[69]

Formation of the mixture of acetates **30/31** was straightforward. Subsequent generation of the enolate(s) with LICA as before produced numerous attendant by-products from self-condensation, but it was nevertheless possible to isolate a single carboxylic acid **32** in 29% yield (Scheme 2). Upon deketalization with oxalic

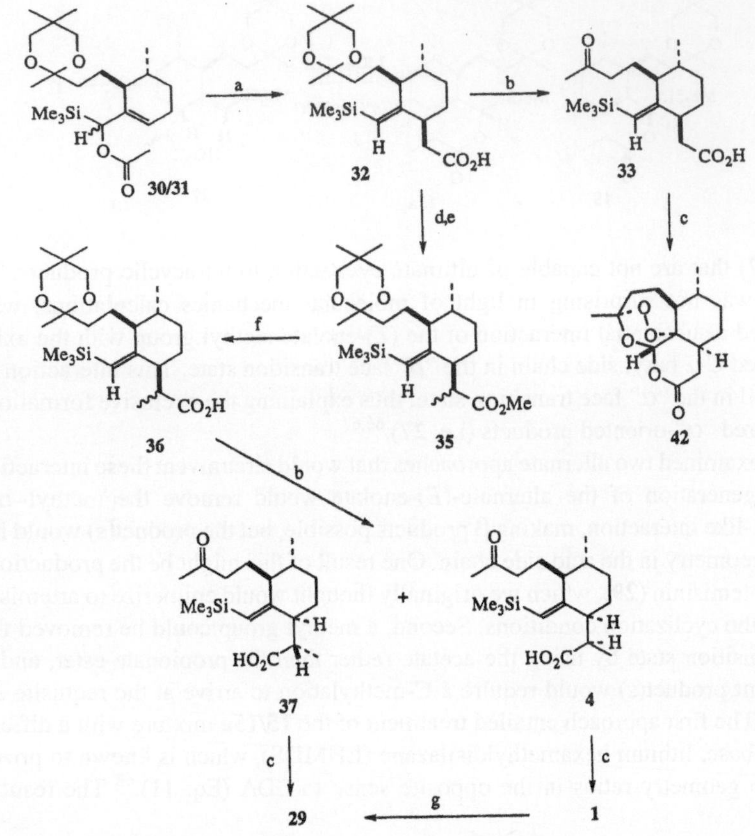

Scheme 2. Key: a) LICA, THF, −78 °C to 23 °C; b) 10% aq. oxalic acid, SiO₂, CH₂Cl₂; c) O₃/O₂, MeOH; then TFA, CHCl₃; d) Me₂SO₄, K₂CO₃; e) LDA, THF; MeI; f) KOH, MeOH; g) LDA, THF, −40 °C; then HOAc.

acid-impregnated silica gel,[70] keto acid **33** was obtained in 81% yield. Subsequent ozonolysis in methanol afforded a complex product mixture that could be treated without purification with trifluoroacetic acid in chloroform to furnish (+)-9-des-methylartemisinin **42** in 56% yield.[69] At this stage, it seemed apparent that desired tetracyclic products were being produced. However, to prove our structural assignments unequivocally, the total synthesis was completed before improving the total synthetic route. Hence acid **32** was converted to the corresponding methyl ester under basic conditions, enolized with LDA, and then alkylated with methyl iodide to provide ester **35** as a 7:3 diastereomeric mixture (83%). Sequential deprotection of ester **37** and ketal functions gave the separable keto-acids **4** (major) and **37** (minor), whose respective stereochemistries were ascertained by conversion of the former to natural product. Minor acid **37**, upon ozonolysis and acid-catalyzed ring closure, did not afford artemisinin but instead a substance identical with 9-epiartemisinin (**29**). Control experiments further supported the identity of synthetic 9-epiartemisinin. Enolization of authentic artemisinin with LDA at low temperature followed by kinetic quench[23] gave material identical to **29** produced either from **37** or the propionate Claisen.[17,69] To finish this preliminary work, the acid tentatively assigned as **4** was submitted to the usual conditions: ozonolysis and acid workup provided material identical in all respects to the natural product (+)-artemisinin (**1**).

With a workable route in hand, we then set about the task of improving the overall sequence leading to the natural product. Clearly, the route suffered from a lack of stereoselectivity in the Brook rearrangement and later in the side-chain alkylation. We felt that it was likely that only one of the diastereomers produced in the Brook rearrangement was leading to "β" targets. This was evident from the Claisen rearrangement: half of the product was acid **32**; the remainder of the acidic component was not transformed to tetracyclic products on ozonolysis/acidification. In fact, examination of possible transition-state geometries for the Claisen rearrangement suggested that a difference might be expected between **30** and **31**. With β- and α-face transition states, as well as boat and chair conformers, there are a minimum of eight possible transition states to consider for the mixture of **30/31**. Excluding the boat conformers as being energetically unfavorable,[66,67,71,72] there are still at least two possible transition states for each diastereomer. Drawing the four most likely transition states derived via MMP2 (Figure 1), we can readily see that **30b** would be preferred over **30a**: a sizable axial–axial trimethylsilyl–C2–butyl interaction is avoided and the trimethylsilyl group is in a pseudo-equatorial relationship.[71,72]

For **31**, the opposite ranking seems likely: **31a** is preferred over **31b**, again because of the equatorial disposition of the trimethylsilyl group. If these arguments are valid, then diastereomerically pure **30** should afford only desired erythro acid upon Claisen rearrangement. Unfortunately, it was not possible to separate diastereomers **30/31** chromatographically, and the steroselective synthesis of **30** was targeted.

Molecular mechanics (MM2) calculations of the unsaturated aldehyde **16** revealed an interesting possibility. A comparison of the lowest energy conformers of **16** demonstrated a clear preference for diaxially oriented **16a** ($\Delta E_{rel} > 3$ Kcal/M). Inspection of this conformer suggested that an incoming nucleophile could approach from one face of the carbonyl to lead to a single product whose relative stereochemistry corresponds to diastereomer **30** (Figure 1), as depicted in Figure 2.

Thus, if trimethylsilyl anion were used as the nucleophile, then requisite diastereomer **30** could become available. Of the various counterions, Li^+, Na^+, and K^+ have been examined by others. None of these species were suitable for direct 1,2-addition to carbonyl compounds. For example, trimethylsilyl lithium (Me_3SiLi) adds nicely 1,4- to enones by a one-electron-transfer process, but does not provide α-silyl alcohols from ketones or aldehydes. In contrast tris(trimethylsilyl)alumi-

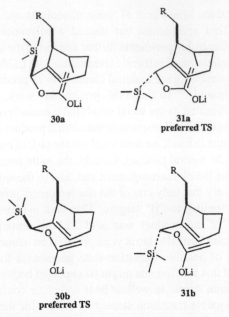

Figure 1. Potential transition states in the Claisen ester–enolate rearrangement of diastereomers **30** and **31**.

num etherate (TTAE) underwent unfettered 1,2-addition to benzaldehydes to furnish α-silyl alcohols. In our hands, this reagent reacted with cyclohexene carboxaldehyde **38** at low temperature in ether to afford the corresponding allylic alcohol in nearly quantitative yield. Furthermore, intermediate aluminate salt **38a** was stable to the Brook rearrangement as compared to the lithium analogue. As a result, the aluminate salt was captured in situ with acetic anhydride (accelerated by 4-(N,N-dimethylamino)pyridine) to give desired silyl acetate **39** in 90% yield (Eq. 12).

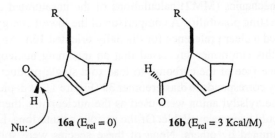

Figure 2. Transoid (**16a**) and cisoid (**16b**) rotomers of the preferred conformer of aldehyde **16**. Nucleophilic attack could be predicted to occur upon the less sterically encumbered α-face of **16a** to provide the C-1′ S diastereomer **30**.

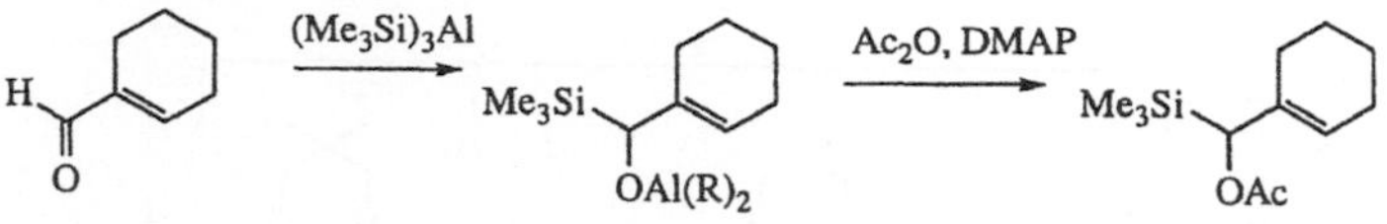

$$ (12) $$

38   38a   39

With these encouraging results in hand, we returned to aldehyde **16**. Upon reaction with TTAE and subsequent quenching with acetic anhydride, **16** was transformed to a single diastereomer **30** in 88% yield (Scheme 3). Although it was not determined which diastereomer (**30** vs. **31**) had actually been produced at this stage, the material underwent ester–enolate rearrangement (2.1 mol-equiv LICA, THF, –78 to 23 °C) to a single acid, identical to **32**, in 51% yield on the first attempt. The balance of the material from base treatment of **30** corresponded to competing Claisen condensation: β-ketoester **40** and silyl ether **25** (together with desilylated alcohol) were obtained in yields of 12 and 28%, respectively (Eq. 13).

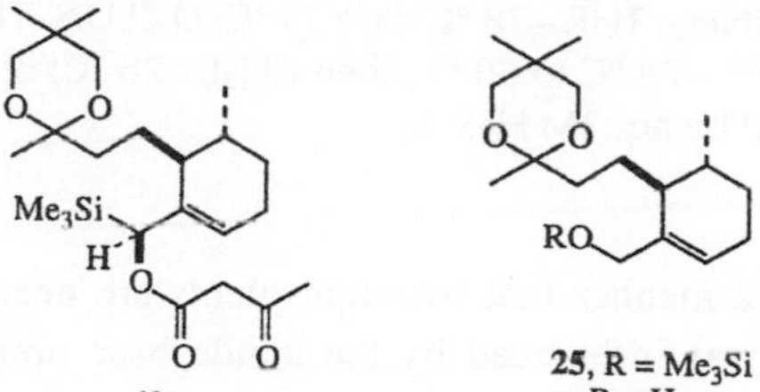

$$ (13) $$

40   **25, R = Me$_3$Si** or R = H

With diastereomerically pure **32** now available, it was possible to determine its stereochemistry. Regiochemistry about the vinyl silane moiety in **32** was ascertained by nuclear Overhauser enhancement difference (DNOE) experiments. Decoupling experiments with either downfield methylene proton adjacent to the carboxylic acid δ 2.62 (dd, 1 H, $J = 9.5$, 15.0 Hz) or δ 2.48 (dd, 1 H, $J = 5.9$, 15.0 Hz) identified the C-6 proton resonance δ 2.78 (m, 1 H). Similarly, the C-2 proton resonance was located [δ 2.11 (m, 1 H)]. Irradiation of the vinylsilane proton singlet at δ 5.38 led to an enhancement of 10% of the C-6 and none of the C-2 proton resonance, thus demonstrating a *syn* relationship between the vinyl proton and the C-6 proton. The fact that acid **32** was converted to the natural product confirmed that a β-oriented side chain had been produced at C-6 and clinched the stereochemistry of **32** as depicted.

With stereocontrol mastered, the preparation of **32** was optimized. The efficiency of the Claisen rearrangement strongly depended upon the conditions employed. For example, the amount of base needed to be strictly controlled in that at least two equivalents of base were required. With a single equivalent of base, only self-condensation was observed and further, excess base did not improve the yield. Perhaps the by-products formed early during enolization react further with remaining

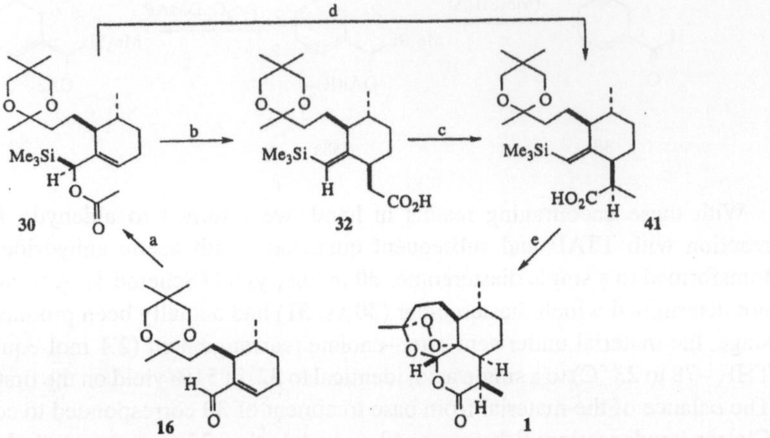

Scheme 3. Key: a) Tris(trimethylsilyl) aluminum etherate, Et₂O, –78 °C; then Ac₂O, DMAP, to 23 °C; b) 2LiNEt₂, THF, –78 °C then 23 °C; c) 2LDA, THF, 50 °C; then CH₃I, –78 °C; d) 4LiNEt₂, THF, –78 °C to 50 °C, then CH₃I, –78 °C; e) O₃/O₂, CH₂Cl₂, –78 °C; then SiO₂ followed by aq. 3M H₂SO₄.

base/enolate in such a manner that two equivalents are needed. In addition, the product distribution was influenced by the amide base employed. With highly hindered lithium bases such as LICA or lithium tetramethylpiperidide (LiTMP), self-condensation products accounted for as much as half of the reaction products. When less bulky amides were used, the rate of deprotonation minimized self-condensation to acetoacetate, but also increased direct displacement by amide anion on the ester resulting in acetamide formation. For example, LDA, lithium diethyl-amide (LDEA), and lithium pyrrolidineamide (LiNC₄H₈) each gave **32** in yields of 25, 63, and 20%, respectively. Thus, LDEA was routinely used to obtain preparative amounts of the desired product **32**.

The cumbersome route in Scheme 2[69] had been prompted by frustrated attempts to prepare the dianion of acid **32** on the microscale. A reexamination of this reaction on a larger scale showed that warming of a THF solution of **32** with two equivalents of LDA at 50 °C for 2 h led to an orange solution of dianion. Addition of methyl iodide then gave rise to a single diastereomerically pure homologous acid, **41**, in nearly quantitative yield (Scheme 3). The stereochemical identity of **41** was reasonably assumed to be erythro from its conversion to the natural product **1**. The possibility of epimerization at some stage in this process was ruled out by the clean conversion of threo acid **37** to 9-epiartemisinin **29**.

This serendipitous result was advantageously applied to the synthesis of (+)-^{14}C-artemisinin[73] using ^{14}C-methyl iodide as well as trideuteroartemisinin from CD₃-

I.[74] This alkylation is of general utility for a wide variety of alkyl halides and can be employed to furnish a myriad of C-9 analogues of the natural product.

Most recently, our total synthesis was streamlined further. Since the Claisen rearrangement which provided **32** required excess base, and was followed in a separate step by dianion formation, it seemed reasonable that the two steps could be combined. For example, treatment of acetate **30** with several equivalents of base should lead directly to the dianion of **32**, which could then be alkylated in situ to provide the homologated acid **41**. Indeed, treatment of **30** with four equivalents of LDEA (−78 to 50 °C) provided the desired dianion of **32**, which upon cooling and admission of methyl iodide, gave the acid **41** in 57% yield.

Finally, the conversion of acid **41** to the natural product **1** was reconsidered. Previously, separate deprotection of ketal **41** to ketoacid **4** (80% yield) was done prior to ozonolysis. The possibility of a one-pot ozonolysis, deprotection, and cyclization sequence was entertained. Thus, ozonolysis of **41** in dichloromethane, when followed by successive addition of aqueous sulfuric acid and silica gel, led in reasonable yield (33–39%) to (+)-artemisinin (**1**), identical in all respects to the authentic natural product. Other solvents were examined for this sequence: while we originally observed the rearrangement of dioxetane **9** to hydroperoxide **10** in methanol, we found that methanol was incompatible with the intramolecular formation of lactone from **41**. Interestingly, most other solvents (hexane, ethyl acetate, etc.) were poor in comparison to dichloromethane. When a fairly dilute solution of **41** in dichloromethane (0.01 M) was subjected to ozone, a higher yield of artemisinin (**1**) was obtained because of a lower ratio of non-peroxidic desoxyartemisinin. Therefore, additives such as *t*-butyl hydroperoxide and *t*-butyl peroxide were used to maintain an oxidative environment during acid treatment after ozone exposure, but they had little effect. In fact, crude products were much cleaner by thin-layer chromatography with the addition of *t*-butylhydroxytoluene (BHT) subsequent to reaction of **41** with ozone, and a slightly higher yield of artemisinin was obtained.

In summary, a stereoselective 10-step total synthetic route to the antimalarial sesquiterpene (+)-artemisinin (**1**) was developed. Crucial elements of the approach included diastereoselective trimethylsilylanion addition to α,β-unsaturated aldehyde **16**, and a tandem Claisen ester–enolate rearrangement-dianion alkylation to afford the diastereomerically pure erythro acid **41**. Finally, acid **41** was converted in a one-pot procedure involving sequential treatment with ozone followed by wet acidic silica gel to effect a complex process of dioxetane formation, ketal deprotection, and multiple cyclization to the natural product (+)-artemisinin (**1**). The route was designed for the late incorporation of a carbon-14 label and the production of a variety of analogues for structure–activity-relationship (SAR) studies. We were successful in preparing two millimoles of ¹⁴C-**1**[73] which was used for conversion to ¹⁴C-arteether for metabolism[75] and mode of action studies.[76,77]

III. ANALOGUE SYNTHESES

We have synthesized a number of analogues which may be divided into the following major groups: (1) optically active, substituted QHS (**42–63, 77–84, 94–99, 108–114, 116–123, 127, 134–140**; see Figures 3–5 and Tables 2–5) and dihydro-QHS (e.g., arteether, **142, 143, 148** Figure 5), analogues that were produced via branches from our total synthesis; (2) racemic analogues derived from bicyclic synthetic intermediates, 6,9-desmethyl QHS (**155**) and truncated system **156** (see Figure 6); (3) seco-analogues of racemic nature with lactone substituents (**166, 174, 176, 184, 185**) and optically active substituted cyclohexanes (**179–182**); (4) highly abbreviated and flexible racemic QHS analogues **177** and **178** (see Figure 8); and (5) 13-carba analogues of artemisinin **240, 242, 258, 259, 262–266** (see Tables 5 and 6, Eq. 5).

A. C-9 Substituted Analogues of Artemisinin

From a historical perspective, the first analogues were semisynthetic derivatives of artemisinin itself, which had been isolated in modest amounts from plants. Despite these early limitations, several of these analogues have now achieved clinical utility because their structural alterations provide improved drug delivery characteristics. These analogues retain the entire tetracyclic structure and incorporate pendant substituents. By comparison to these early analogues, our total

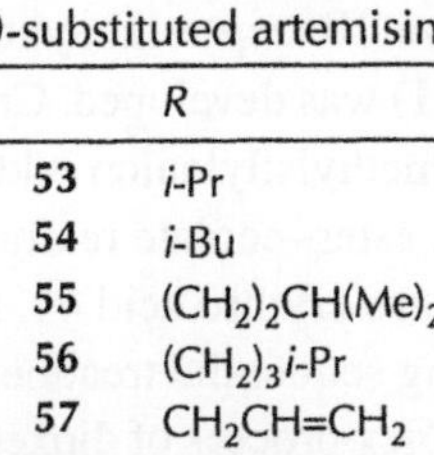

Figure 3. C-9-substituted artemisinin analogues.

R	R′		R	R′
42	H	H	53	i-Pr	H
43	Et	H	54	i-Bu	H
44	Pr	H	55	(CH₂)₂CH(Me)₂	H
45	Bu	H	56	(CH₂)₃i-Pr	H
46	C₅H₁₁	H	57	CH₂CH=CH₂	H
47	C₆H₁₃	H	58	CH₂CO₂H	H
48	C₁₃H₂₇	H	59	H	CH₂CH=CH₂
49	PhCH₂	H	60	H	E-CH₂CH=CHMe
50	PhCH₂CH₂	H	61	H	Z-CH₂CH=CHMe
51	Ph(CH₂)₃	H	62	H	Me
52	Ph(CH₂)₄	H	63	Me	Me

synthesis gave us a unique opportunity to attach groups onto novel sites on artemisinin otherwise inaccessible from the natural product. Several of our C-9 substituted analogues are depicted in Figure 3. Acid 32 (see Scheme 3) was available in fairly large amounts from our optimized total synthesis and served as a versatile synthetic intermediate. For example, 9-desmethyl analogue 42 was originally obtained upon ozonolysis of the corresponding keto-acid derived from 32.[69] But acid 32 was later converted in one pot directly to desmethylartemisinin 42 without prior removal of the ketal (Eq. 14).[17]

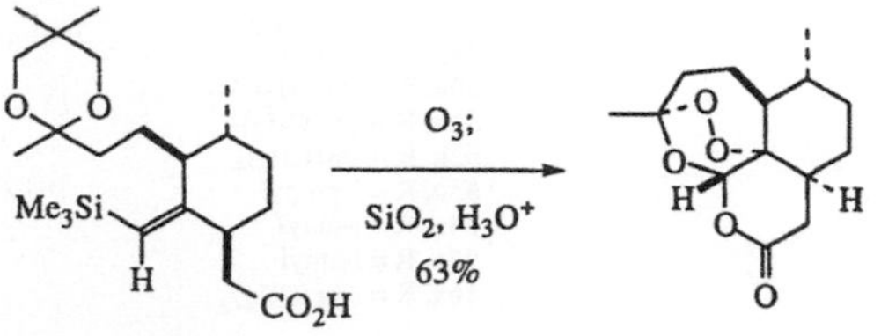

$$(14)$$

Alkylation of the corresponding dianion of acid 32 was very convenient and led to numerous 9-alkyl products. For example, analogue 58 was prepared via the LDA-generated dianion of 32, which was alkylated with *t*-butyl bromoacetate to provide acid-ester 64. Crude vinylsilane 64 was submitted to successive ozone addition and acidification. The resultant tetracyclic peroxide 65 was subsequently treated with trifluoroacetic acid to cleave the *t*-butyl ester to the free the acetic acid appendage of target 58 in 20% overall yield from 64 (Eq. 15).

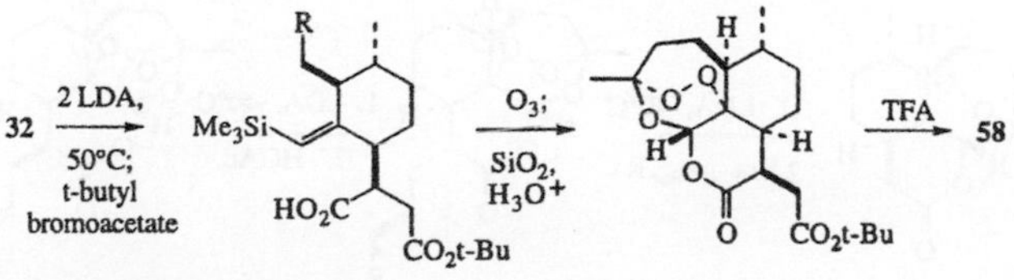

$$(15)$$

In the same straightforward manner, alkylation of 32 provided homologous acids 43a through 56a. Subsequent exposure of the resultant alkylation products to ozone, followed by acidification, gave targets 43–56 (Eq. 16).

Attachment of substituents incompatible to ozone required a different general synthetic route. For example, allyl acid 59 displayed poor chemoselectivity on exposure to ozone. Contrary to reports of alkaline decomposition of the lactone–acetal–peroxide functionality,[16] the enolate of the prefabricated tetracycle 42 could be generated at low temperature and alkylated. As a notable example, this methodology produced a single allylated tetracycle, the α-epimer 59. This initial alkylation product was unchanged after prolonged acid treatment, and structure 57 was

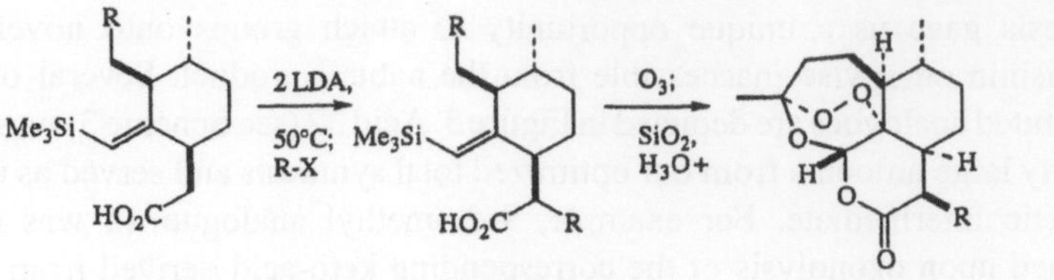

(16)

32

43a, R = Et
44a, R = n-Pr
45a, R = n-Bu
46a, R = n-C$_5$H$_{11}$
47a, R = n-C$_6$H$_{13}$
48a, R = n-C$_{13}$H$_{27}$
49a, R = benzyl
50a, R = Ph CH$_2$CH$_2$
51a, R = Ph(CH$_2$)$_3$
52a, R = Ph(CH$_2$)$_4$
53a, R = i-propyl
54a, R = i-butyl
55a, R = i-amyl
56a, R = i-Pr(CH$_2$)$_3$

43-56

tentatively assigned on the assumption that the axial C-9 substituent would epimerize.

However, when a mixture of both crotyl regioisomers **60** and **61** were obtained and separated, unambiguous assignment of C-9 to an α orientation could be ascertained from ^{1}H NMR coupling constants (Eq. 17). After this comparison, it was clear C-9α allyl analogue **59** had been produced earlier. However, **59** underwent kinetic enolization with LDA at low temperature and subsequent acid quench afforded desired C-9β allyl analogue **57**.

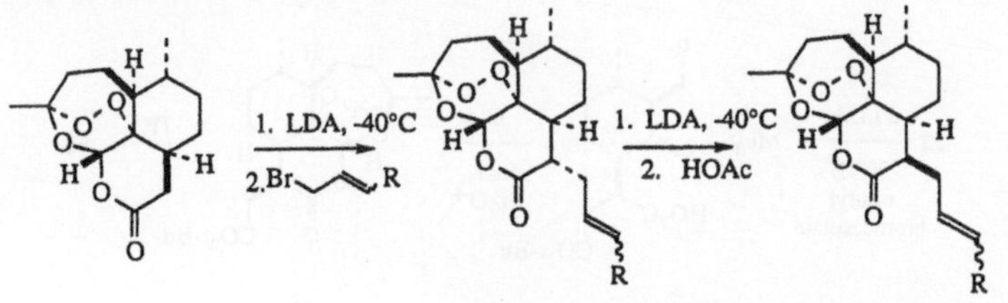

(17)

42

59, R = H
60, R = E-Me
61, R = Z-Me

57, R = H

The same enolate methodology also produced C-9 epiartemisinin **62** from artemisinin.[23] The enolate could also be intercepted with methyl iodide to provide *gem*-dimethyl analogue **63**. Alternative approaches to C-9-substituted artemisinin derivatives have been reported starting from artemisitene.[78,79]

B. C-3 Substituted Analogues of Artemisinin

Because intermediate **32** could be readily produced in large quantities, investigation into the chemistry of the ketal-bearing side-chain seemed a worthwhile goal as it could in principle provide tetracyclic analogues substituted at the C-3, C-4, and C-5 positions. The effect on antimalarial potency of modification to the C-3 and C-4 positions of tricyclic analogues of artemisinin have been investigated.[80] Previously, we had demonstrated that a similar 5,5-dimethyl-1,3-dioxane ketal, differing only in the remote carboxylate side chain, could be selectively deketalized with aqueous oxalic acid treated silica gel. Similarly, ketal **32** could be readily converted into keto-acid **33** (Scheme 4). Yields in this reaction were found to be sensitive to conditions: at higher acid concentrations, protodesilylation led to formation of unwanted exomethylene by-product **66**. When conducted carefully, yields were typically around 80%. In these circumstances there appeared to be very little protodesilylation, the yields being less than theoretical due to facile adsorption of keto-acid **33** onto silica gel.

Initial attempts to alkylate the kinetic enolate of keto-acid **33** (generated upon treatment with 2 mol equiv of LDA) with methyl iodide were unsuccessful due to formation of mono-, di-, and trimethylated ketones. This problem would presumably have worsened with less reactive alkylating agents and therefore an alternate methodology was sought. Hydrazone chemistry appeared to offer the desired regioselectivity as it has been reported that the less substituted side of *N,N*-dimethylhydrazones are preferentially deprotonated, regardless of the hydrazone geometry (*E* or *Z*).[81,82]

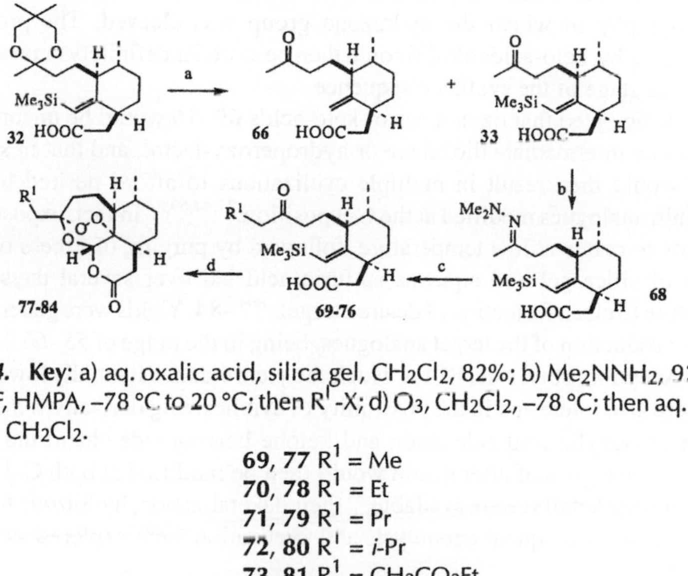

Scheme 4. Key: a) aq. oxalic acid, silica gel, CH_2Cl_2, 82%; b) Me_2NNH_2, 92%; c) 2 LDA, THF, HMPA, $-78\,°C$ to $20\,°C$; then R^1-X; d) O_3, CH_2Cl_2, $-78\,°C$; then aq. H_2SO_4, silica gel, CH_2Cl_2.

69, 77 R^1 = Me
70, 78 R^1 = Et
71, 79 R^1 = Pr
72, 80 R^1 = *i*-Pr
73, 81 R^1 = CH_2CO_2Et
74, 82 R^1 = CH_2Ph
75, 83 R^1 = $(CH_2)_2(4$-Cl Ph)
76, 84 R^1 = $(CH_2)_3Ph$

N,N-Dimethylhydrazone **68**, furnished from keto-acid **33** upon treatment with *N,N*-dimethylhydrazine, was found to be extremely water sensitive. Attempts to form the hydrazone were thwarted by low yields under a number of conditions in which solvents were present. Azeotropic removal of water, with or without molecular sieves, was also unsatisfactory. Eventually, it was found most convenient to simply dissolve the keto-acid in neat dimethylhydrazine without desiccant. After heating for a number of hours, followed by cooling and removal of excess dimethylhydrazine, formation of the desired hydrazone was apparent by NMR due to loss of the methyl ketone resonance at δ 2.14. This initially formed hydrazone existed as a dimethylhydrazonium carboxylate, but it was found that reversion to free carboxylic acid **68** occurred in vacuo, as evidenced by the proton NMR run in dry CDCl$_3$.

With simple methodology in hand for formation of requisite hydrazone **68**, we returned to our alkylation studies. Treatment of hydrazone **68** in tetrahydrofuran (THF) with 2 mol-equiv of LDA at low temperature led to formation of hydrazone enolate as was evidenced by the slow appearance of alkylated products by TLC. However, these reactions were reluctant to go to completion and it was found that if hexamethyl phosphoric triamide (HMPA) were added to the intermediate metalloenamine, that alkylation would reach completion in a matter of hours. While the preliminary alkylation product(s) were now relatively stable regioisomeric hydrazones, alkylated exclusively on the methyl group, and could be used crude in the ensuing ozonolysis reaction, products were easier to characterize after silica gel chromatography in which the hydrazone group was cleaved. The products of chromatography, keto-acids **69–76** could then be stored indefinitely or used directly for the final stage of the synthetic sequence.

It was anticipated that ozonolysis of keto-acids **69–76** would be uncomplicated leading to an intermediate dioxetane or hydroperoxy-lactol, and that in situ acidification would then result in multiple cyclizations to afford desired tetracyclic artemisinin analogues modified at the C-3 position.[27,44,69,83] In fact, exposure of the keto-acids to ozone at low temperature, followed by purging of excess ozone and addition of silica gel and aqueous sulfuric acid led over several days at room temperature to clean formation of desired targets **77–84**. Yields were generally quite good for production of the target analogues, being in the range of 55–66% with one notable exception, ethyl ester **81** for which the yield was substantially lower at 26%.

We were also interested in the possibility of synthesizing derivatives modified in both the carboxylic acid side chain and ketone bearing side chain; the resulting tetracyclic analogues of artemisinin would then be modified at both C-3 and C-9. As the requisite ketal(s) were available,[83] their deketalization, hydrazone formation, alkylation, and subsequent ozonolysis and cyclization were explored as shown in Scheme 5.

Starting with known "butylated" ketal-acid **85**, readily synthesized in high yield directly from ketal-acid **32**, deketalization as before with oxalic acid treated silica gel gave comparable results to before, with keto-acid **86** being produced in 85%

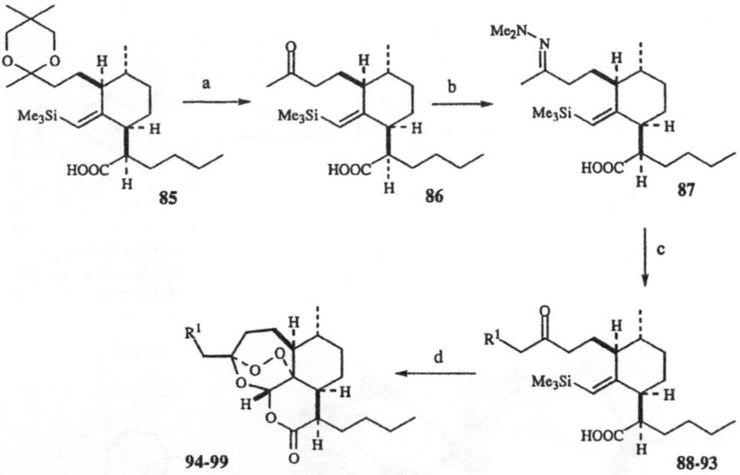

Scheme 5. [a]**Key:** a) aq. oxalic acid, silica gel, CH$_2$Cl$_2$, 85%; b) Me$_2$NNH$_2$, 93%; c) 2 LDA, THF, HMPA, –78 °C to 20 °C; then R^1-X; d) O$_3$, CH$_2$Cl$_2$, –78 °C; then aq. H$_2$SO$_4$, silica gel, CH$_2$Cl$_2$.

yield. Derivatization of this ketone as the hydrazone provided, in high yield, expected hydrazone **87**, once again as an apparently pure regioisomer as inferred by the appearance in the NMR spectra of a singlet for the vinyl silane proton of **87** at δ 5.30. Alkylation of **87** by treatment first with 2 mol equiv of LDA followed by addition of HMPA and the alkylating agent gave intermediate hydrazones which upon chromatography underwent hydrolysis and afforded the penultimate intermediate keto-acids **88–93** in reasonable yields ranging from 42–66%. Finally, low-temperature ozonolysis followed by treatment with silica gel and aqueous acid furnished target tetracyclic analogues **94–99** substituted both at C-9 and C-3. Unfortunately, yields for targets **94–99** were roughly half of those obtained for tetracycles not substituted at C-9 (**77–84**).

It is noteworthy (Scheme 6) that cyclization of intermediate hydroperoxy-aldehyde equivalents such as **103**, whether from total synthetic studies,[52] analogue work,[83,84] or this work, bearing a substituent in the carboxylate side chain at C-1′ (e.g. **103**, R = alkyl) typically occur in a range of 15–35%. On the other hand, those lacking a substituent (e.g. **102** or **104**, R = H) are more efficiently converted to tetracycles (**42** and **107**) with yields roughly from 50–65%. We have noted that this effect is unrelated to the efficiency with which initial oxidative addition occurs providing oxetanes **101**. Molecular modeling studies[85] indicate that increased torsional strain from the C-9 R substituent to the C-8 methylene in product **106** is reflected in less rapid cyclization of intermediate **103**, resulting in an increase in competing decomposition of **103** and thus lower yields of desired products such as

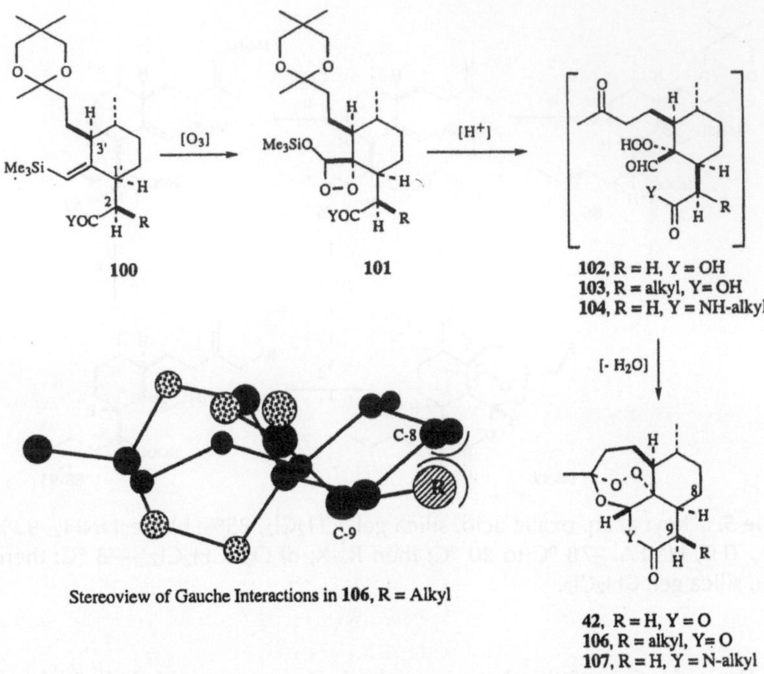

100 [O₃] → **101** [H⁺] →

102, R = H, Y = OH
103, R = alkyl, Y= OH
104, R = H, Y = NH-alkyl

[- H₂O]

Stereoview of Gauche Interactions in **106**, R = Alkyl

42, R = H, Y = O
106, R = alkyl, Y= O
107, R = H, Y = N-alkyl

Scheme 6.

106. When the R substituent is H, as in **102**, the energetically unfavorable C-9/C-8 interaction in product **42** is relieved and thus the reaction is more efficient.

C. 10-Deoxoartemisinin Modification to Artemisinin Analogues

Upon treatment of the natural product (+)-artemisinin **1** with NaBH₄ and BF₃-etherate in methanol-THF by Jung's procedure (1–100 g),[31] the known product **108**, 10-deoxoartemisinin, was obtained in 68% yield. When this procedure was applied to **1** or previously reported lactones substituted at C-9 (**42–46, 51**) or C-3 (**77–84**) on small scales (20–100 mg), yields of **108–114** and **116–123** varied dramatically from run to run (Scheme 7). It was found that without suitable modification to the reported procedure, small-scale reactions would frequently result in extensive decomposition. In fact, even with suitable monitoring of the reaction for completion, deoxoanalogues **115** (Scheme 8) could not be prepared by this approach. Thus, it was discovered that this reduction could only be safely achieved on small scale by periodic interruption for TLC monitoring. In this manner, the C-9 substituted analogues of artemisinin **43–46** were reduced in yields ranging from 35 to 58%, while C-3 substituted (C-9 = desmethyl) analogues **116–123** were furnished in 32 to 64% yield (see Scheme 1).

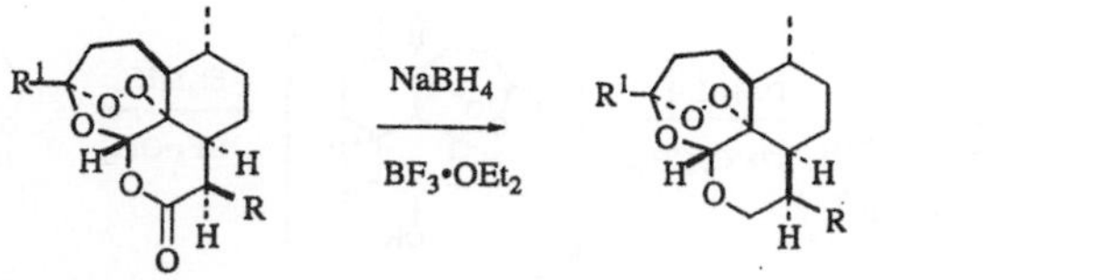

1, 42, 51, 77–84

1, 108, R = Me; R^1 = Me;
109, 42, R = H; R^1 = Me
110, 43, R = Et; R^1 = Me
111, 44, R = nPr; R^1 = Me
112, 45, R = nBu; R^1 = Me
113, 46, R = Bu; R^1 = Me
114, 51, R = (CH$_2$)$_3$Ph; R^1 = Me
116, 77, R = H; R^1 = Et

108–114, 116–123 (32–68%)

117, 78, R = H; R^1 = Pr
118, 79, R = H; R^1 = nBu
119, 80, R = H; R^1 = CH$_2$CH(CH$_3$)$_2$
120, 84, R = H; R^1 = (CH$_2$)$_4$Ph
121, 82, R = H; R^1 = (CH$_2$)$_2$Ph
122, 83, R = H; R^1 = (CH$_2$)$_3$(4-ClPh)
123, 81, R = H; R^1 = (CH$_2$)$_2$CO$_2$Et

Scheme 7.

We also found that ester functionality was compatible with this reduction sequence. Thus, treatment of the ester-lactone **81** with sodium borohydride and boron trifluoride etherate provided the tetrahydropyran **123** in 55% purified yield. Upon reaction with sodium hydroxide, **81** underwent simple ester hydrolysis to furnish the carboxylic acid **127** (see Table 3).

In order to access the elusive haloaromatic species **115**, alternative approaches were examined. As shown in Scheme 8 alkylation of the synthetic intermediate **32** with 3-(*p*-chlorophenyl)propyl bromide led as expected to clean production of erythro-acid **124**. As previously described,[17] low-temperature ozonolysis of **124**

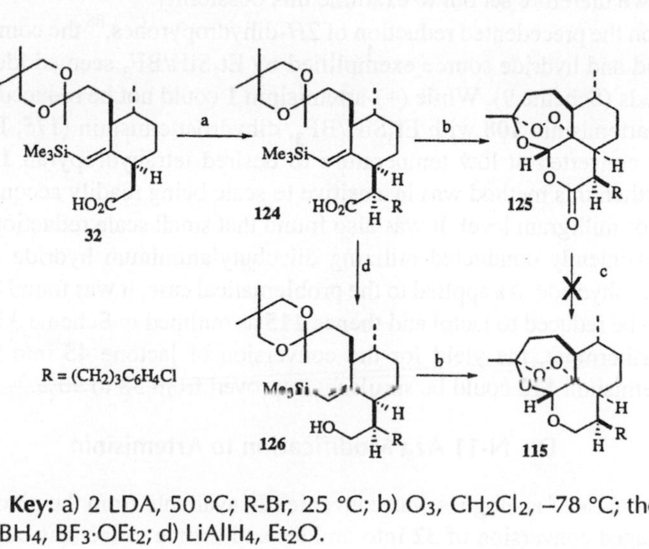

R = (CH$_2$)$_3$C$_6$H$_4$Cl

Scheme 8. Key: a) 2 LDA, 50 °C; R-Br, 25 °C; b) O$_3$, CH$_2$Cl$_2$, −78 °C; then H$_3$O$^+$, SiO$_2$; c) NaBH$_4$, BF$_3$·OEt$_2$; d) LiAlH$_4$, Et$_2$O.

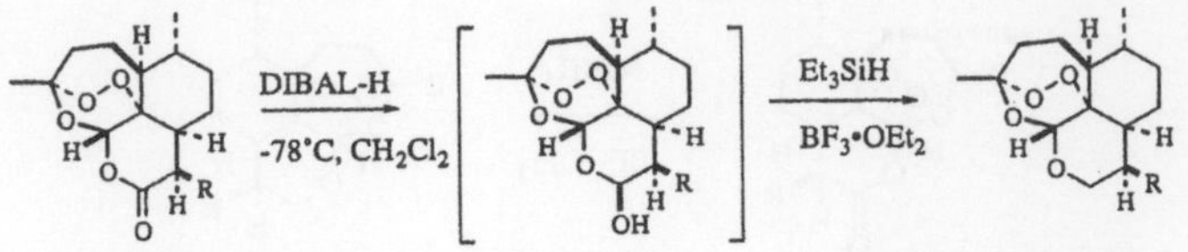

1, R = Me   **175**, R = Me   **108**, R = Me (96%)
125, R = (CH₂)₃C₆H₄Cl   **115**, R = (CH₂)₃C₆H₄Cl (92%)
45, R = Bu   **112**, R = Bu (90%)

Scheme 9.

followed by in situ acidification proceeded, through a series of rearrangements, to desired lactone **125**. Attempted reduction of **125** using the Jung/McChesney methodology[31] (borohydride/Lewis acid) was unsuccessful in providing **115**. Rather than examine this reaction in detail, the short term goal of rapidly furnishing material for in vitro antimalarial bioassay was accomplished by alteration of the oxidation state prior to ozonolysis. Thus, reduction of **124** with LiAlH₄ in ether gave alcohol **126** in good yield. Surprisingly, its ozonolysis/cyclization to **115** was accomplished in only 7% yield.

While immediate gratification was thus obtained, the overall approach to these compounds seemed less than satisfactory, particularly in light of the enhanced biological activity of analogue **115**. Apparently, the problem of reproducibility in these reductions was related to some aggregate involving the heterogeneous nature of the reaction, the inherent difficulties in accurately transferring minute quantities of hygroscopic reductant, and the need to employ refluxing conditions to complete the reduction. Lower reaction temperatures combined with homogeneous conditions could provide a means to reproducibly access 10-deoxo analogues of artemisinin, and we therefore set out to examine this possibility.

Based on the precedented reduction of 2H-dihydropyrones,[86] the combination of Lewis acid and hydride source exemplified by Et₃SiH/BF₃ seemed ideally suited to our needs (Scheme 9). While (+)-artemisinin **1** could not be reduced directly to 10-deoxoartemisinin **108** with Et₃SiH/BF₃, dihydroartemisinin (**175**, R = H) was smoothly converted at low temperature to desired tetrahydropyran **108** in 96% yield. Further, this method was insensitive to scale being readily accomplished on the gram or milligram level. It was also found that small scale reductions could be more conveniently conducted utilizing diisobutylaluminum hydride in place of sodium borohydride. As applied to the problematical case, it was found that lactone **125** could be reduced to lactol and thence **115** as outlined in Scheme 3 in excellent yield. Furthermore, the yield for the conversion of lactone **45** into 9-butyl-10-deoxoartemisinin **112** could be similarly improved from 58 to 90%.

D. N-11 Aza Modification to Artemisinin

Another class of analogues was conveniently available from the pivotal acid **32**. We envisaged conversion of **32** into amides and thence to lactams on addition of

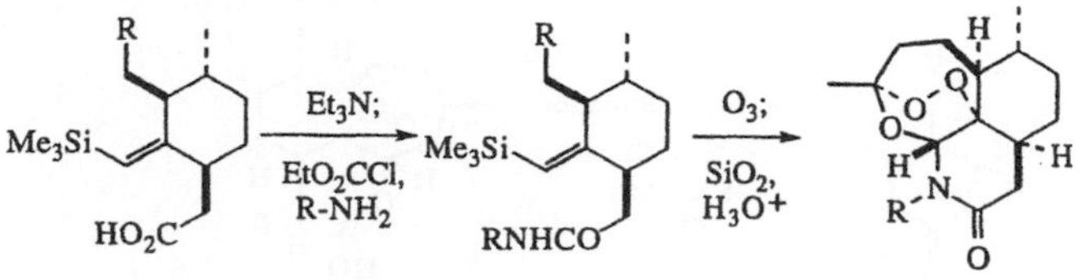

32

128, R = Me
129, R = Pr
130, R = benzyl
131, R = CH_2CO_2(*t*-Bu)
132, R = $CH_2CH_2NMe_2$
133, R = $(CH_2)_5CO_2H$

134, R = Me
135, R = Pr
136, R = benzyl
137, R = CH_2CO_2(*t*-Bu)
138, R = $CH_2CH_2NMe_2$
139, R = $(CH_2)_5CO_2H$
140, R = CH_2CO_2H

Figure 4. Aza-modification to artemisinin.

ozone followed by acid-catalyzed cyclization. Thus, a cold solution of the triethylammonium carboxylate salt of **32** was treated with ethyl chloroformate and the resultant mixed anhydride reacted with various primary amines to give the amides **128–133**, which proved excellent substrates for reaction with ozone and subsequent acidification to afford the lactam analogues **134–139** (Figure 4). A straightforward approach has been reported in which artemisinin is reacted under catalysis with various amines.[87]

In certain cases, further transformations were warranted for deprotection or derivatization: the *N*-(2-acetic acid) analogue **140** was provided upon hydrolysis of ester **137** with trifluoroacetic acid in dichloromethane.

E. Dihydroartemisinin Analogues

Very early studies by other workers[19–22,88] described the higher potency of arteether relative to artemisinin (**1**). This had forecast the selective reduction of the lactone of our novel analogues as a routine method, with the likely potential to increase antimalarial activity. Previous transformations of artemisinin are summarized in Scheme 10.

Artemisinin (**1**) can be selectively reduced with $NaBH_4$ to afford dihydroartemisinin (**175**, DHQHS), which has been shown to be roughly ten times as potent as parent material **1**. The lactol DHQHS (**175**) was converted to various ethers by treatment in alcohols with boron trifluoride (acetal formation conditions). Notably artemether (**144**) and arteether (**145**) were prepared in this manner. Alternatively, DHQHS can be acylated to yield esters or carbonates, as shown for sodium artesunate (**146**) as an example.

Our contribution to this area is also included in Scheme 10. The relatively hindered lactone of C-9 alkyl analogues of artemisinin are selectively and efficiently reduced with diisobutyl aluminum hydride (DIBALH) to the β-lactols **141**, for example. From the prior workers, the configuration at the C-10 center of

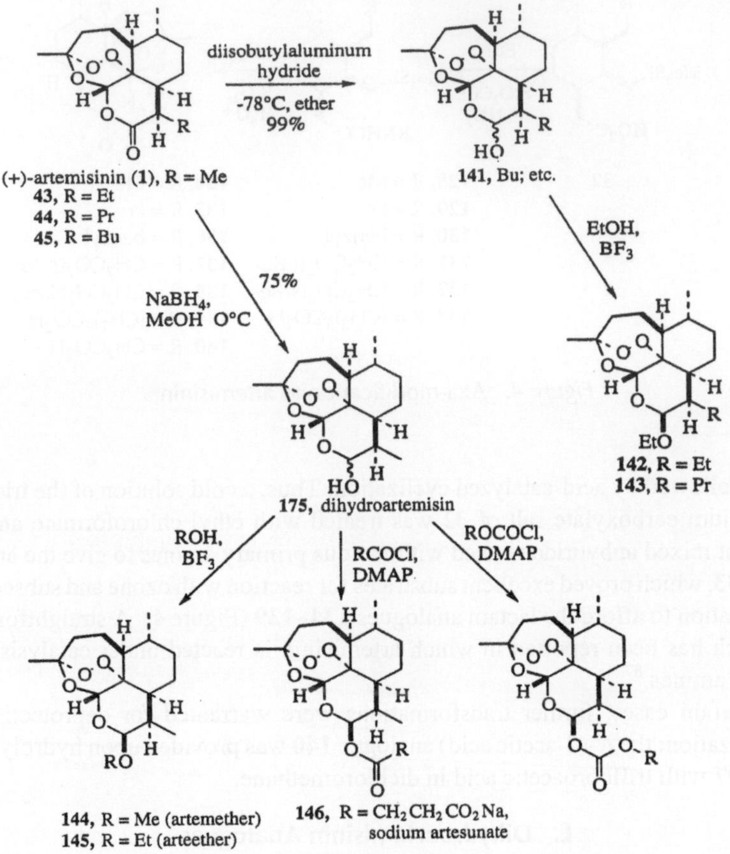

diisobutylaluminum
hydride
-78°C, ether
99%

(+)-artemisinin (1), R = Me
43, R = Et
44, R = Pr
45, R = Bu

141, Bu; etc.

EtOH,
BF₃

NaBH₄,
MeOH 0°C

75%

175, dihydroartemisin

142, R = Et
143, R = Pr

ROH,
BF₃

RCOCl,
DMAP

ROCOCl,
DMAP

144, R = Me (artemether)
145, R = Et (arteether)

146, R = CH₂CH₂CO₂Na,
sodium artesunate

Scheme 10.

dihydroartemisinin was determined easily from diagnostic coupling patterns in the
^{1}H NMR.[9,10] Scheme 10 depicts the customary lactol etherification to the novel C-9
alkyl analogues of arteether **142** and **143**.

In total, we prepared six new dihydroartemisinin analogues as shown in Figure
5 from the corresponding lactones and three new analogues from artemisinin itself.
9-Desmethylartemisinin **42** was transformed into the lactol **147**, which was then
converted to the β-ether **148**. Treatment of DHQHS under anhydrous conditions
with acid and *tert*-butylhydroperoxide gave perether **150** in 62% yield, which
contains two peroxide moieties.

We also prepared carbonate analogues by acylation of dihydro-QHS, consistent
with Scheme 10. Although numerous analogues derived from DHQHS have appeared in the literature, linkage of DHQHS to cell–membrane components has not

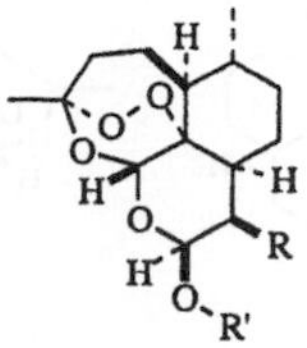

147, R = H, R' = H
148, R = H, R' = Et
150, R(β) = Me, R'(β) = -O-Bu
151, R(β) = Me, R'(β) = -CH$_2$CH(OCOC$_{15}$H$_{31}$)CH$_2$(OCOC$_{15}$H$_{31}$)
152, R(β) = Me, R'(α) = -COO-cholestanyl
141, R(β) = Bu, R' = H

Figure 5. Dihydroartemisinin analogues.

been disclosed. Thus, analogues **151** and **152** were made from DHQHS via linkage to either cholesterol for **152** or a diglyceride such as dipalmitin for **151**. If these drugs interact with the parasite cell membrane, lipophilic **151** or **152** hopefully would be actively delivered and/or incorporated into the site of action. The plasma half-life might also be extended by compartmentalization into fatty tissue and forestalling catabolic processing. Recently, the 1,2,4-trioxane ring has been incorporated into the A-ring of a steroid.[89]

For another dihydro analogue, we wondered about conceptually tying down the ethyl group of arteether (**145**) by connection to the C-9 methyl as an intriguing assessment of the rotational freedom of the C-10 substituent on activity. To achieve this, an ω-alcohol was connected to C-9 in analogue **153**, which was in turn cyclized to novel pentacycle **154** (Eq. 18).

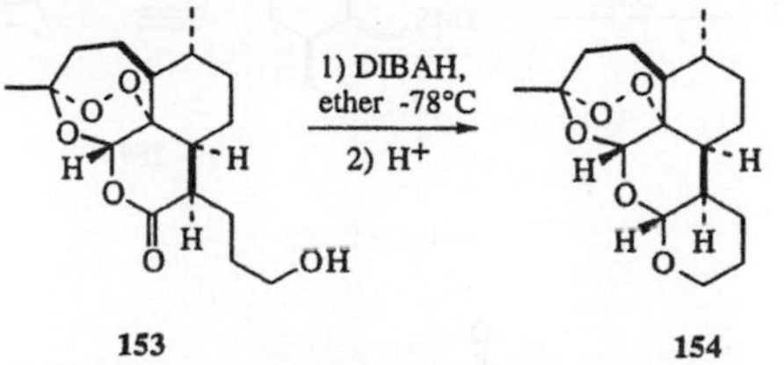

$$\text{(18)}$$

153   154

F. Nor-Analogues of Artemisinin

Racemic (±)-6,9-desmethylartemisinin (**155**) (Figure 6) was prepared according to the synthetic route shown in Scheme 11,[27] as discussed previously as an approach to the total synthesis of QHS via process development of **13 → 14**.

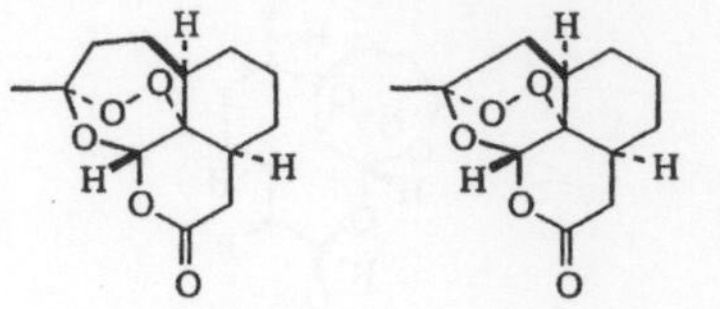

155   **156**

Figure 6. Analogues accessed through the bicyclodecene route.

Known bicyclo[4.3.1]enone **157**[58] was converted into vinylsilane **158** with bis(trimethylsilyl)methyl lithium.[55] Diene **158** underwent selective ozonolysis at the cis-olefin under conditions to produce differentially oxidized termini:[90] aldehydo-ester **159** was homologated with a phosphine oxide anion[91] to enol **160**. Subsequent hydrolysis of **161** provided substrate **162**, which after tandem ozonolysis–acidification gave racemic 6,9-desmethyl analogue **155**. Unfortunately, initial efforts failed to resolve **155** into its two optical isomers with cellulose triacetate.[92] However, the antimalarial activity of racemate **155** is intriguing, as discussed in a later section.

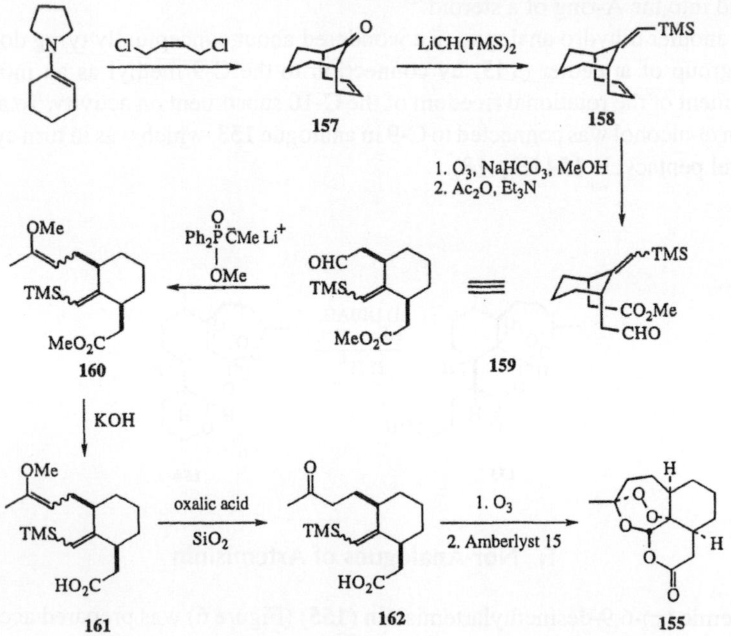

Scheme 11.

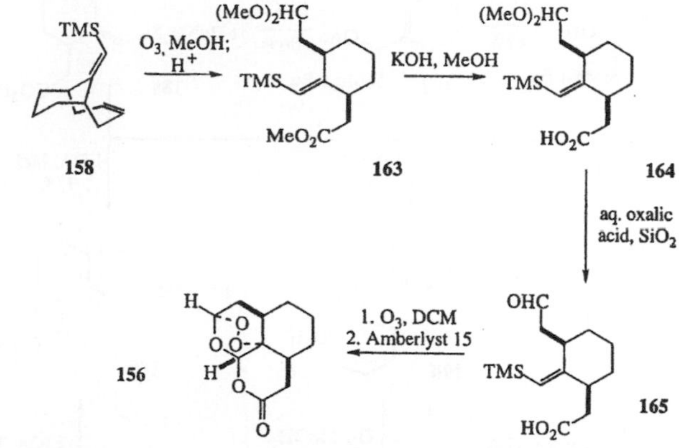

Scheme 12.

In the preparation of **155**, a Wittig-type reaction of **159** to **160** served to incorporate carbons needed to build the tetracyclic system of artemisinin. As seen in Scheme 12, we took a portion of diene **158** and bypassed the introduction of any other carbons. After sequential deprotection of **163** via **164** and **165**, closure to a new, more compact tetracyclic peroxide **156** was accomplished with our existing methodology.

G. Tricyclic Analogues of Artemisinin

In the past, we have synthesized tricyclic analogues of artemisinin, such as **184**, from simple precursors such as **183** (Eq. 19). The fact that **184** has about 20% of the activity of **1** and is somewhat easier to synthesize has stimulated further efforts in this area.

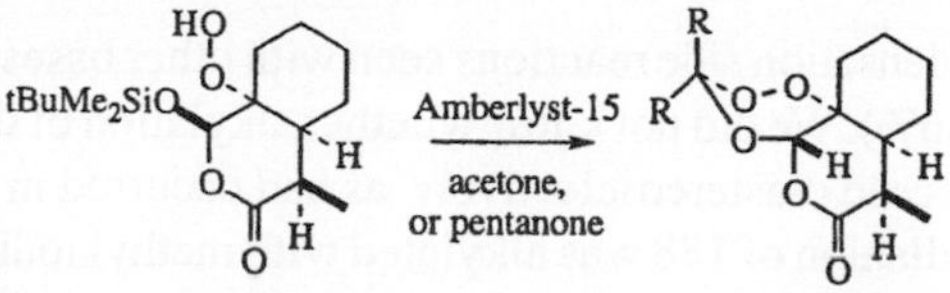

(19)

SAR data indicated that additional alkyl groups in the vicinity of the peroxy group (**184** vs. **185**) reduced activity substantially. Because we wished, for similar reasons, to vary the lactone ring of **184**, we prepared the butyl derivative **168** as shown in Scheme 13.

Acetylation of alcohol **186** gave the ester **187** in 87% yield after distillation. Claisen rearrangement of **187** with lithium diethylamide as base avoided the usual

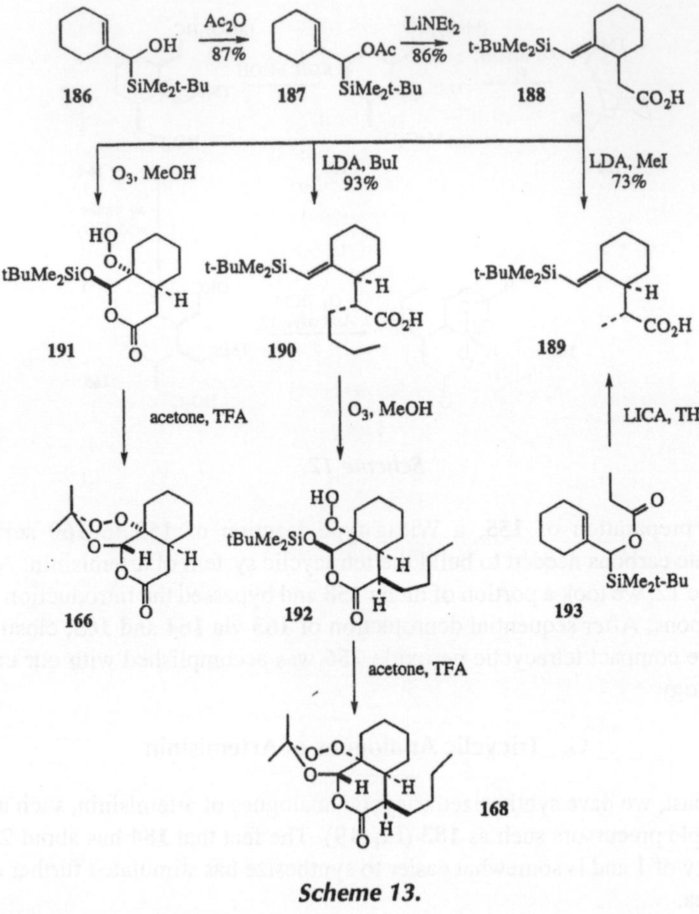

Scheme 13.

competing self-condensation side reactions seen with other bases and gave acid **188** in excellent yield (86%). We did not know whether alkylation of the dianion derived from **188** would proceed diastereoselectively, as had occurred in the total synthesis **32 → 41**. Thus, the dianion of **188** was alkylated with methyl iodide and proceeded, quite unexpectedly, with complete diastereoselection to give **189** in 73% yield. Structure **189** was correlated with the known propionate Claisen product, **193 →** **189**. We were therefore confident that the alkylation of **188** with butyl iodide would give the desired diastereomer, and thus prepared acid **190** in 93% yield. On ozonolysis, **190** was transformed to hydroperoxide **192** (55%) as expected. Finally, **192** was treated in acetone with TFA to give the desired analogue **168**.

Other racemic analogues **166, 167,** and **169** were similarly prepared as shown in Scheme 13, to examine the effects of systematic variation on the lactone ring in these readily made analogues. For example, vinylsilane acid **188** was carefully treated with ozone/oxygen to provide the hydroperoxy lactone **191** in 17% yield; subsequent treatment of **191** in acetone with trifluoroacetic acid gave the desired tricyclic analogue **166** in 38% yield.

Access to analogues with higher substitution was easily achieved. From either the propionate ester **193** or acetic acid **188**, we previously made the propionic acid appendaged **189**, which was in turn alkylated to the *gem*-dimethyl acid **194** in 76% yield (93% based on recycled starting material). The vinylsilane of **194** underwent addition of ozone to eventually afford hydroperoxide **195**, and final ring closure was accomplished with trifluoroacetic acid and acetone to afford *gem*-dimethyl analogue **169** in 19% overall yield from **194** (Eq. 20).

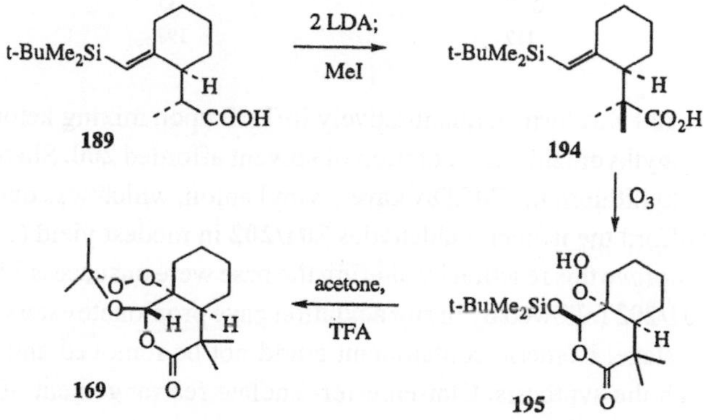

(20)

In a parallel synthesis, alcohol **186** was esterified to hemisuccinate **196** in 28% yield (64% based on recyclable starting material). The unprecedented Ireland–Claisen rearrangement of a hemisuccinate was effected by excess LDEA in THF. Upon warming overnight from −78 °C, diacid **197** was produced in 76% yield. The geometry depicted for **197** was expected by analogy and confirmed by DNOE experiments. Treatment of diacid **197** with ozone led to production of a very labile hydroperoxide **198**, which was therefore treated immediately with acid and acetone to give carboxyl analogue **167** in 6% overall unoptimized yield from **197** (Eq. 21).

Upon inspection of the ^{1}H NMR spectra of seco analogues **166**, **167**, and **169**, temperature-dependent behavior was observed for **166** and **167**. By contrast, *gem*-dimethyl analogue **169** had a sharply resolved ^{1}H NMR spectrum at room temperature.

With this class of tricyclic analogues of **1**, a major area of interest has been the synthesis of an optically active substance, as previously submitted analogues were racemates. Accordingly, we designed a synthesis with a homochiral starting material possessing the correct absolute configuration. Thus, analogue **172** was synthesized from 3R-methylcyclohexanone (**199**, commercially available) as outlined in Scheme 14.

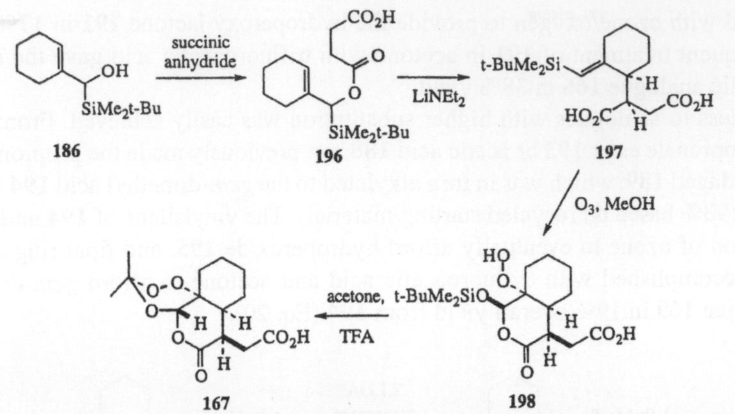

(21)

Hydrazone **200** was formed quantitatively in THF upon mixing ketone **199** with *p*-toluenesulfonylhydrazide. Evaporation of solvent afforded **200**. Shapiro reaction of **200** with alkyllithium in TMEDA gave a vinyl anion, which was quenched with dry DMF to afford the isomeric aldehydes **201/202** in modest yield (1:1 mixture). Attempts to improve this reaction by altering the base were unsuccessful. Silylanion addition to **201/202** followed by in situ acylation gave propionate esters **203** in good yield. At this stage, isomeric contaminant could not be removed and was simply carried through the synthesis. Claisen ester–enolate rearrangement of **203** gave a

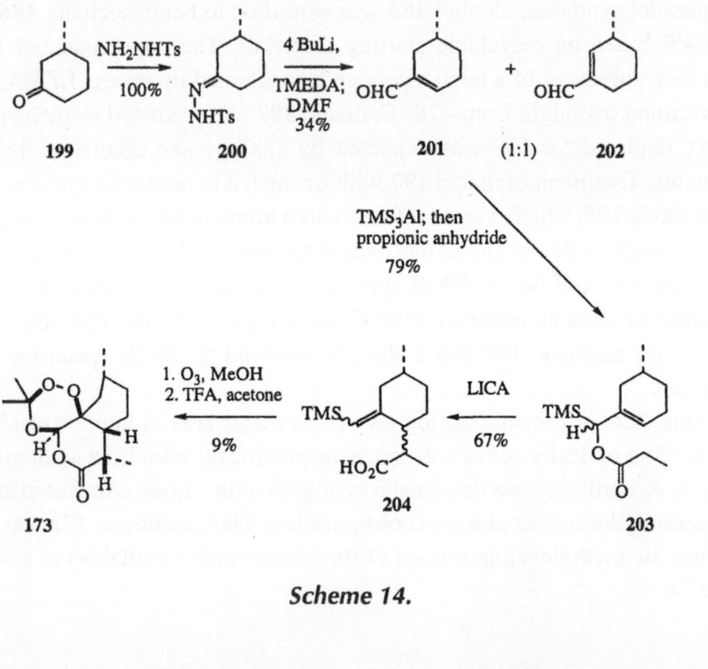

Scheme 14.

complex mixture of acids **204**. At this point some chromatographic separation was possible, and **204** had less isomeric contamination than did **203**. Ozonolysis of **204** followed by cyclization in acetone afforded only one discernible product: isomeric analogue **173**. The fact that **173** and not **170** had been produced by the sequence in Scheme 14 was determined by independent synthesis of **170**, as shown in Scheme 15.

Chiral ketone **205**, prepared from isopulegol, was reacted with (methoxydimethylsilyl)trimethylsilylmethyllithium[57] to afford **206** as an *E/Z* mixture. Simple deprotection/oxidation served to convert **206** to the acid **207**. Upon ozonolysis of **207** in methanol, removal of solvent, and addition of either acetone or acetaldehyde and acid catalyst, tricycle **170** and known **171**[50] were produced. The fact that a known product was produced from a common precursor was sufficient proof of structure **170**. Peroxide **170** (Scheme 9) was slightly different from **173** by NMR, although melting points and optical rotations ($[\alpha]^D$) were quite similar. To synthesize the methyl homologue, 4,5-secoartemisinin **172**, alkylation of starting ketone **205** was performed, providing **209**. Carrying out the synthesis as before furnished

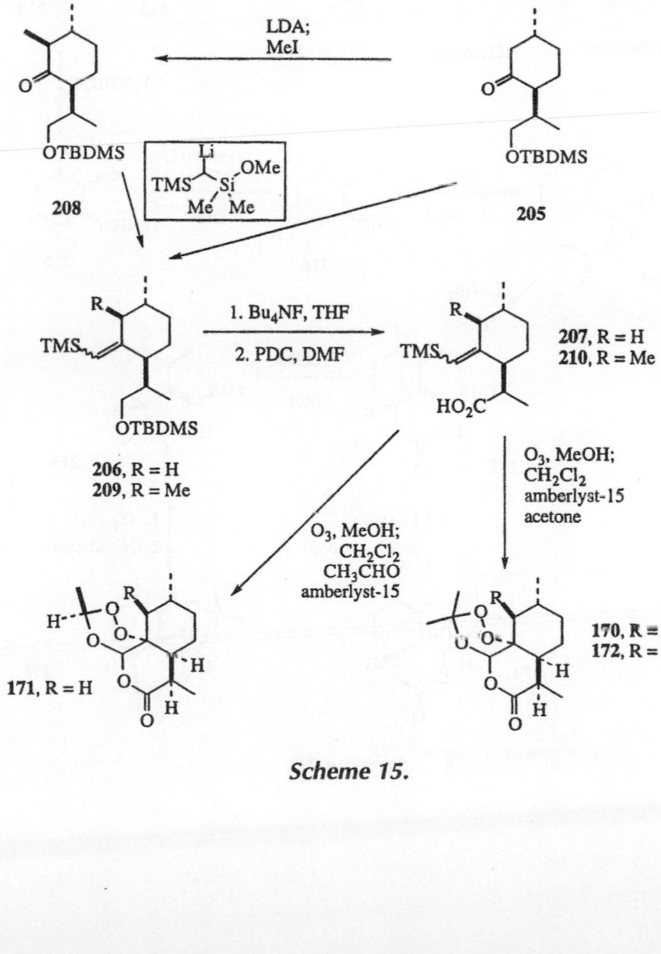

Scheme 15.

the desired seco analogue of artemisinin, **172**. To unambiguously prove the correct stereochemistry was obtained, an X-ray crystallographic analysis of **172** was performed.

The earlier regioisomeric problem upon fragmentation of the tosylhydrazone (i.e. production of mix **201/202**) was overcome by increased substitution for selectivity. Therefore, 3R-pulegone was used as a starting material for synthesis (Scheme 16). The enolate of pulegone was generated with lithium isopropylcyclohexylamide (LICA) and alkylated with methyl iodide to furnish mainly 2,3-dimethyl-6-iso-propylidene cyclohexanone **211** along with by-product **212**, which was previously observed by Reusch et al.,[93] but **212** did not react in the following conversion of crude material. When the mixture containing isopropylidene **211** was placed in acid and submitted to prolonged heating, acetone distilled prior to the water azeotrope of the epimeric mix of **213:214** (1:1.96, as determined by NMR) in 72% yield. The

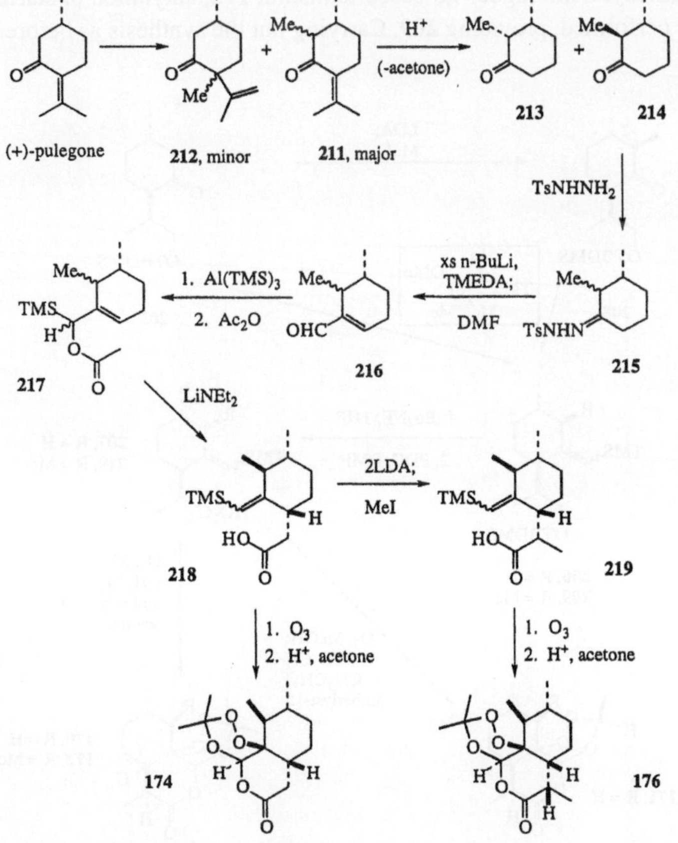

Scheme 16.

corresponding tosylhydrazone mix **215** from the mixture of ketones was made as before and underwent *n*-butyllithium-effected fragmentation in TMEDA to a regioisomerically pure cyclohexenyl anion, which was capped with dimethylformamide to afford isomeric aldehydes **216** in 76% yield. The mixture was treated with tris(trimethylsilyl)aluminum (III) etherate,[17,94] and followed by acetylation to provide a mixture of all possible diastereomers of **217**. The lack of selectivity was surprising in contrast to the total synthesis of QHS, in which synthetic intermediate **41** differs in the presence of a 6'-methyl instead of a larger alkyl chain. Upon exposure to lithium diethylamide, mixture **217** rearranged to a mixture of diastereomeric cyclohexylacetic acids, which after rigorous chromatographic separation furnished geometric isomers of acid **218** in a 1:1 ratio by NMR and 28% yield. Acids **218** were submitted to ozone and acidification in a single pot to give trioxane **174** in 22% yield. Alternatively, acids **218** were methylated via the corresponding LDA-generated dianion to propionic acid **219**, which was subsequently reacted with ozone and acidified to provide trioxane **176**. As discussed above, these particular analogues are isomeric about all but the starting 3R chiral center. Close inspection reveals that they are antipodal (except for the 3-methyl) to the materials prepared from Scheme 9. Two optically active trioxanes **174** and **176** also display temperature-dependent NMR behavior, like other aforementioned seco-analogues.

The ramifications of the different conformational natures of **166–176** and other similar analogues on biological activity may be a promising area of future exploration. As a preliminary approach, a comparison of the low-energy conformers predicted by molecular mechanics and two-dimensional NMR data are under analysis. A possible correlation with conformational families of these flexible analogues and the antimalarial activity is under examination.

Another class of analogues which we felt would be of interest were the ring-D seco analogues of artemisinin (viz. **180**), derived conceptually from scission of the 8a, 9 bond of **1** (Eq. 22).

$$\tag{22}$$

These compounds were expected to possess useful antimalarial activity because the crucial peroxy moiety is held in, what we believe to be, the requisite relative orientation for maximal activity, and the carbonyl group is capable of rotation into novel orientations unavailable to the natural product. A further advantage of this class is that the carbonyl substituent is readily introduced by simple acylation reactions. Other virtues of this class of compounds are: (1) their synthetic accessibility; (2) the wide variety of analogues available; and (3) enantiomeric purity of the products is assured.

As shown in Scheme 17, **180** was available from the common, totally synthetic, intermediate **17**. Using reported methodology for the introduction of vinylsilanes,[57] **17** was reacted smoothly with (methoxy dimethylsilyl) trimethylsilyl methyllithium in pentane to afford (*E/Z*) vinyl silane **220** in 54% yield. The main by-product in this reaction was ketone **17**, which could be recycled; thus, based on recovered **17**, the yield of **220** was 93%. Hydrolysis of ketal **220** occurred without protodesilylation upon exposure to aqueous oxalic acid absorbed onto silica gel to give ketone **221** in 80% yield. Upon low-temperature ozonolysis of **221** in methanol, a remarkably stable dioxetane **223** was produced, as evidenced in the ^{1}H NMR

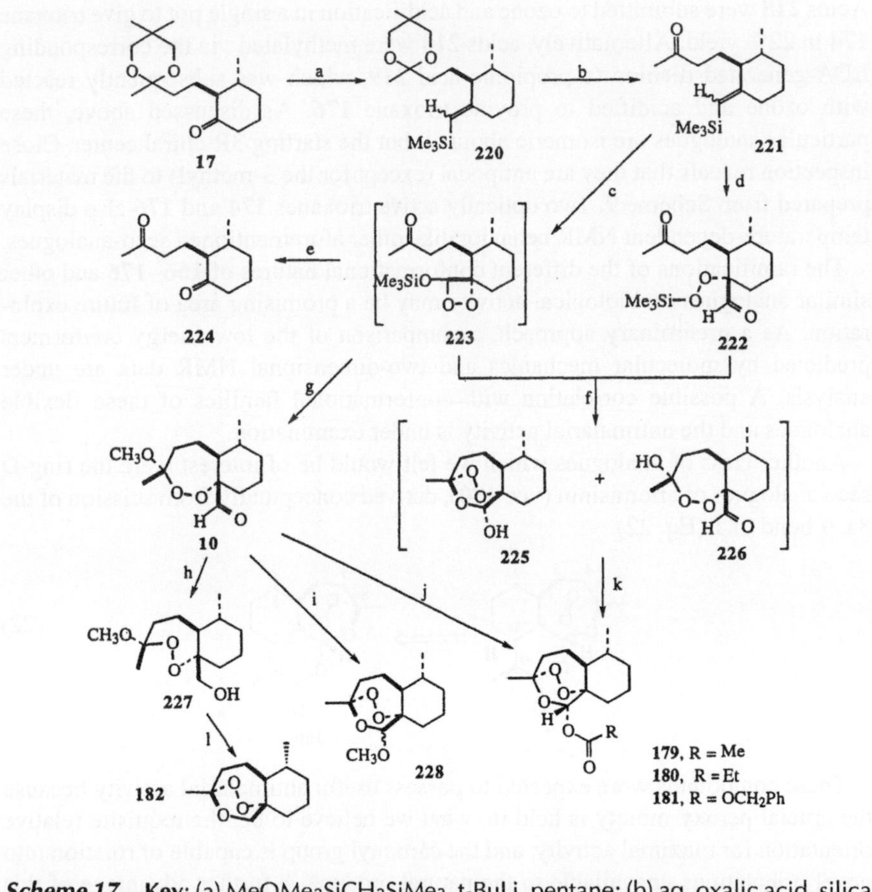

Scheme 17. Key: (a) MeOMe2SiCH2SiMe3, t-BuLi, pentane; (b) aq. oxalic acid, silica gel, CH2Cl2; (c) O3/O2, MeOH, −78 °C; (d) O3/O2, CH2Cl2, −78 °C; (e) CDCl3, 23°C; (f) moist CHCl3, TFA; (g) c followed by BF3 etherate; (h) NaBH4, MeOH, 0 °C; (i) BF3 etherate, MeOH, (MeO)3CCH3; (j) RCOX, Amberlyst-15, solvent; (k) RCOX, DMAP, CH2Cl2. (l) p-TsOH, CH2Cl2.

spectrum (δ 6.1, s). On prolonged standing, **223** underwent [2 + 2] cycloreversion to mainly afford diketone **224**. By contrast, when dioxetane **223** was intercepted with Lewis acid (BF₃), crystalline aldehyde-ketal **10** was produced in good yield (69%).

Aldehyde **10** was a useful intermediate due to its chemical stability in storage and ready conversion to artemisinin analogues. On treatment of **10** in propionic anhydride with protic acids (HClO₄ or H₂SO₄) or more conveniently with polymer-bound acid (Amberlyst-15), with or without co-solvent (CH₂Cl₂), the 8a,9-seco analogue of artemisinin **180** was obtained in 22% yield. It was also possible to treat dioxetane **223** under the same conditions to arrive at **180** or **179** by substituting acetic anhydride or propionic anhydride, respectively, and in this fashion analogue **179** was obtained in 30% yield.

Hydrolysis of dioxetane **223**, or of ketal **10**, led to an inseparable mixture of expected product **225**, as well as bicyclic isomer **226**. The mixture was 1:2 (**225:226**), and underwent standard acylation reactions to give, for example, **179** on treatment with Ac₂O/pyridine/CH₂Cl₂. Carbonates such as **181** were available from **225** upon treatment with various chloroformates in pyridine/CH₂Cl₂. In other words, the alcohol **225** could be funneled away from the mixture by reaction with electrophiles, providing the desired tricyclic products. The bicyclic aldehyde **10** was isomerized to obtain tricyclic ketal **228** under dehydrating conditions in the presence of an alcohol.

Finally, facile reduction of aldehyde **10** to alcohol **227** occurred with NaBH₄ in MeOH at 0 °C. Exposure of **227** to acid in CH₂Cl₂ led to the expected *trans*-ketalization product **182** in 79% overall yield. Seco-analogue **182** has, of course, the A, B, and C rings of artemisinin.

In connection with our quest to identify the minimal structural requirements and the design of increased flexibility for antimalarial activity among the artemisinin class of compounds, we conceptually cleaved all the rings in artemisinin (Scheme 18) and targeted peroxide esters **177** and **178**, which were prepared as rapidly as hoped. Commercially available 2-ethyl-1-butene was epoxidized with *m*-chloroperoxybenzoic acid to give 2,2-diethyloxirane, which was ring-opened in situ upon addition of *t*-butyl hydroperoxide and *p*-toluenesulfonic acid. Crude peroxide alcohol **229** was divided into equal portions and acylated to either propionate **177** or butyrate **178**, respectively.

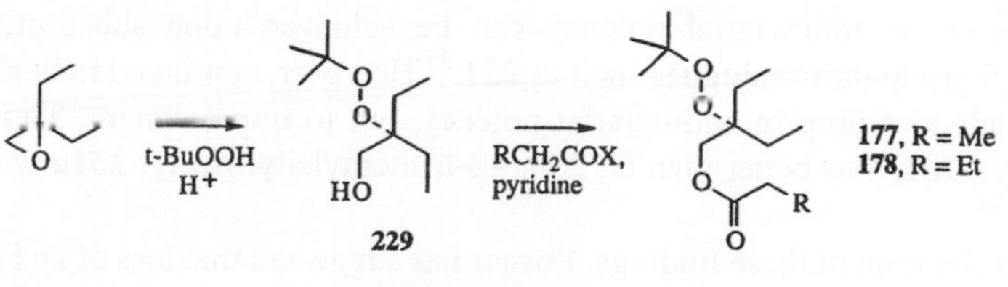

Scheme 18.

IV. MODE OF ACTION

A. Reactive Intermediates

Considerable effort has been expended by many researchers to understand how structural features of peroxides and trioxanes related to artemisinin effect antimalarial activity. As for many bioactive substances, a knowledge of their mode of action is very helpful in the design of more potent congeners. However, it is the exception rather than the rule that efficacy at the most basic mechanistic level, e.g. binding affinity to a protein, translates directly to potency in man. Indeed, how efficiently a drug reaches its target (drug accumulation) and how a drug is metabolized and eliminated can become far more important than intrinsic potency in defining an overall in vivo activity.

A proposed mode of action for the artemisinin class has been described by Meshnick and collaborators.[95-97] As *P. falciparum* is rich in hemin, hemozoin,[98] and free iron, artemisinin can readily react with hemin, a ubiquitous cellular component of malaria, to produce damaging radicals[99] and a structurally unidentified covalent adduct between hemin and artemisinin.[75] Meshnick has found that [^{14}C]-artemisinin reacts with intercellular hemin found in the parasite to generate reactive radicals that alkylate specific parasite proteins of 25, 50, 65, and > 200 kDa.[20] Since protein alkylation is highly specific, Meshnick has suggested that one of them might be the target for the artemisinin class. Addition of iron sequestering agents, in vitro, negated the antimalarial activity of artemisinin. The extended π system of hemin might be expected to interact with certain analogues of artemisinin resulting in enhanced binding relative to **1**. Yuthavong et al. have recently demonstrated a correlation of peroxide binding to hemin with antimalarial activity.[79] Computer modeling[100] of the interaction of hemin with artemisinin is consistent with our 3-D QSAR studies employing comparative molecular field analysis (CoMFA)[101,102] which will be discussed in the following section in detail. Our pharmacophore model shows that certain substituents could prevent complexation with the peroxide-containing face and thereby reduce potency.

Posner has provided information regarding   the reactive intermediates: Fe(II) promotes peroxide bond homolysis to oxyradicals (**232** and/or **233**) which then rapidly undergo 1,5-H atom transfer to generate the less reactive carbon radical **234** (Scheme 19).[22] Posner has suggested that stabilization of an intermediate radical like **234** should lead to more potent analogues of artemisinin. Indeed, modest improvement in antimalarial potency can be achieved upon substitution of a 4β-methyl group into analogues such as **231**.[23] However, continued radical stabilization leads to a drop in antimalarial potency, not to improvement. Thus, while 4β-methyl, **231**, was better than H, **230**, 4β-trimethylsilylmethyl **231a** was worse (Eq. 23).[24]

In consideration of these findings, Posner has suggested that loss of an Fe(IV)=O species from the intermediate carbon radical **234** can effect reoxidation of enolic

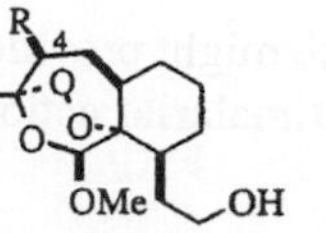

$$230, R = H$$
$$231, R = Me$$
$$231a, R = CH_2SiMe_3$$

(23)

intermediates thereby providing epoxides (**235**).[26] These epoxides have also been promoted as active species by Posner, that might alkylate parasite proteins. Jefford has debated the Fe(IV)=O hypothesis and has provided evidence that an iron–oxo intermediate is not formed.[27] Instead, **237** is ascribed by these researchers as the biogenic radical responsible for the damaging properties of **1**. Haynes has suggested that the mechanism of action of **1** involves Lewis acid catalyzed elimination of the peroxide bond to form a bulky version of hydrogen peroxide, hydroperoxide **239**.

The Fe(II) stimulated rearrangement of several tricyclic artemisinin analogues has been studied with the hope of defining a mechanistic feature that could be exploited in analogue design.[47,103,104] Based on studies with labeled tricyclic analogues of **1**, it has been suggested that upon Fe(II)-promoted fragmentation, two possible radicals result (Scheme 19): **232** and **233**. Of these, only **233** possesses optimal structure for 1,5-H atom transfer and more stable carbon-based radical formation, i.e. **234**. It has been suggested that the antimalarial activity of structures such as **1** were dependent on formation of radicals such as **234**. It has been further

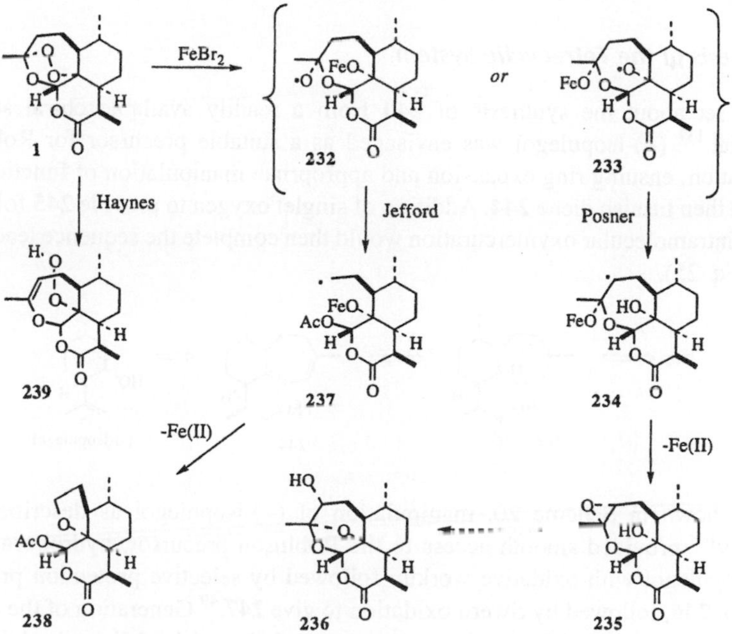

Scheme 19.

suggested that collapse of **234** to an epoxide might provide an active alkylator (e.g. **235**) that is in fact responsible for the antimalarial action and protein alkylating properties of this class of drugs.

B. 13-Carba Modification to Artemisinin

Due to the rigid nature of the artemisinin ring system, carba modification (**240** for **1**) would not be expected to significantly modify the shape of the resultant homologue. In fact, simple energy minimization and overlap of **1** and **240** revealed very little difference in shape.[105] Thus, the antimalarial activity of **240** relative to **1** can be ascribed in a general sense to the importance of the 1,2,4-trioxane substructure within the artemisinin tetracycle. Also of interest is the deoxo modification at C-10, 13-carba-10-deoxoartemisinin **241**. Other carbamodified, topologically familiar but nonperoxidic artemisinin-like ring systems **242** and **243** have been reported to be inactive.[106,107]

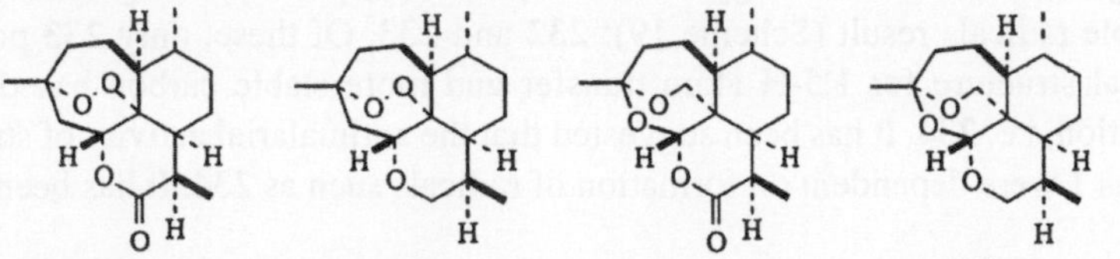

$$(24)$$

240   241   242   243

Synthesis of the Tetracyclic System

We set about the synthesis of **240** from a readily available chiral starting material.[137] (−)-Isopulegol was envisaged as a suitable precursor for Robinson annelation; ensuing ring expansion and appropriate manipulation of functionality would then furnish diene **244**. Addition of singlet oxygen to provide **245** followed by an intramolecular oxymercuration would then complete the sequence leading to **240** (Eq. 25).

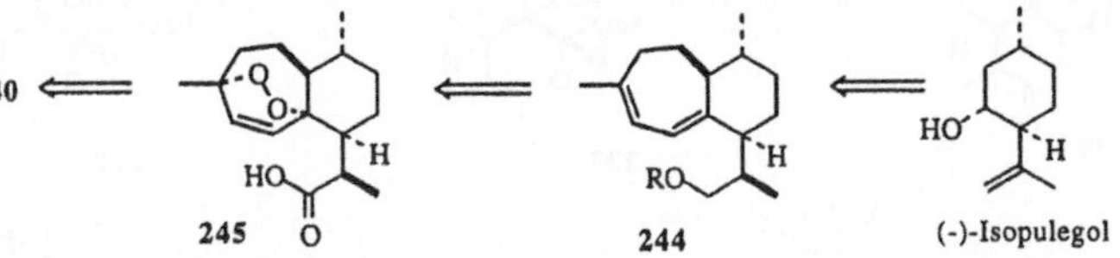

$$(25)$$

240 ⟸   245   O   244   (−)-Isopulegol

As shown in Scheme 20, manipulation of (−)-isopulegol as described previously[108] provided smooth access to the Robinson precursor: hydroboration of (−)-isopulegol with oxidative workup followed by selective protection provided alcohol **246**, followed by Swern oxidation to give **247**.[49] Generation of the kinetic enolate of **247** with LDA at low temperature followed by Michael addition to 3-trimethylsilyl-3-buten-2-one gave, after mild in situ hydrolysis of the α-silyl

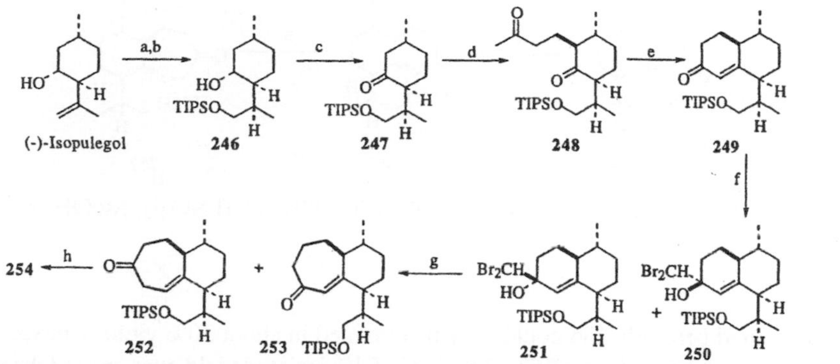

Scheme 20. Key: a) BH₃-THF, then H₂O₂, NaOH, 40%; b) (i-Pr)₃SiCl, DMAP, Et₃N, 96%; c) (COCl)₂, DMSO; Et₃N, 95%; d) LDA, 3-trimethylsilyl-3-butene-2-one, then pH 2, 73%; e) *l*-proline, MeOH, 43%; f) Br₂CH₂, LDA, –95 °C, 88%; g) BuLi, THF, –95 °C, 53% of **252**, h) KH, DME, Tf₂NPh, 89%.

group, diketone **248**. Aldol cyclization and elimination of water from **248** to generate bicyclic enone **249** was much more difficult than anticipated due to facile epimerization of the propanol side chain. It was found that this epimerization could be inhibited to some extent upon treatment of **248** with 1.5 mol equiv of *l*-proline, leading directly to the bicyclic enone **249**. Addition of dibromomethyllithium at low temperature afforded a mixture of diastereomers **250** and **251**. The mixture of **250** and **251** was then ring expanded upon treatment with butyl lithium at –95 °C and furnished a mixture of β,γ-unsaturated enone **252** along with α,β-unsaturated enone **253** (3:1 mixture of **252**:**253**). It was later found that only isomer **251** underwent the desired ring expansion. Dienol triflate **254** was then regiospecifically formed upon treatment of enone **252** with KH in THF followed by addition of Tf₂NPh. Triflate **254** was just stable enough to be chromatographed rapidly over silica gel; it had to be utilized directly for the next reaction.

While this approach to ring expansion yielded the desired product **252** in an overall yield of 47%, photochemical (room lights) and thermal sensitivity (room temperature) of the dibromomethyl intermediates inspired the development of an alternate approach. As shown in Scheme 21, decalenone **249** could be smoothly methylenated with Tebbe reagent to provide diene **255** in excellent yield. Clean and reproducible oxidative ring expansion could then be effected with thallium trinitrate to furnish enone **252** in 62% overall yield (unoptimized).

The final steps in the overall sequence were then carried out as demonstrated in Scheme 22. Coupling of the enol triflate with dimethylcopper lithium provided the diene **244** in excellent yield. Singlet oxygenation of **244** was then investigated with a variety of dyes in different solvents. We were surprised to find that a mixture of diastereomeric peroxides were obtained, only varying slightly with conditions, and

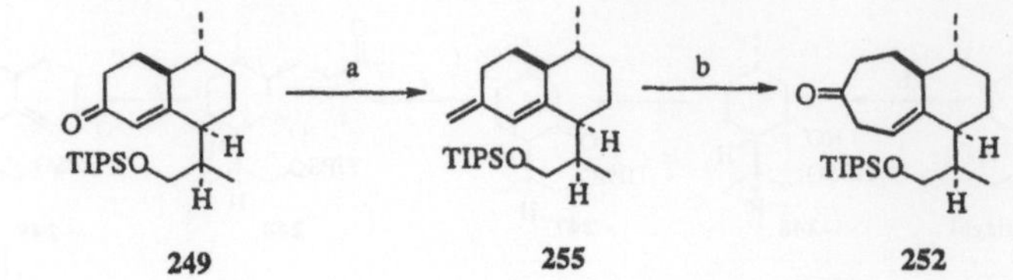

Scheme 21. Key: a) Tebbe Reagent, THF, 0 °C, 95%; b) Tl(NO₃)₃, MeOH, CHCl₃, 72%.

that desired peroxide **256** could only be obtained in about 30% yield. Unexpected predominant diastereomer **257**, obtained in 54% isolated yield, was carried through the sequence of silyl group deprotection and Hg(II) cyclization (Scheme 22) to arrive at diastereomeric peroxide **258**. Desired peroxide **256** was then deprotected with tetrabutylammonium fluoride in THF to give alcohol **259**.

Oxymercuration of alcohol **259** was readily accomplished using mercuric trifluoroacetate in THF; the organometallic product could then be carefully demercurated with sodium borohydride to provide (+)-10-deoxy-13-carbaartemisinin **241**. Attempted oxidation of alcohol **259** directly to acid **261** using pyridinium dichromate in DMF was not successful as the intermediate aldehyde could not be further oxidized with this reagent. However, alcohol **259** could be oxidized directly to the acid **261** using chromium trioxide in acetic acid. Oxymercuration–demercuration then gave target compound **240**. The structures of **258** and **241** were unambiguously

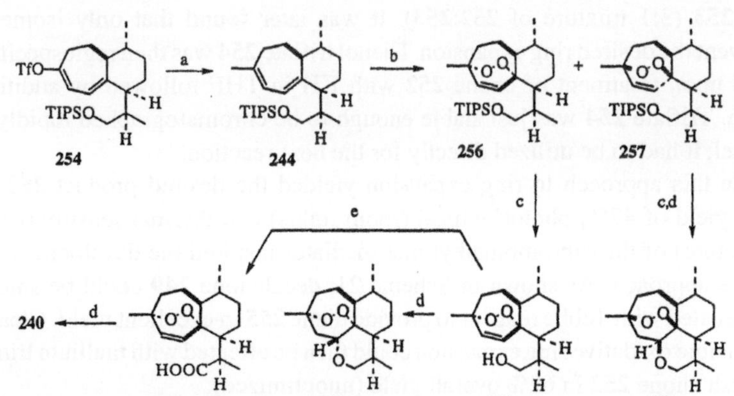

Scheme 22. Key: a) Me₂CuLi, Et₂O, 94%; b) ¹O₂, EtOAc, Rose Bengal Salt, 30% **256**, 54% **257**; c) Bu₄NF, THF: 89% **259**; 70% alcohol from **257**; d) Hg(OTFA)₂, THF; then NaBH₄/NaOH: 76% **241**; 78% **258**; 60% **240**; e) CrO₃, AcOH, H₂O, 64%.

determined by single crystal X-ray crystallographic analysis, while the structure of **240** was reasonably extrapolated from **241**.

Tricyclic Analogue Syntheses

Having successfully assembled the desired tetracyclic system, examination of corresponding tricyclic systems derived from synthetic precursor **259** were relatively straightforward. Smooth access to lipophilic methyl and benzyl ethers **262** and **263**, respectively, was accomplished via S_n2 reaction of the sodium alkoxide, derived from alcohol **259** and NaH with either methyl iodide or benzyl bromide (Scheme 23).

It seemed prudent that the same ethers be examined in the absence of potentially labile functionality, thus removal of unsaturation in **262** and **263** was considered. Hydrogenation of **259** over Pd/C or Pt was unsuccessful; in either case reduction of the peroxide group was problematical. Hydrogenation over Wilkinson's catalyst gave a new product, but with the unsaturation retained. While selective alkene hydrogenation can sometimes be achieved in the presence of a peroxide bond, the double bond of **259** was apparently too hindered in this case. Diimide, on the other hand, worked reasonably well for this reduction. Thus, treatment of **259** in dichloromethane solution with potassium azodicarboxylate followed by addition of acetic acid led, after several days, to roughly 60% conversion of **259** to the saturated version, **264**. Now, ether formation as before provided the saturated methyl and benzyl ethers **265** and **266**, respectively, in good yields.

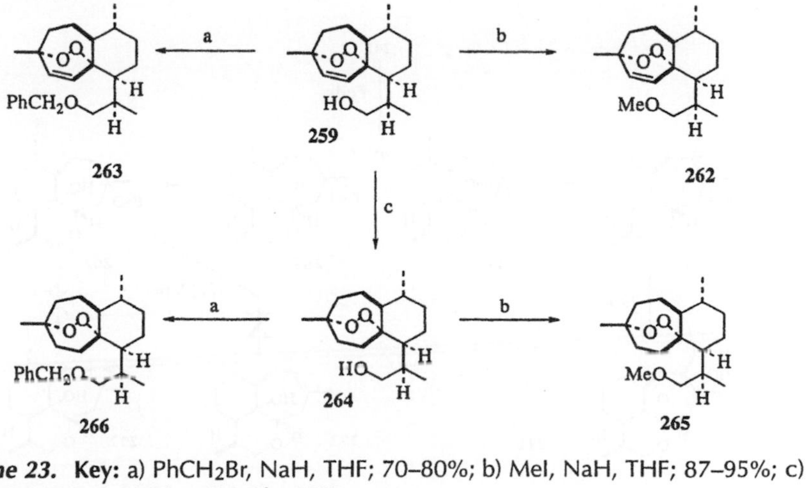

Scheme 23. **Key:** a) PhCH₂Br, NaH, THF; 70–80%; b) MeI, NaH, THF; 87–95%; c) KO₂CN=NCO₂K, HOAc, CH₂Cl₂; 60%.

Rearrangement Chemistry

The intrinsic instability of **235** makes its existence difficult to prove; intramolecular closure to 4-hydroxy-1-deoxyartemisinin **236** or simple aqueous hydrolysis
leads to its destruction. If **235** (and related structures) were the bottleneck through
which the artemisinin class exerted its antimalarial effect, then replacement of O-13
by CH_2 might give an isolable intermediate epoxide of greater stability, but also
with lesser activity.

Under reported conditions,[104] the only tractable product from rearrangement of
241 was alcohol **271**, isolated in 79% yield as shown in Scheme 24. From a
mechanistic standpoint, it would seem that radical **267** is a probable precursor to
271, while it is less clear how **270** is derived from **241**. Homolysis of the peroxide
bond of **241** could lead to either **267** or **268**. If C-4 H-atom abstraction from **268**
were to occur giving radical **269**, then it would be more difficult to explain the
product **271**. In fact, a more reasonable product of radical **268** might be the epoxide
272 which was, however, not formed in this reaction.

If proton transfer occurs from C-4 to radical **267**, then **270** would result directly.
The later, more direct pathway, unfortunately proceeds through a four-membered
transition state which would be predicted to occur at a much slower rate than the
six-membered TS leading to **269**. However, no evidence for unreacted alkene **273**
resulting from expulsion of Fe(IV)=O from **269** was evidenced,[109] suggesting that
the path through radical **268** was not followed in this example.

Rearrangement of the isomeric β-oriented peroxide **258** under identical conditions to those used for **241** provided **279** and **281** (Scheme 25) in 39 and 20% yields,
respectively. Cycloheptene **282** was not detected in this rearrangement while
epoxide **281** was, suggesting that elimination of Fe(IV)=O does not occur and that

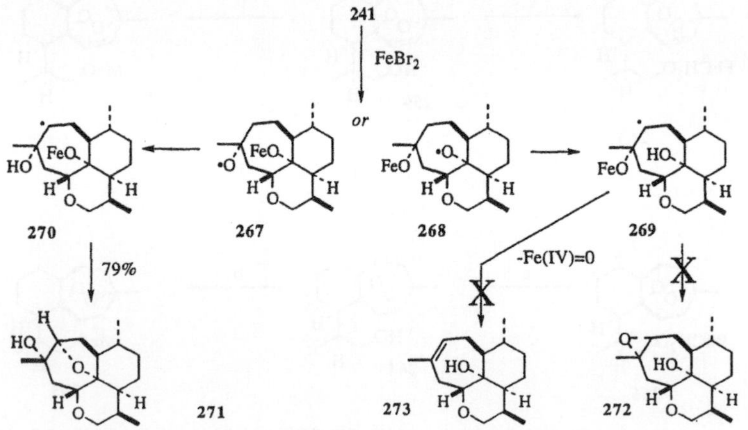

Scheme 24.

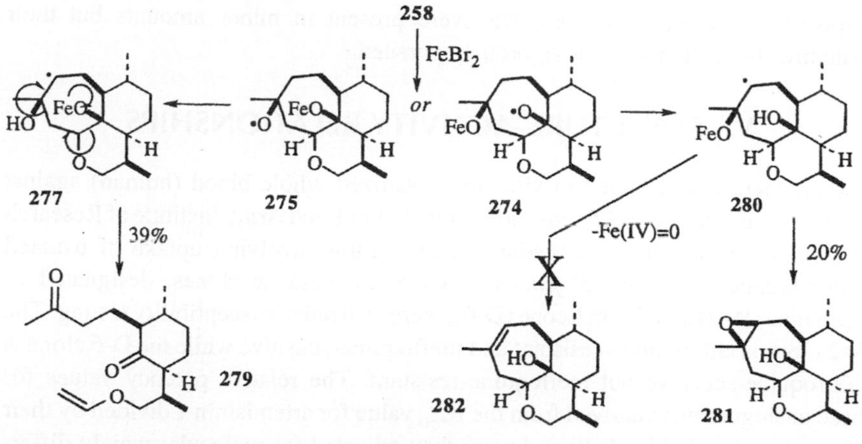

Scheme 25.

epoxide formation follows another pathway presumably involving loss of Fe(II) from **280**.

As the rearrangement products of the carba modified 10-deoxo analogues (D-ring tetrahydropyran) were somewhat different from the products from rearrangement of artemisinin **1** (D-ring lactone), we felt that it would be worthwhile to explore the rearrangement of 10-deoxoartemisinin **108**. Upon exposure of **108** with 2.5 equiv of FeBr$_2$ in THF at room temperature (Scheme 26), reaction was complete within minutes and gave, after purification, products **288** and **286** in 79 and 8% yield,

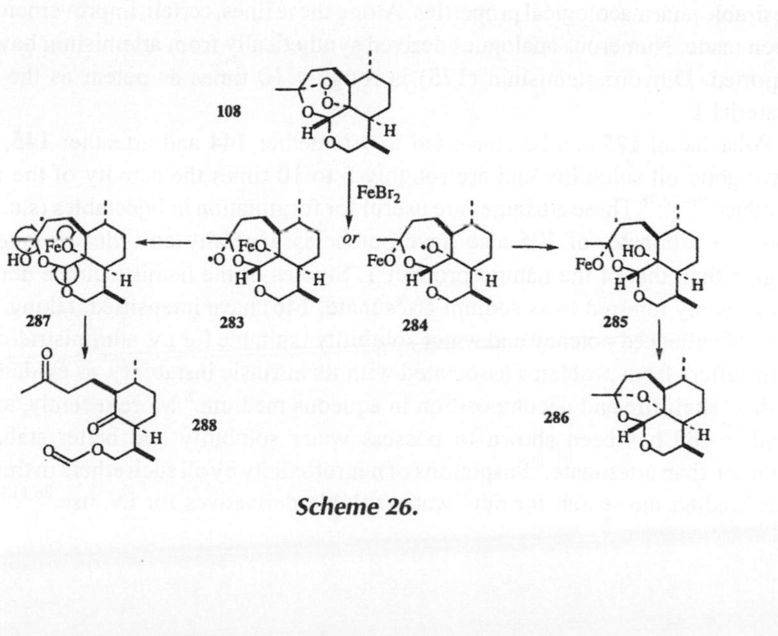

Scheme 26.

respectively. Several other products were present in minor amounts but their structure elucidation was not vigorously pursued.

V. STRUCTURE–ACTIVITY RELATIONSHIPS

The analogues were tested in vitro in parasitized whole blood (human) against drug-resistant strains of *P. falciparum* at the Walter Reed Army Institute of Research by a modification of the procedure of Desjardins involving uptake of tritiated hypoxanthine.[110,111] Two *P. falciparum* malaria parasite clones, designated as Indochina (W-2) and Sierra Leone (D-6), were utilized in susceptibility testing. The W-2 clone is chloroquine-resistant and mefloquine-sensitive while the D-6 clone is chloroquine-sensitive but mefloquine-resistant. The relative potency values for these analogues were derived from the IC_{50} value for artemisinin 1 divided by their IC_{50} values (see Tables 1–8) and were then adjusted for molecular weight differences by multiplication of the ratio of the molecular weight of the analogue divided by the molecular weight of artemisinin. This approach to reporting activity was based in part on the fact that the analogues were tested on different occasions in which the IC_{50} for the control, artemisinin, had varied anywhere from 0.1 to 4 ng/mL based on parasitemia levels and other factors.

Before the results of our synthetic analogues are presented, it is relevant to first discuss the state of knowledge concerning SAR and current philosophies of designing improved artemisinin analogues. The astounding antiparasitic profile exhibited by artemisinin combined with certain drawbacks to its use—such as its limited availability (< 0.5% from *Artemisia annua L.*), relatively low potency in man (0.9–1.2 g/3 days), lack of substantial oral activity, poor oil or water solubility, short half-life, and fetotoxicity—prompted the search for analogues with more desirable pharmacological properties. Along these lines, certain improvements have been made. Numerous analogues derived synthetically from artemisinin have been reported. Dihydroartemisinin (175) is roughly 10 times as potent as the parent material 1.

Polar lactol 175 can be converted to artemether 144 and arteether 145, which have good oil solubility and are roughly 5 to 10 times the activity of the natural product.[19–22,88] These attributes are useful for formulation in injectables (s.c. or i.m. route). Carbonates of 175 also have potencies generally an order of magnitude higher than that of the natural product 1. Studies of the hemisuccinate derivative (commonly referred to as sodium artesunate, 146) have intensified, taking advantage of enhanced potency and water solubility (suitable for i.v. administration), but 146 suffers from problems associated with its intrinsic instability, as evidenced by a short shelf life and decomposition in aqueous medium.[8] More recently, artelinic acid[112] 289 has been shown to possess water solubility and better stability in solution than artesunate.[8] Suspicions of neurotoxicity by all such ethers in this series has kindled the search for new water-soluble derivatives for i.v. use.[88,113] While

10-ethers such as arteether **145** are impotent neurotoxins in vitro,[114–117] in vivo conversion of **145** to dihydroartemisinin **175** is a facile process[20,76,77,118] and further, **175** is the most neurotoxic artemisinin analogue reported. Support for the sometimes lethal toxicity of 10-lactol ethers in vivo is provided by animal studies.[119]

289, R = CH$_2$PhCO$_2$H (artelinic acid)

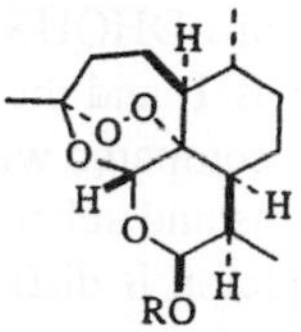

(26)

Although the aforementioned analogues contribute to the knowledge of structure–activity relationships (SAR) of **1**, the goal of obtaining ideal analogues of artemisinin for the treatment of malaria requires much more SAR information. Compounds related to **1** are scattered throughout the literature and are interesting in light of our desire to construct an SAR framework as the basis for the rational design of analogues of **1** (for related summaries, see refs. 120–122). It is noteworthy in this regard that deoxyartemisinin **290**, which lacks the peroxide linkage of **1**, is topologically similar to **1** but devoid of antimalarial activity.[123] This deoxy analogue **290** is a human urinary metabolite of **1**, but can also be synthesized from **1** by treatment with triphenylphosphine. Additional urinary metabolites such as dihydrodeoxyartemisinin and the so-called crystal-7 are also devoid of antimalarial activity.[123]

1 $\xrightarrow[\text{or } H_2, \text{ Pd/C; } H^+]{\text{Ph}_3\text{P}}$ **290**

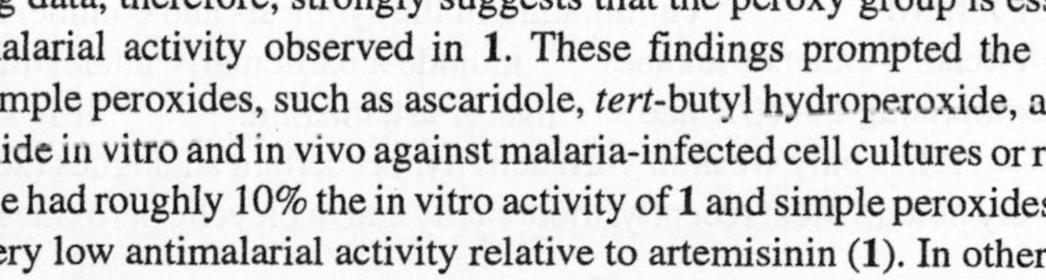

(27)

Existing data, therefore, strongly suggests that the peroxy group is essential for the antimalarial activity observed in **1**. These findings prompted the testing of various simple peroxides, such as ascaridole, *tert*-butyl hydroperoxide, and hydrogen peroxide in vitro and in vivo against malaria-infected cell cultures or rodents.[124] Ascaridole had roughly 10% the in vitro activity of **1** and simple peroxides typically display very low antimalarial activity relative to artemisinin (**1**). In other laboratories using dissimilar in vitro techniques, *tert*-butyl hydroperoxide was found to be (by extrapolation) roughly 10,000 times less active than **1** and similar to hydrogen peroxide in potency. In in vivo testing, ascaridole was orally inactive, and simple hydroperoxides showed protective effects against malaria infestation when administered interperotoneally (i.p.). The results presented for a range of peroxides tend to support the hypothesis that the peroxy group in **1** is essential for activity but extremely dependent upon complementary topology (shape, size, and polar-group arrangement). Recent studies also suggest that the details of intramolecular radical rearrangement reactions initiated by Fe(II) and/or hemin are also important variables in defining antimalarial potency.[47,80]

Two closely related but unsaturated versions of peroxide **1** are known; artemisitene **291** and anhydro-DHQHS **292**. The *exo*-methylenelactone **291** is about one-quarter as active as **1**, and the olefin **292** is entirely inactive. When the various structures above are compared with **291**, the thought arises that the C-9 and C-10 positions of **1** can withstand substantial variation without loss of biological activity. However, this suspicion is difficult to reconcile with the lack of activity

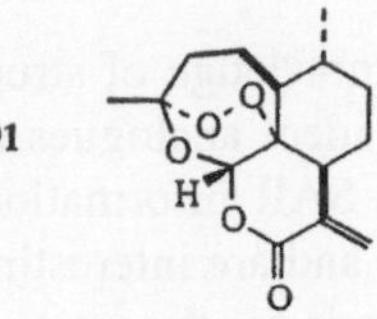

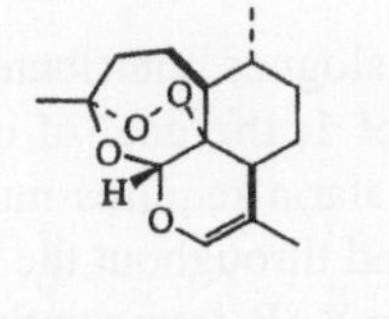

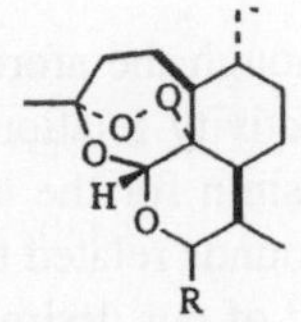

(28)

displayed by olefin **292**. Molecular models (MM2) constructed to permit comparison of **1** versus **291** and **292** are not particularly illuminating. We believe that the explanation for the difference in activity of **291** versus **292** does not reside in the presence or absence of an oxygen substituent at C-10 or in minor differences in the ring conformations between **291** and **292**. It seems likely that a somewhat more complex rationale is involved, since in contrast to the above, 10-deoxoartemisinin **108** is four to eight times more potent than artemisinin,[31] and other 10-deoxo analogues of artemisinin can be quite potent.[33] Also, 10-substituted-10-deoxo analogues **293** are known that in some cases possess good in vitro antimalarial activities.[32,35]

A single ring-cleaved derivative, called "desethanoartemisinin," **171**, was reported[50] but not tested for activity. We resynthesized **171** by our own route (Scheme 9)[49] and found very weak activity (3% of **1**). Other synthetic trioxanes of the type **294** also showed weak activity.[40] A wide range of 1,2,4-trioxanes such as **295** were prepared by Jefford[36–38,125,126] culminating in the synthesis and commercial evaluation of arteflene.[95] Others trioxanes[39,41] include a particularly interesting class of tricyclics comprising the A/B and C rings of artemisinin.[46–48,127] It is significant that **230** (R = H) has only weak in vitro activity, yet certain analogues (**231**) in this series are very potent. Most recently, moderately active bicyclic peroxides **296** have

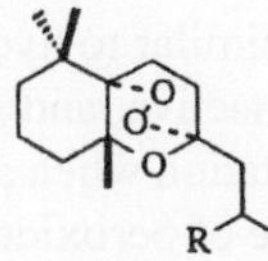

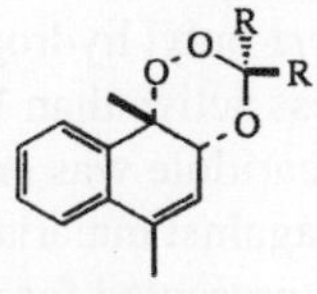

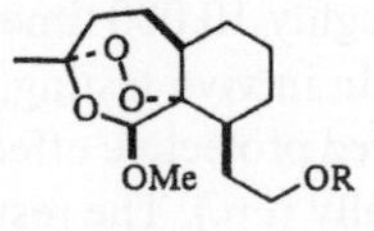

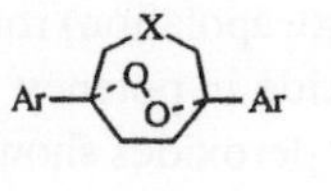

(29)

been easily prepared by cycloaddition of singlet oxygen to the corresponding dienes.[128]

1,2,4,5-tetraoxanes have been examined and display modest activity,[43] and even ozonides can display antimalarial activity, again of modest potency.[129] Therefore, although the 1,2,4-trioxane ring moiety in artemisinin is required for activity, a specific lipophilic arrangement of atoms is also important for potency. There are specific examples in which these simplified analogues possess excellent potency in vitro. At the current time however, most highly potent analogues are more closely related to the artemisinin tetracyclic architecture.

Racemic desmethyl tetracyclic analogue **155** has roughly 50% the activity of artemisinin. This is an interesting finding that raises several issues. First, if optical activity is a criterion for antimalarial potency, the fact that **155** is a racemate should be reflected in a lower relative activity. Alternatively, it may be that the C-6 and/or the C-9 methyl groups contribute to the activity of **1** and that optical activity is not important. Finally, of course, it is possible that some combination of effects is operative in that one enantiomer of **155** is less active than the other and that one or the other (or both) of the methyl groups affects the activity of **1**. If it is assumed that only the (+)-enantiomer of **1** is active, then a comparison of (+)-9-desmethyl-artemisinin **42** with (±)-6,9-desmethylartemisinin (**155**) offers some clues regarding SAR at the C-6 position. Because removal of the C-9 methyl leads to a six-fold enhancement in potency and removal of both C-6 and C-9 methyls (along with a 50% reduction due to the compound being a racemate) leads to a 50% reduction, it may be true that the C-6 methyl contributes a six-fold enhancement to the activity of the basic tetracyclic structure of **1**. As yet, the effect of substitution at the C-6 position has not been fully explored.

To partially address a solution to the SAR information provided by **42** and **155**, it would be useful to carry out an enantioselective synthesis of the antipode (mirror image) of artemisinin **1** starting from (3*S*)-(–)-pulegone instead of (3*R*)-(+)-pule-gone (Schemes 2 and 3). While it has been claimed in a few cases that both enantiomers of simple trioxanes have identical activities, the antimalarial effect of more complex systems resembling artemisinin may be more sensitive to absolute configuration. A successful synthesis of the enantiomer of naturally derived artemisinin would accomplish several important goals. First, it would help establish whether the mode of action of **1** involves interaction with a specific receptor, an optically active site, a membrane-associated transport protein, or was not dependent on chirality. If the enantiomer of **1** proved to be totally inactive, then one might safely say that receptor or receptor-like specificity is involved in the mechanism of action of **1**. However, if the enantiomer of **1** were found to have identical activity, it might well be argued that a nonspecific mechanism is involved. Currently, the interaction of peroxides with either free Fe(II) or hemin (both achiral) is generally accepted as the first step in the mechanism of action. This argument would depend to some extent on the degree of activity of the enantiomer of **1**. Second, depending on the outcome of the biological testing of the enantiomer of **1**, further SAR studies would be more directed in focus. For example, one would know whether to carry out syntheses of analogues enantioselectively or merely stereospecifically. Finally, the results of testing of **42** and **155** could be more readily explained, and this would provide further valuable insight into SAR in this family of drugs. For example, if it is true that the C-6 methyl is important to activity, then varying this position of the molecule would be a fruitful area of research.

We have examined the C-9 position of **1** in some detail because these analogues are readily available by alkylation of the dianion derived from acid **32**, produced in the total synthesis of **1**. Of interest is the effect of hydrophobic/hydrophilic groups at C-9 (see Figure 7). Accordingly, ethyl **43** through *n*-hexyl (**44**, **45**, **46**, and **47**, respectively) and tetradecyl **48** were examined. An amazing enhancement in potency was gained for ethyl **43** and propyl **44** (~13). The intermediate butyl **45** and pentyl **46** analogues begin to lose activity and by *n*-hexyl **47**, activity drops (~7). Upon extension to tetradecyl **48**, activity is almost nil. Log P values did not correlate with our findings despite a reported correlation between lipophilicity and antimalarial activity for arteether-like derivatives.[130]

It is possible that as the carbon-chain length increases, the chain begins to fold back on the tetracycle, thus blocking some key binding interaction. This idea could be tested by introduction of *trans* unsaturation into the hexyl chain (or the longer homologues). We have also investigated the effects of branching in the C-9 substituent (see Table 1). Isopropyl **53**, isoamyl **55**, isobutyl **54**, and isohexyl **56**

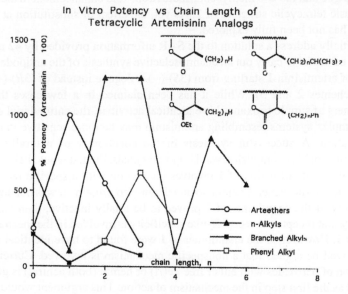

Figure 7. Plot of in vitro potency vs. chain length of four major classes of artemisinin analogues.

Table 1. In Vitro Antimalarial Activities of C-9 Substituted Analogues of Artemisinin

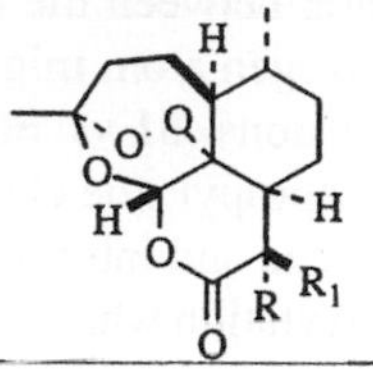

Structure	R	R_1	Relative Activity[a]	
			W-2 Clone	D-6 Clone
1	H	Me	1.00	1.00
42	H	H	6.50	1.80
62	Me	H	0.67	0.14
63	Me	Me	0.006	0.02
43	H	Et	12.26	6.42
44	H	Pr	12.25	5.50
53	H	i-Pr	0.84	2.03
45	H	Bu	1.28	0.96
54	H	i-Bu	0.24	0.20
46	H	C_5H_{11}	8.67	3.67
55	H	$i\text{-}C_5H_{11}(i)$	0.98	1.57
47	H	C_6H_{13}	5.50	4.67
56	H	$i\text{-}C_6H_{13}$	0.74	0.31
48	H	$(CH_2)_{13}CH_3$	0.001	0.001
50	H	CH_2CH_2Ph	1.00	1.69
51	H	$(CH_2)_3Ph$	4.45	6.11
52	H	$(CH_2)_4Ph$	3.00	1.37
58	H	CH_2COOH	0.003	0.002
57	H	$CH_2CH=CH_2$	0.72	0.31
60	E - $MeCH=CHCH_2$	H	0.22	0.38
61	Z - $MeCH=CHCH_2$	H	0.07	0.09

Note: [a]IC_{50} of analogue/IC_{50} of artemisinin = relative activity

branched homologues have interesting relative activity. Isopropyl **53** is about equipotent with **1**, whereas *n*-propyl **44** is about 13 times as potent as **1**.

These findings might tempt one to predict that 3-phenylpropyl analogue **51** would be about equipotent with **1**. Surprisingly, **51** was about 8 times more potent than **1**. Furthermore, there is also a "length" dependency with phenyl terminators, as shown in Figure 7. These results suggest that a hydrophobic pocket exists in a putative artemisinin "receptor" with well-defined dimensions. A Pi acceptor in this pocket could provide additional binding energy and might explain the enhanced potency of phenylpropyl **51**. For example, if the acceptor were a tryptophan residue, then hydrophobic–hydrophilic repulsion between the branched hydrocarbon C-9 analogues (e.g. **56**) and the indole nitrogen atom might explain the disparity between side chains of similar steric dimensions (**51** vs. **56**). Another possible explanation for the enhanced potency of phenylpropyl side chains versus alkyl side chains may be related to a dual effect of both size and inhibited metabolism. Alkyl chains can undergo facile ω- or $(\omega - 1)$ hydroxylation which would be blocked by the presence of a phenyl ring.

Further support for a hydrophobic binding domain in the southeast quadrant of the artemisinin pharmacophore comes from the carboxylic acid **58** (originally designed as a chemically stable replacement for sodium artesunate), which was found to be nearly devoid of antimalarial activity. Of course, bioavailability of the acid **58** could be poor in contrast to hydrophobic analogues.

Simple homologous olefins at C-9, such as the allyl analogue **57** were planned to examine the effect of donors. Allyl analogue **57** was surprisingly only 30–70% as potent as artemisinin. Compared to the propyl or aryl analogues, this was an unexpected result.

A number of 11-aza-9-desmethylartemisinin analogues (lactam) were synthesized and tested in vitro: N-methyl **134**, $N–CH_2CO_2H$ **140**, $N–CH_2CH_2N(Me)_2$ **138**, N-benzyl **136**, and $N–(CH_2)_5CO_2H$ **139**. Interestingly, N-methyl **134** was about as active as the natural product. However, N-benzyl **136** was amazingly more potent than **134** and roughly twice as active as **1**. Carboxylic acids **139** and **140** were quite inactive, which is consistent with the inactivity of the "southeast" acid **58**. If the mode of action of this class of drugs involves binding of (a) the α-face peroxide grouping with the iron core of hemin, and (b) the N-11 and/or C-9 alkyl substituents with the somewhat flat hydrophobic face of the porphyrin ring system, then these test results could be reasonably explained. A series of 11-azartemisinin analogues, prepared directly from artemisinin, have been reported.[131]

A continued examination of both C-9 and N-11 substituted analogues of **1** might allow one to more accurately define the "southwest" and "southeast" regions of the artemisinin binding site. Optimization of substituents with either the C-9 or N-11 class of analogues would then lead naturally to hybrid structures which integrate these findings into combined C-9, N-11 analogues which might be expected to have greatly enhanced activity.

That this additivity cannot be taken for granted with the artemisinin molecule can be gleaned from another study that has been carried out. For example, both arteether (10 times more potent than the parent molecule) and ethyl analogue **43** ($\times$ 13) were separately quite potent. It was hoped, simplistically, that conversion of **43** to corresponding ethyl ether derivative **142** would result in an analogue which was substantially more potent than either **43** or **142**. This simple reasoning did not hold and, in fact, **142** was only 3–6 times more potent than **1** and was about half as active as **43**. The ethyl ether derivative of propyl analogue **143** was equally disappointing.

Molecular mechanics calculations of **142** and **143** reveal that steric interactions between adjacent bulky alkyl chains force the C-9 and C-10 groups into conformations that are unlike artether **145**. One might argue that these new conformations are unacceptable and can account for the drop in activity. However, the intermediate lactols should not suffer from these intramolecular interactions and should retain activity. Surprisingly, however, butyl-substituted lactol **141** (reduction product of **45**) was found to be less active than dihydroartemisinin.

It was felt that this potential conformational problem could be circumvented by joining the two ends of the alkyl substituents, at C-9 and C-10 in arteether, to form a tetrahydropyran ring. Pentacyclic ether **154** was synthesized, but found to be less active than arteether.

Another compound of interest is ring-contracted analogue **156** which was synthesized to examine the effect of increased ring strain on antimalarial activity. A more reactive peroxy group incorporated into an analogue topologically similar to the natural product (the minimized structure of **156** has a reasonable overlap with **1**) might lead to enhanced activity.

Analogue **156** was found to be thermally unstable compared to artemisinin, which is an indication of ring strain. However, **156** has poor activity suggesting that rapid metabolism or chemical decomposition of the reactive peroxy group of **156** results in deactivation prior to reaching the site of action.

Substitution solely at C-3 and the effect of dual substitution at C-3 and C-9 were examined for their influence on antimalarial potency in vitro (see Table 2). Addition of another carbon at C-3 giving ethyl analogue **77** (R^1 = Me, but lacking the C-9 methyl group, R = H) results in little effect on activity, while next homologue **78** (R^1 = ethyl) is quite potent compared to artemisinin. Interestingly, 3-butyl homologue **79** and its branched counterpart **80** have diminished antimalarial activity relative to control. The overall trend for this series is qualitatively similar to that observed for C-9 substitution: a peak and trough in activity is seen upon substitution of *n*-alkyl groups at C-9 of 2 to 6 carbons in chain length. For C-9, maximal activity was achieved at ethyl and propyl (12 times activity of artemisinin), but had dropped by hexyl to around 5 times the potency of artemisinin. In this C-3 substituted series, the activity profile is attenuated with a peak occurring at propyl **78** (7–21 times activity of artemisinin) but with lower activity on either side such as ethyl **77** and butyl **79**. Interestingly, placement of an ester group at C-3 as in the propionate ethyl ester **81** results in an analogue with better activity than artemisinin (2 times artemisinin) while the closest homologue to **81** is relatively inactive isobutyl derivative **80**.

Next, the effect of aryl substitution was considered. The series substituted at C-3: 2-phenylethyl **82**; 3-*p*-chlorophenylpropyl **83**; and 4-phenylbutyl **84**; displayed roughly 1, 100, and 300% the activity of artemisinin, respectively. Compared to the analogous series at C-9 (2-phenylethyl, 100%; 3-phenylpropyl, 500%; 4-phenylbutyl, 300%), these results are only substantially at variance for phenylethyl analogue **82**. Furthermore, as was the case for C-9 substituted aryl–alkyls, the steric

Table 2. Relative In Vitro Antimalarial Activity of 3-Substituted Analogues of Artemisinin against *Plasmodium falciparum*

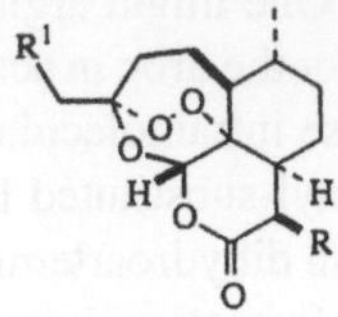

Structure	R^1	R	Relative Activity[a] D-6	W-2
1	H	Me	100	100
77	Me	H	88	112
78	Et	H	2102	673
79	Bu	H	20	18
80	i-Pr	H	53	45
81	EtO_2CCH_2	H	232	232
82	$PhCH_2$	H	3	1
83	p-ClPh$(CH_2)_2$	H	114	127
84	Ph$(CH_2)_3$	H	220	281
94	Me	Bu	184	257
95	Pr	Bu	28	33
96	$PhCH_2$	Bu	1	1
97	p-ClPh$(CH_2)_2$	Bu	43	53
98	Ph$(CH_2)_3$	Bu	39	48
99	EtO_2CCH_2	Bu	1382	2285

Note: [a]Relative activity = 100 × [IC$_{50}$ artemisinin (control value)/IC$_{50}$ analogue] MW analogue/MW artemisinin.

bulk typified by a aryl ring did not adversely effect activity but branched hydrocarbons did lower potency appreciably.

On the whole, analogues of artemisinin substituted at C-3 were found to be less active than those substituted at C-9. Because it was a simple matter to combine methodologies for synthesis of analogues at both C-3 and C-9, we thought that dual substitution would provide additional interesting SAR information. The activities of analogues **95** through **98** were unimpressive while the activity of **94** and **99** were good to excellent. The following trends were observed: For increasing alkyl bulk at C-3 a drop in antimalarial efficacy was noted (**77**, relative activity of 1.1; **79**, relative activity of 0.2). Upon butyl substitution at C-9, the corresponding dual substituted analogues (3-alkyl, 9-butyl) showed a doubling of activity (**94**, relative activity of 2.6; **95**, relative activity of 0.33). For the C-3 aryl–alkyl substituted analogues alone, an increase in activity was observed with increasing chain length between ring system and the aryl ring (2 carbons for **82**, relative activity of 0.01; 3 carbons for **83**, relative activity of 1.3, and 4 carbons for 84, relative activity of 2.8). Dual substituted aryl–alkyl analogues (3-arylalkyl,9-butyl) were generally less active than 3-substituted aryl–alkyl analogues alone (e.g. **83**, relative activity of 1.3; **97**, relative activity of 0.53). Thus, a number of opposing trends were evident.

The high potency of dual-substituted analogue **99**, on the other hand, is somewhat of a surprise in light of the foregoing discussion. By comparison to **81** (ester side chain), **99** should be roughly no better than two-fold more potent than artemisinin. However, one is faced with a 14–23-fold *enhancement* in activity for **99** relative to artemisinin.

These findings suggest the following qualitative statements: (a) a binding site or acceptor for artemisinin and analogues exists with limited dimensions at both C-9 and C-3, more tolerant of aryl and ester substitution on *n*-alkyl chains than of branched alkanes of any length; and (b) dual substitution at C-3 and C-9, explored for only 9-butyl analogues, was on the whole detrimental to activity with one interesting exception, **99**. As it happens, **99** would be predicted to dock with hemin better than other members of this series. Perhaps the efficacy of hydrogen bonding of **99** to hemin is reflected in the enhanced activity of this analogue. The interaction energies of artemisinin and dihydroartemisinin with hemin have been calculated in Sybyl using the dock routine.[132] We have applied this technique to the interaction of **99** with hemin in the Fe(II) oxidation state. The lowest E alignment of **99** with hemin places the ester group in the vicinity of the carboxyl groups of hemin, as shown in Figure 8, and allows for hydrogen bonding between hemin and **99**. If this hypothesis is true, then other hydrogen bond acceptors with appropriate tethers at the C-3 position of artemisinin might be expected to have enhanced activity.

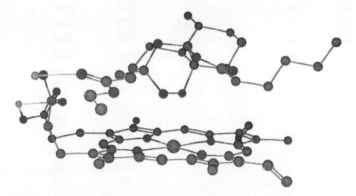

Figure 8. Docking interaction of hemin with artemisinin analogue **99** in Sybyl showing hydrogen bonding between hemin side chains and the C-3 ester group of analogue **99**.

For the analogues of 10-deoxoartemisinin 108 substituted at C-9 (110–115), or desmethyl analogue 109, certain qualitative relationships emerge upon examination of Table 3. In the W-2 clone, the homologous series ranging from H to propyl shows a relatively steady potency at 5 times the activity of control (artemisinin, 1). For butyl 112, a dramatic improvement in potency was seen at about 21 times control, while at pentyl, 113, activity had dropped off somewhat to around 1.5 times the potency of artemisinin. We have found that activity in the W-2 clone is generally paralleled by the more sensitive D-6 clone. Thus, potency ranged from 2 times control for analogue 109, up to 58 times control for butyl derivative 112, and finally decreased as expected for pentyl homologue 113. Profound potency enhancement

Table 3. Relative In Vitro Antimalarial Activity of 3 and 9-Substituted Analogues of 10-Deoxoartemisinin against *Plasmodium falciparum*

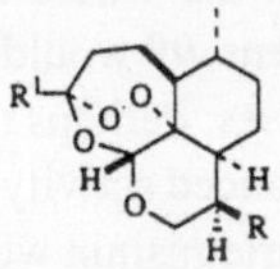

Structure	R^1	R	Relative Activity[a]	
			D-6	*W-2*
108	Me	Me	659	567
109	Me	H	237	190
110	Me	Et	914	466
111	Me	Pr	473	550
112	Me	Bu	5826	2090
113	Me	$CH_3(CH_2)_4$	170	145
114	Me	$Ph(CH_2)_3$	5073	2506
115	Me	p-ClPh$(CH_2)_3$	6991	3317
116	Et	H	10	10
117	Pr	H	722	685
118	Bu	H	653	556
119	$(CH_3)_2CHCH_2$	H	183	250
120	$Ph(CH_2)_4$	H	336	380
121	$Ph(CH_2)_2$	H	6	2
122	p-ClPh$(CH_2)_3$	H	13	28
123	$(CH_2)_2CO_2Et$	H	422	506
127	$(CH_2)_2CO_2H$	H	0.09	0.09

Note: [a]Relative activity = $100 \times$ [IC$_{50}$ artemisinin (control value)/IC$_{50}$ analogue] MW analogue/MW artemisinin, i.e. relative activity of artemisinin = 100.

was observed for 3-aryl propanes **114** and **115** in both W-2 and D-6 clones, being in the range of 25–70 times more potent than control. *To our knowledge, analogues* **112, 114,** *and* **115** *are the most potent analogues of artemisinin yet reported in vitro which lack the 10-acetal linkage.*

It is interesting to note that lactones **42–46, 51** corresponding to the above tetrahydropyrans **109–114** (with the exception of **125**, the precursor to **115**) showed a similar but attenuated trend in activities. For example, lactones **43** and **44** were most potent in this series (ethyl and propyl) at about 12 times the activity of artemisinin, and had dropped significantly in potency by butyl (**45**) and pentyl (**46**). Similarly, 3-phenylpropyl lactone **51** was reasonably potent at 4–6 times control. It is clear that removal of the lactone carbonyl in the series **108–115** provides excellent potency enhancement and that overall trends relative to lactones **42–46, 51, 125** are not radically altered, but do seem to be offset somewhat (e.g. ethyl most potent in lactone series; butyl most potent in tetrahydropyran series).

For C-3 ethyl-substituted analogue **116**, a mere one-carbon homologation of compound **109**, leads to a significant drop in potency. However, homologation by two carbons (C-3 propyl analogue **117**) results in roughly a 70-fold improvement over **116**. For C-3 butyl **118**, activity had dropped only slightly. However, the effect of branching (i.e. C-3 isobutyl, **119**) is apparently detrimental towards antimalarial potency. This same effect was noted for C-9 substituted lactones: *n*-alkanes were usually significantly more active than the homologous iso-alkanes. Interestingly, the effect on potency of aryl–alkyl substitution at C-3 is completely different than noted in the C-9 series.

Aralkyl analogue **122**, isomeric with highly potent 9-*p*-chlorophenylpropyl analogue **115**, is practically devoid of antimalarial activity. While ethylphenyl derivative **121** is impotent, 4-phenylbutyl homologue **120** is about 3 times more active than control. These results support the notion that *antimalarial potency among artemisinin analogues cannot be explained on the sole basis of hydrophobicity.*

Calculated log P data (CLogP)[133] for artemisinin analogues, previously shown to correlate reasonably well with measured log P,[134] has been obtained for these analogues as shown in Table 4. For 3-(*p*-chlorophenyl)propyl-substituted analogues **122** and **115** illustrated above, log P data show that **122** is roughly as lipophilic as **115**. Subtraction of the fragment constant π for a methyl group of about 0.56 brings the CLogP for **115** of 5.21 (derived by 5.77–0.56) into parity with the calculated CLogP of **122** of 5.25. Furthermore, using the entire dataset and attempting simple regression analysis in Tsar[135] with bulk independent variables such as log P or molar refractivity was fruitless, yielding r^2 values approaching zero. Polynomial regression analysis after deletion of **121** provided the "best" statistical correlation with $r^2 = 0.348$, where "log relative activity" (LogRA) = $-8.68 + 3.93$ log P $- 0.396$ (log P)^2. Verloop B_1 (length) parameters and log P were correlated loosely with LogRA, giving LogRA = 0.65(CLogP) $- 0.16B_1 - 0.14$; $r^2 = 0.366$. More extensive multiple regression analyses in Tsar combining other parameters

Table 4. Calculated Logarithmic Partition Coefficients (Log P) for 10-Deoxoartemisinin Analogues

	Calculated Log P	
Structure	CLogP[133]	Tsar[135]
108	2.58	3.26
109	2.07	2.86
110	3.11	3.65
111	3.64	4.05
112	4.17	4.45
113	4.70	4.84
114	5.06	5.66
115	5.77	6.18
116	2.59	3.42
117	3.12	3.82
118	3.65	4.22
119	3.52	4.15
120	5.07	5.83
121	4.01	5.04
122	5.25	5.95
123	2.08	2.96
127	1.14	2.59

such as moment of inertia, lipole, total dipole, bond dipole, charges, topological, connectivity, and shape indices were likewise without statistical significance.

At a glance, the removal of the lactone carbonyl might not be expected to alter the postulated primary mode of antimalarial action put forth by Posner and Meshnick. While it is true that 10-deoxoartemisinin **108** undergoes Fe(II)-promoted rearrangement to give different products than those provided by artemisinin **1**, the biochemical significance is not clear.[136,137] The existence of a nontrivial structure–activity correlation for these analogues suggests the possibility of selective transport-mediated phenomena. Some evidence for this thesis is provided by the observation of enhanced uptake of dihydroartemisinin into parasitized, over nonparasitized, red blood cells.[138] Along these lines, we found it interesting to note that while ester **123** demonstrated activity comparable to 10-deoxoartemisinin **108**, carboxylic acid **127** was virtually devoid of antimalarial activity. Further, free carboxylic acids attached at either C-9 or N-11 positions demonstrated a thorough lack of activity.[83,134,139]

Based on Posner's proposed mechanism of action, we hoped synthesis of 13-carbaartemisinin **240** would allow us to address issues such as: can the process whereby radical intermediates such as **232/233** undergo intramolecular collapse, be inhibited by replacement of the O-13 atom by CH_2? Perhaps this modification would lead to longer lived intermediate radicals that would have greater activity. The facility with which this process occurs could presumably be determined by observation of the Fe(II)-catalyzed fragmentation pathway of **240**. In light of an earlier assertion that 4β-methyl-substituted tricyclic artemisinin analogues were more potent antimalarials than artemisinin due to C-4 radical stabilization,[47] replacement of C for O at the C-13 position should lead to a more stable C-4 radical because of the destabilizing effect of the β-oxygen would be removed. Although not an immediate goal of this research, replacement of O-13 by CH_2 might eventually furnish analogues whose hydrolytic stability relative to artemisinin could provide access to congeners with improved oral activity.

As can be seen in Table 5, neither target **240** (~4%) nor **241** (~16%) displayed substantial antimalarial potency relative to artemisinin (100%) in vitro against the W-2 clone of *Plasmodium falciparum*, yet the isomeric peroxide **258** was found to possess good antimalarial activity (~60%). The fourfold enhancement in activity of **241** relative to **240** is analogous to the ranking of activities between 10-deoxoartemisinin **108** and artemisinin **1**.

The relative inactivity of **240**, in vitro, in comparison to artemisinin is somewhat surprising in that the geometry of these two molecules are very similar. Molecular mechanics calculations of **240** provide a structure which overlaps nicely with the reported crystal structure of artemisinin **1**.[7,140] That the crystallographic coordinates

Table 5. Relative In Vitro Antimalarial Activity of 13-Carbaartemisinin Analogues Against *Plasmodium falciparum*

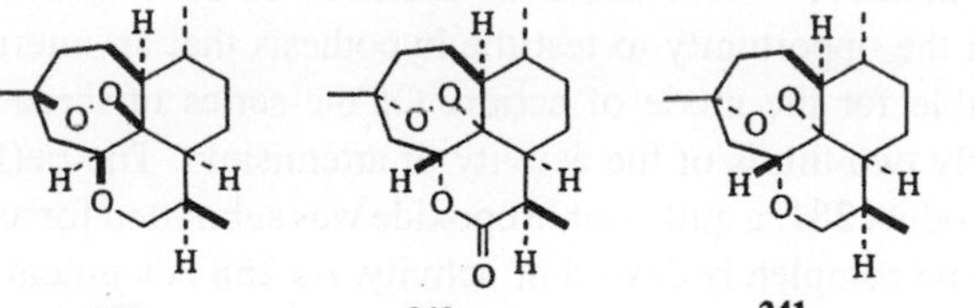

	IC$_{50}$ (ng/mL)		Relative Activity[a]	
Structure	D-6	W-2	D-6	W-2
1 (Artemisinin)	0.97	0.48	100	100
240	38,8	12.63	2.5	3.8
258	1.72	0.78	53	58
241	2.87	2.79	32	16

Note: [a]Relative activity = 100 × [IC$_{50}$ artemisinin (control value)/IC$_{50}$ analogue] MW analogue/MW artemisinin.

of these analogues correspond to calculated structures can also be seen for **241** whose X-ray structure is essentially the same as the MM2 calculated structure. Thus, while natural product-like architecture is maintained in analogues **240** and **241**, activity is diminished. This would seem to present clear proof that a 1,2,4-trioxane moiety is essential for good antimalarial potency. This argument is mitigated to some extent by the finding of substantial activity for novel, β-fused diastereomer **258**. Also, it has been reported that a steroidal analogue of **258** having a β-fused peroxy group was more effective than artemisinin in vivo.[89]

It has been suggested that the inactivity of C-4α substituted artemisinin analogues could be correlated to their inability to undergo C-4 hydrogen atom abstraction. While C-4 radical stability would be predicted to have an impact on antimalarial potency, these results would contradict the notion that improved C-4 radical stability is related, a priori, to improved antimalarial activity because the C-4 radical derived from **230** should be more stable than radical **233**. Indeed, more recent studies have shown that within a homologous series, continued stabilization of an intermediate radical at C-4 leads to decreased antimalarial activity, not increased potency as originally hypothesized.[103] The C-4 radicals derived from **230, 258**, or **241** would be expected to be more stable than their oxa-counterparts (β-destabilization by O), but it is difficult to assess the exact degree of reactivity required for antimalarial potency. Apparently, a careful balance between C-4 radical half-life and C-4 radical reactivity is essential for potency and this balance is not generally achieved in the 13-carba modification.

An additional step in the radical mechanism has been suggested; namely, that collapse of C-4 radical intermediate **234** to a neutral but highly reactive alkoxy-epoxide **235** occurs (Scheme 5).[104] Protein alkylation then presumably occurs via **235** and not radical intermediates such as **234**. Unfortunately, epoxide **235** is probably too unstable to be handled or identified. In the case of **258** however, we were granted the opportunity to test the hypothesis that an intermediate epoxide was responsible for the mode of action. Of the series of three tetracycles, **258** retained nearly two-thirds of the activity of artemisinin. The Fe(II)-induced rearrangement product **281**, a quite stable epoxide was submitted for antimalarial assay and found to be completely devoid of activity. As **258** is a potent antimalarial but the epoxide **281** is not, it seems reasonable to suggest that the antimalarial activity of **258** is unrelated to epoxy intermediates.

A good deal of importance has been placed on a detailed analysis of the products obtained from the Fe(II)-mediated rearrangement of artemisinin and its analogues. Due to the novelty of our findings in regards to the rearrangement of not only the 13-carba analogues **258** and **241**, but also 10-deoxoartemisinin **108**, some discussion of these results relative to these reports seems warranted.

Typical products from FeBr$_2$ treatment of artemisinin are 4-hydroxy-1-deoxoartemisinin **236** (Scheme 5), 1-deoxoartemisinin **290** and a ring-contracted product **238**, occurring in a 1:6:3 ratio, respectively.[104]

$$\text{1} \xrightarrow{\text{FeBr}_2} \quad \textbf{236} \quad + \quad \textbf{238} \quad + \quad \textbf{290} \tag{30}$$

Exclusive formation of ring-contracted products such as **238** in tricyclic analogues appears to correlate with low antimalarial activity in that secondary C-4 radical formation (e.g. **234**) does not occur on the pathway to **238** but instead radical **237** is formed.[47] How is the rearrangement chemistry of 10-deoxoartemisinin and the 13-carbaartemisinins reconciled with these findings? No products from cleavage of the C-3/C-4 bond (e.g. **237**) leading to **238**-like products are evidenced in any of these analogues, or in 10-deoxoartemisinin **108**. In fact, novel cleavage to monocyclic cyclohexanones occurs as a major pathway for **108** and **258**, but not for **241**.

It could be argued that products from **258** and **108**, vinyl ether **279** and formate ester **288**, were produced via C-4 radicals, but that ensuing intramolecular rearrangement pathways are altered relative to artemisinin and certain tricyclic analogues. β-Fused peroxide **258** would seem to follow two pathways: radical **275** does not fragment to form a tetrahydrofuran such as **238**, but instead undergoes unraveling to produce cyclohexanone **279**. While **281** might be formed via alternate radical **280** via β-scission to **282** with concomitant reoxidation by Fe(IV)=O,[104] the absence of **282** in the reaction mixture despite production of epoxide **281** argue against such a mechanism. 10-Deoxoartemisinin **108**, like **258**, appears to follow a major decomposition pathway via radicals **283** and **284** (Scheme 8), that furnishes a cyclohexanone (**288**). Finally, for **241** the major product likely occurs from C-4 radical, **270** (Scheme 6).

These rearrangement results indicate that minor modifications to structure can lead to major changes in decomposition pathway and therefore should be viewed from a standpoint of structure–activity relationships with some skepticism.

A detailed analysis of the products of annihilation of these intermediate radicals does not appear to be of use in the prediction of antimalarial potency. The presence of unraveled products such as **279** and **288** in the Fe(II)-promoted rearrangement of bioactive analogues indicates that a completely unique decomposition pathway is followed after minor structural modification (lactones **1** versus pyrans **108**). Production for the first time of an inactive epoxide intermediate (**281**) in the biomimetic rearrangement of active analogue **258** argues against the importance of intermediate epoxides in the mode of action. The carba modification, wherein the nonperoxidic trioxane oxygen atom is replaced by carbon, leads to a substantial loss in antimalarial potency in the series of compounds with artemisinin-like stereochemistry. Paradoxically, abnormal 13-carba analogue **258** retains most of the antimalarial activity of artemisinin. Finally, the tricyclic derivatives of poorly active 13-carba analogue **241**, analogues **259** and **262–266** were likewise relatively impotent in the in vitro antimalarial screen (Table 6). What effect this modification will have on in vivo activity is currently under investigation.

Seco derivatives **166** through **182** are important: they have attractive synthetic availability and provide general SAR information quickly. There are several points of interest regarding the structural features of these analogues relative to artemisinin **1**. From molecular modeling, **1** is rigid and the minimum energy conformation matches the solid-state structure determined by X-ray crystallography. In sharp contrast to **1**, the seco analogues displayed a vast array of conformations in solution, evident from their peculiar NMR behavior. A homologous series was studied in detail: **170**, **171**, **172**, and **184**. A mixture of conformers of **184** were observed as broadened patterns in the proton NMR. The NMR profile of **184** was temperature-dependent, exhibiting different details when examined at −10, 3, 20, and 60 °C. X-ray analyses of **184** and **172**[49] afforded additional conformational information. The solid-state structure for **171** is known.[50,51] Analogue **170** is very similar to **184** by NMR and MM2. Detailed energy minimizations of possible conformers (16 per compound) were undertaken and low-temperature 2-D NMR experiments (NOESY, ROESY, COSY) conducted.[141] Together with the X-ray information, we were able to determine that at −10 °C in chloroform solution, there were two

Table 6. Relative In Vitro Antimalarial Activity of Tricyclic-13-carbaartemisinin Analogues Against *Plasmodium falciparum*

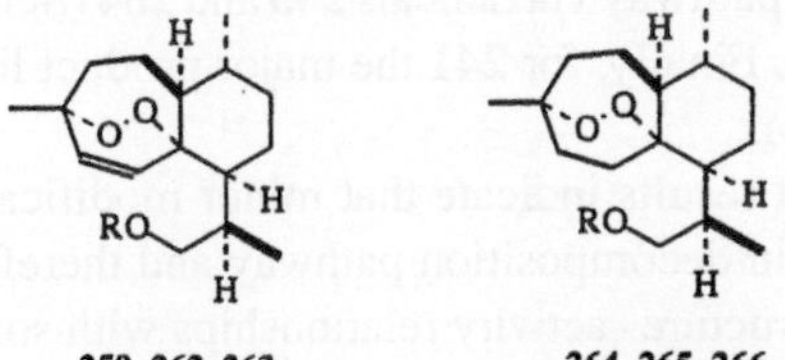

259, 262, 263   264, 265, 266

Structure	R	IC$_{50}$ (ng/mL) D-6	IC$_{50}$ (ng/mL) W-2
1 (Artemisinin)		1.4	2.2
259	H	1000	NA
262	Me	920	NA
263	CH$_2$Ph	NA	NA
264	H	300	1500
265	Me	310	700
266	CH$_2$Ph	290	660

Note: [a]Relative activity = 100 × [IC$_{50}$ artemisinin (control value)/IC$_{50}$ analogue] MW analogue/MW Artemisinin.

conformers (7:3 ratio) of **184**. The major conformer corresponded to the all-chair arrangement found in the X-ray structure. The minor conformer was in a chair–chair–boat arrangement where the boat trioxane ring was not the same boat conformer as found in artemisinin. Neither conformer overlapped perfectly with artemisinin **1** and might explain why **184** has only modest potency compared to the natural product.

The situation for **172** was quite similar, but trends between conformers were reversed: the solid-state conformer (chair–chair–boat) corresponded to the minor solution isomer of **184**, whereas the minor solution isomer of **172** corresponded to the major all-chair conformer of **184**. Imakura's "desethano" artemisinin **171** was found to exist in the all-chair arrangement by X-ray and exhibited a nontemperature-dependent proton NMR spectrum. MM2 studies of **171** showed that the all-chair conformer was 4–5 kcal/M more stable than the next lowest energy conformer, a chair–chair–boat, and were thus consistent with experimental observations. Studies to correlate conformer populations and energetics with antimalarial activities have not provided clear answers.[141]

In vitro bioassay results for **170–172** and **184** can be found in Table 7. Relative potencies in W-2 and D-6 clones for **184** were 6 and 20%, respectively, while for **170**, values of 75 and 108% were found, and the relative potencies for **172** were 14 and 7%. The analogue **171** was very nearly inactive (ca. 3%).

In this series, both **171** and **172** have both high barriers to ring inversion and the adoption of an artemisinin-like conformer, and anomalously low relative activity. The relative activity could be related to the ability of the seco analogue to adopt an artemisinin-like conformer or "active conformer", and was borne out by molecular mechanics calculations in the following way: analogues **170–172**, and **184** were restricted to the "active conformer" by holding the peroxide dihedral angle to 49° and setting the interatomic distance between atoms located at C-3 and C-5a (artemisinin numbering system) to 3 Å for minimization. There is an apparent correlation because **171** and **172** have low relative potencies (high barriers to the active conformer) while higher potencies are found for **170** and **184** (low barriers to the active conformer).

Little improvement in activity of seco analogues was evident with structural changes that were beneficial in the tetracyclic series. For example, replacement of the lactone ring methyl group in **184** with a butyl group (**168**) resulted in relative potencies in the range of 5–7% of artemisinin, even less than those for **184**. Replacement of the C-9 methyl group of artemisinin with a pentyl group in **46** led to a 4–9-fold enhancement.

An alternative rationale for the general inactivity of the seco analogues may be related to the proposed mechanism of action. If carbon-centered radical formation is important for the mechanism of action, then it may be argued that the change in structure somehow adversely effects this chemistry by affecting the ability of the initially formed oxy-radical to undergo intramolecular H-atom transfer to furnish **298** (path B, Scheme 27). If a 1,5-diketone generated by rearrangement of the

Table 7. In Vitro Antimalarial Activities of Seco Analogues of Artemisinin

166, 168, 170-2, 184-5   173, 174, 176   179-182

Structure	R_1	R_2	R_3	R_4	R_5	R_6	W-2	D-6
							Relative Activity	
Artemisinin, 1							100	100
166	Me	Me	H	H	H		0.5	0.4
184	Me	Me	H	H	Me		6	20
171	H	Me	H	Me	Me		NA	NA
185	Et	Et	H	H	Me		3	15
168	Me	Me	H	H	Bu		7	5
170	Me	Me	H	Me	Me		75	108
172	Me	Me	Me	Me	Me		14	7
173	Me	Me	H	Me	Me		9	8
174	Me	Me	Me	Me	H		—c	58
176	Me	Me	Me	Me	Me		24	23
179						OAc	—a	
180						OOCOEt	0.5a	—c
181						OCOBz	24b	13
182						H	—c	37b
Random Seco Structures								
177							NAd	NA
178							NA	NA
10							NA	NA

Notes: aUnstable oil.
bStable solid.
cNo data available in this clone.
dNA = poorly active.

peroxide via radical rearrangement is an active species, as has been recently proposed,[80] then the lack of a connecting bridge in **184** could lead to loss of acetone from intermediate **297** (path A), and to the generation of nontoxic mono-keto products such as **299** or **300**. It is interesting to note that **300** is a by-product in the synthesis of the hydroperoxide **183** as well as the trioxanes **184** and **185**.

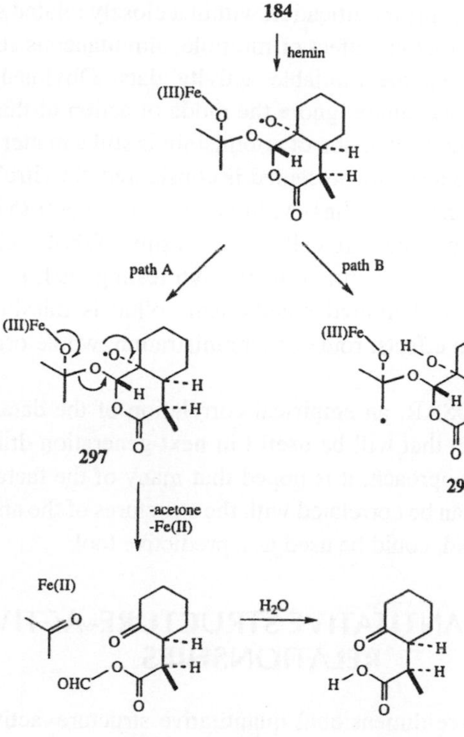

184

| hemin

(III)Fe

path A   path B

297   **298**

-acetone
-Fe(II)

Fe(II)   H₂O

299   **300**

Scheme 27.

Considering these findings, the 8a,9-seco analogues (Table 7) were expected to demonstrate better activity than the 4,5-seco analogues because they more closely resemble the natural product in the crucial peroxy region of the molecule. For example, **180** overlaps very nicely with all but the lactone ring of **1** from MM2 and an X-ray analysis, yet **180** was found to be poorly active. Ester **180** is an unstable oil and could have decomposed prior to testing. This seems reasonable because closely related carbonate derivative **181** is a stable solid and is fairly active (about 30% of artemisinin). On the other hand, perhaps these analogues are not active because the ester side chains point in unusual directions and might not occupy a putative hydrophobic pocket. Analogue **182**, which lacks the entire lactone ring, is more potent than carbonate **181** and contradicts the lack of activity by dehydro-derivative **292**.

From the foregoing observations, one can appreciate the usefulness of this structure–activity relationship data in the design of antimalarial agents. We can, for example, understand simple modification within a closely related series. However, it is not possible to predict the effect of multiple, simultaneous structural modifications on activity using the available activity data. Obviously, a reasonable interpretation of this data cannot ignore the mode of action of this class of drugs, but unfortunately the mode of action of artemisinin is still a matter of some debate. Furthermore, while the test data presented is considered "in vitro" data, in reality the results obtained in the Desjardins–Milhous assay are closer to in vivo data than an enzyme assay because they are cell culture results. Whole cells incorporate a variety of secondary effects common to in vivo testing such as cellular uptake, accumulation, export, and limited metabolism. What is missing in culture, of course, are distribution effects, route of administration, whole organ metabolism, and so on.

It is possible that QSAR, an empirical correlation of the data, may provide a pharmacophoric model that will be useful in next-generation drug design within this class. In such an approach, it is hoped that many of the factors which define antimalarial potency can be correlated with the structures of the analogues, and the resulting model, if valid, could be used as a predictive tool.

VI. QUANTITATIVE STRUCTURE–ACTIVITY RELATIONSHIPS

Development of a three-dimensional quantitative structure–activity relationship (3-D QSAR) model representing the in-vitro (cell culture) antimalarial activity of a dataset of artemisinin analogues is a complex endeavor. How specific structural features effect the sequence of events leading to accumulation of drug at its site of action as well as the mode of action itself define the in vitro potency of a drug. Peroxidic antimalarial activity might be disrupted by metabolic inactivation or premature deactivation by iron salts,[142] effected by its accumulation into the cell,[143,144] or how the molecule interacts (dock) with hemin.[75] However, by judicious selection of analogues, and careful model construction (alignment, conformational predictions), a 3-D QSAR study can lead to a predictive model providing useful SAR information. Using comparative molecular field analysis (CoMFA),[101] a protocol introduced by Cramer et al. in 1982, we have developed artemisinin pharmacophores based on two distinctly different conformational determinations.

A critical issue involved in the creation of these pharmacophores is the determination of the "active" conformation each compound will retain. Without structural data supporting a specific "active" conformation, one may be forced to assume that a compound is active in an energetically minimized conformation. In these cases, the molecule's X-ray structure can serve as a template upon which other analogues can be overlaid. Treatment of a database of 202 artemisinin analogues based on the "lowest energy" conformation assumption has led to a statistically excellent model.

A second, equally robust model was constructed using a mechanism of action hypotheses to affect conformation choice. Here, docking interaction with hemin alters the analogue conformation, and the final structure included in creation of the pharmacophore now represents the result of a force-field calculation incorporating the influence of the hemin molecule.

A. Biological Activity

In creating our QSAR models we have collected compounds that have a variety of structural features, but that are otherwise similar enough to presume that they act by the same mechanism of action. Developing these models entailed acquisition of 202 compounds shown in Tables 8–16 found primarily in scientific literature.[9,19–21,31,33,46,47,69,78,83,87,112,134,136,137,139,141,145–153] Only analogues that had been tested in vitro (Milhous–Desjardins assay) against the chloroquine-resistant, mefloquine-sensitive *P. falciparum* W-2 clone were included. Relative activity was used as a logarithmic term taken from the experimentally derived IC_{50} of artemisinin (IC_{50} values in ng/mL) divided by the IC_{50} of the analogue, corrected for molecular weight:

$$\text{LogRA} = \text{Log of Relative Activity} =$$
$$\text{Log}[(IC_{50} \text{ of Artemisinin}/LC_{50} \text{ of Analogue})][(\text{Molecular Weight of Analogue}/\text{Molecular Weight of Artemisinin})] \quad (31)$$

B. Computer Methodology

2-D QSAR Analysis

Initial attempts to uncover a relationship between antimalarial activity and molecular structure focused on calculated physical properties such as dipole moment, log P, and molecular volume inherent in the structure. The $n = 202$ database was analyzed using the software package Tsar (Oxford Molecular, Oxford, England).[135] Most of the properties were determined without difficulty, though some of the molar refractivity, surface area, and log P values could not be calculated. In the case of molar refractivity and surface area, the program was allowed to ignore compounds whose values could not be determined. Additional log P values were calculated using Daylight CLogP (Daylight Chemical Information Systems, Mission Viejo, CA)[133] software. A regression analysis was run using 75 columns of data produced in the predict properties area. A stepwise regression was run using 75 variables; the final regression, based on intermediate results was run using only six of the data columns, including LogRA (the dependent variable), log P, molar refractivity, moment of inertia, Balaban Index, and 1 Å and 2 Å 3D autocorrelograms as the independent variables. Cross-validation[102] was carried out in three groups with a fixed deletion pattern.

Table 8. Relative Activities of Artemisinin Analogues

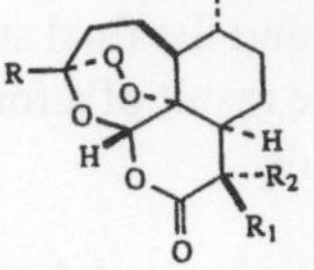

Structure	R	R_1	R_2	LogRA
1	Me	Me	H	1.00
84	(CH$_2$)$_4$Ph	H	H	0.45
61	Me	H	Z-crotyl	-1.10
42	Me	H	H	0.79
62	Me	H	Me	-0.17
60	Me	H	E-crotyl	-0.60
57	Me	allyl	H	-0.10
45	Me	Bu	H	0.17
98	(CH$_2$)$_4$Ph	Bu	H	-0.32
97	(CH$_2$)$_3$(p-Cl-Ph)	Bu	H	-0.28
99	CH$_2$CH$_2$CO$_2$Et	Bu	H	1.36
95	n-Bu	Bu	H	-0.48
43	Me	Et	H	1.40
47	Me	C$_6$H$_{13}$	H	0.86
54	Me	i-C$_4$H$_9$	H	-0.55
56	Me	i-C$_6$H$_{13}$	H	-0.04
53	Me	i-Pr	H	-0.04
55	Me	i-C$_5$H$_{11}$	H	0.07
83	(CH$_2$)$_3$(p-Cl-Ph)	H	H	0.10
82	(CH$_2$)$_4$Ph	H	H	-2.00
97	BuC$_4$H$_9$	H	H	-0.74
81	CH$_2$CH$_2$CO$_2$Et	H	H	0.37
77	Et	H	H	0.05
69	H	H	H	-2.76
80	i-C$_4$H$_9$	H	H	-0.35
78	Pr	H	H	0.83
297	Me	(CH$_2$)$_3$(p-Cl-Ph)	H	1.37
298	Me	Br	CH$_2$Br	-1.64
299	Me	OH	CH$_2$OH	-4.00
300	Me		-OCH$_2$-	-2.09
301	Me		-CH$_2$O-	-4.00
302	Me		-CH$_2$-	-0.89
303	Me		-N=N(CH$_2$)$_2$-	-2.70
304	Me	Et	Et	-0.36
305	Me		-CH$_2$CH$_2$-	-0.94
306	Me	=0		-2.47
46	Me	C$_5$H$_{11}$	H	1.02
52	Me	(CH$_2$)$_4$Ph	H	0.63
50	Me	(CH$_2$)$_2$Ph	H	0.12
51	Me	(CH$_2$)$_3$Ph	H	0.78
44	Me	Pr	H	1.13

Table 9. Deoxyartemisinin Analogues

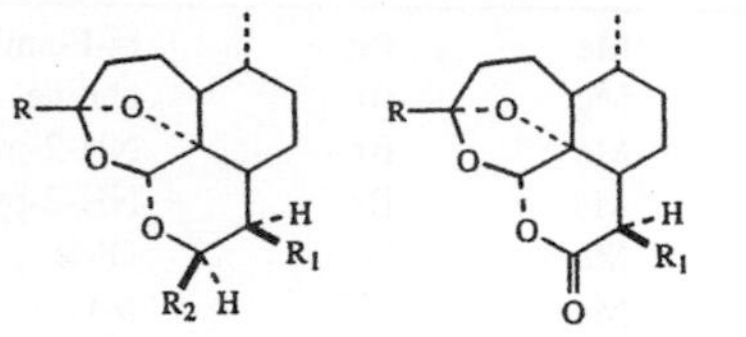

Structure	R	R_1	R_2	LogRA
307	Me	Me	OEt	−4.00
308	Me	Me	OH	−4.00
309	Me	$(CH_2)_4Ph$	—	−4.00
310	Me	n-Pr	—	−4.00
311	Me	C_6H_{13}	—	−4.00
286	Me	Me	H	−4.00
312	Me	Bu	H	−4.00
313	Me	i-C_5H_{11}	—	−4.00
314	$CH_2CH_2CO_2Et$	H	H	−4.00
315	$(CH_2)_4Ph$	H	—	−4.00
316	Et	H	—	−4.00
317	i-C_4H_9	H	—	−4.00
318	i-C_4H_9	H	H	−4.00
319	Me	$(CH_2)_4Ph$	—	−4.00
320	Me	$(CH_2)_3Ph$	—	−4.00
290	Me	Me	—	−4.00

Table 10. 10-Substituted Artemisinin Derivatives

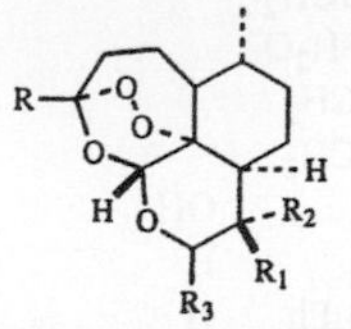

Structure	R	R_1	R_2	$R_3{}^a$	LogRA
108	Me	Me	H	H	0.75
175	Me	Me	H	OH	0.55
145	Me	Me	H	OEt	0.34
141	Me	Bu	H	OH	0.96
321	Me	Me	H	OEt	−1.08
109	Me	H	H	H	0.28
322	Me	Me	Br	NH-2-(1,3-thiazole)	0.66

(*continued*)

Table 10. Continued

Structure	R	R_1	R_2	$R_e{}^a$	LogRA
323	Me	Me	Br	m-F-aniline	0.79
324	Me	Me	Br	aniline	0.18
325	Me	Me	Br	NH-2-pyridyl	−0.09
326	Me	Me	Br	NH-2-pyrimidinyl	−0.77
144	Me	Me	H	OMe	0.28
327	Me	Me	H	α-OEt	0.32
112	Me	Bu	H	H	1.32
110	Me	Et	H	H	0.67
328	Me	Pr	H	OEt	−0.04
329	Me	H	H	OEt	0.43
330	Me	Et	H	OEt	0.50
331	Me	Me	H	$(CH_2)_3CH$	0.78
332	Me	Me	H	Bu	0.06
333	Me	Me	H	OCH_2CO_2Et	0.52
334	Me	Me	H	$O(CH_2)_2CO_2Me$	0.10
335	Me	Me	H	$O(CH_2)_3CO_2Me$	−0.03
336	Me	Me	H	$OCH_2(4\text{-}PhCO_2Me)$	−0.07
337	Me	Me	H	$(R)\text{-}OCH_2CH(Me)CO_2Me$	1.79
338	Me	Me	H	$(S)\text{-}OCH_2CH(Me)CO_2Me$	2.25
339	Me	Me	H	$(R)\text{-}OCH(Me)CH_2CO_2Me$	0.87
340	Me	Me	H	$(S)\text{-}OCH(Me)CH_2CO_2Me$	1.70
341	Me	Me	H	OCH_2-adamantyl	0.28
123	$CH_2CH_2CO_2Et$	H	H	H	0.70
122	$(CH_2)_3$(p-Cl-Ph)	H	H	H	−0.55
127	propionic acid	H	H	H	−3.00
118	Bu	H	H	H	0.75
116	Et	H	H	H	−1.00
119	$i\text{-}C_4H_9$	H	H	H	0.40
117	Pr	H	H	H	0.84
120	$(CH_2)_4Ph$	H	H	H	0.58
121	$(CH_2)_2Ph$	H	H	H	−1.7
342	Me	-OCH_2-		OOH	−0.62
343	Me	-CH_2O-		OOH	−0.57
344	Me	=CH_2		OOH	−0.990
345	Me	=CH_2		OH	−2.39
346	Me	Me	OH	α-OH	−0.89
113	Me	C_5H_{11}	H	H	0.16
114	Me	$(CH_2)_3Ph$	H	H	1.40
111	Me	Pr	H	H	0.74
347	Me	Me	H	CH_2CHF_2	0.11
348	Me	Me	H	$CH_2CF_2CH_3$	−0.17
349	Me	Me	OH	OCH_2CF_3	0.33
350	Me	OH	Me	OCH_2CF_3	−0.70
351	Me	Me	OH	OEt	−0.44
352	Me	OH	Me	OEt	−1.13
353	Me	Me	H	OOt–Bu	0.92

Note: [a]All substituants β unless otherwise noted.

Table 11. Secoartemisinin Derivatives

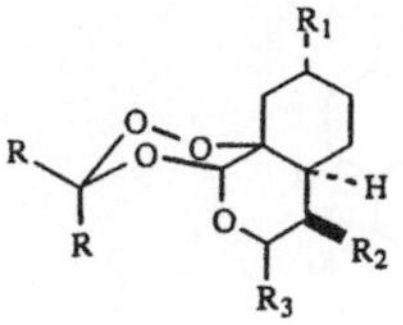

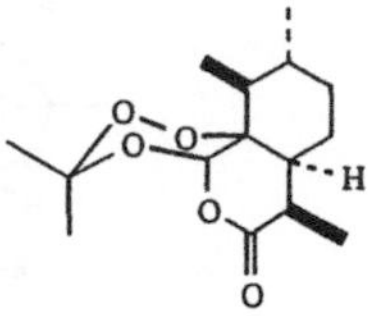

172

Structure	R	R_1	R_2	R_3	LogRA
166	Me	H	H	=0	−2.37
168	Me	H	Bu	=0	−1.13
185	Et	H	Me	=0	−0.60
170	Me	Me	Me	=0	−0.15
172	—	—	—	—	−0.86
184	Me	H	Me	=0	−1.27
354	−(CH₂)₄−	Me	Me	H	−0.26
355	Me	Me	Me	H	−0.80

Table 12. Various Derivatives of Artemisinin and Arteether

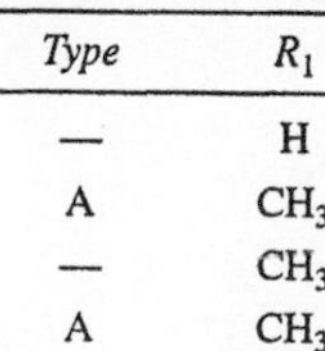

155 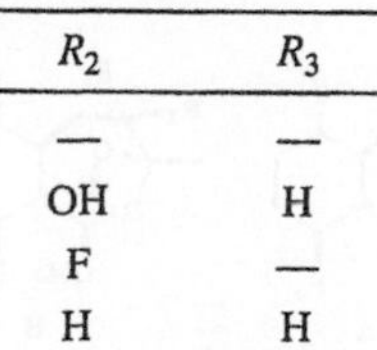   **A** 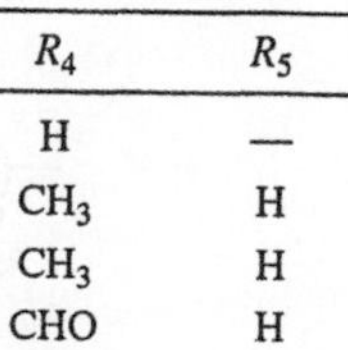   **357**

Structure	Type	R_1	R_2	R_3	R_4	R_5	Log RA
155	—	H	—	—	H	—	−0.36
356	A	CH₃	OH	H	CH₃	H	0.34
357	—	CH₃	F	—	CH₃	H	−0.12
358	A	CH₃	H	H	CHO	H	0.21
359	A	CH₃	=O		CH₃	H	0.16
360	A	CH₃	H	H	CHF₂	H	0.41
119	A	CH₃	F	F	CH₃	H	0.19
131	A	CH₃	H	H	CH₃	OH	−1.27

Table 13. 11-Azaartemisinin Derivatives

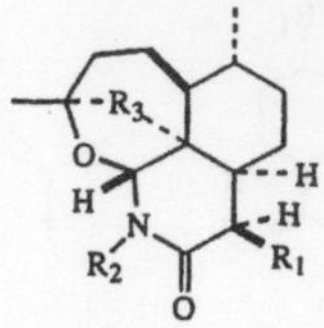

Structure	R_1	R_2	R_3	LogRA
361	H	$(CH_2)_3Ph$	O–O	0.02
362	H	$(CH_2)_4Ph$	O–O	0.16
363	H	C_5H_{11}	O–O	−0.20
364	H	$i\text{-}C_5H_{11}$	O–O	−0.04
365	H	$CH_2(P\text{-}Cl\text{-}Ph)$	O–O	−0.16
366	H	$i\text{-}C_4H_9$	O–O	0.02
77	H	CH_2Ph	O–O	0.34
78	H	Me	O–O	0.70
367	Me	H	O–O	0.00
368	Me	allyl	O–O	−0.04
369	Me	$i\text{-}C_3H_7$	O–O	1.03
370	Me	Me	O–O	0.43
135	H	Pr	O–O	0.05
371	Me	$CH_2\text{-}(2\text{-}C_5H_4N)$	O–O	1.46
372	Me	2-thiophene	O–O	0.17
373	Me	acetaldehyde	O–O	1.47
374	Me	2-furan	O–O	0.11
375	Me	$i\text{-}C_4H_9$	O	−4.00
376	Me	$CH_2\text{-}(2\text{-}C_5H_4N)$	O	−4.00
377	Me	H	O	−4.00

Table 14. Artemisinin Derivatives Lacking the D-Ring

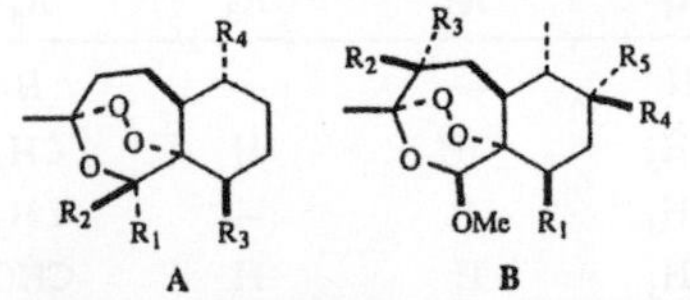

Structure A	R_1	R_2	R_3	R_4	LogRA
181	$-O_2CCH_2Ph$	H	H	Me	−0.51
182	H	H	H	Me	−0.32
378	H	OMe	H	H	−0.31

(*continued*)

Table 14. Continued

Structure A	R_1	R_2	R_3	R_4	LogRA
379	OMe	H	H	H	−1.04
380	OCH$_2$Ph	H	H	H	−0.09
381	OMe	H	(CH$_2$)$_2$O$_2$CNMe$_2$	H	0.33
382	OMe	H	(CH$_2$)$_2$O$_2$CNEt$_2$	H	0.65
383	OMe	H	(CH$_2$)O$_2$CNPh$_2$	H	1.06
384	H	OMe	(CH$_2$)$_2$OP(O)(OPh)$_2$	H	1.10
385	H	OME	(CH$_2$)$_2$OP(O)(OEt)$_2$	H	0.37
386	H	OMe	(CH$_2$)$_2$OP(=S)(OEt)$_2$	H	0.72
387	H	OMe	(CH$_2$)$_2$OS(O)$_2$ (p-toluene)	H	0.20
388	H	OMe	(CH$_2$)$_2$OS(O)$_2$C$_6$H$_4$ (o-CO$_2$Me)	H	−0.87
389	H	OMe	(CH$_2$)$_2$OS(O)$_2$ (5-NMe$_2$napthalen-1-yl)	H	0.33
390	H	OMe	(CH$_2$)$_2$OCH$_3$	H	−0.39
391	H	OMe	(CH$_2$)$_2$OCH$_2$Ph	H	0.75
392	H	OMe	(CH$_2$)$_2$O-allyl	H	0.40
393	H	OME	(CH$_2$)$_2$OCH$_2$- (3,5-dimethyl-1,2-oxazole)	H	0.58
394	H	OME	(CH$_2$)$_2$O$_2$CPh	H	−0.59
395	H	OMe	(CH$_2$)$_2$O$_2$CPh (p-CO$_2$Me)	H	0.27
396	H	OMe	(CH$_2$)$_2$O$_2$CPh (p-CO$_2$H)	H	−0.81
397	H	OMe	(CH$_2$)$_2$O$_2$CPh (p-CONEt$_2$)	H	0.23
398	H	OMe	(CH$_2$)$_2$O$_2$C-C$_6$H$_4$CO$_2$(CH$_2$)$_2$NMe$_2$	H	−0.60
399	H	OMe	(CH$_2$)$_2$O$_2$CCH$_2$-NCO$_2$-(t-Bu)	H	−0.04
400	H	OMe	(CH$_2$)$_2$OCH$_2$(4-Ph-Ph)	H	0.32
401	H	OMe	(CH$_2$)$_2$OCH$_2$(4-F-Ph)	H	0.38
402	H	OMe	(CH$_2$)$_2$OCH$_2$(2-pyr)	H	0.17
403	H	OMe	(CH$_2$)$_2$OCH$_2$(4-pyr)	H	0.14
404	H	OMe	(CH$_2$)$_2$OCH$_2$ (4-N-Me-pyr)	H	−0.90

Structure B	R_1	R_2	R_3	R_4	R_5	LogRA
405[a]	H	H	H	Me	Me	−0.47
406	(CH$_2$)$_2$OH	H	Me	H	H	−1.80
230	(CH$_2$)OH	Me	H	H	H	0.23
407	(CH$_2$)$_2$OH	Me	Me	H	H	−1.80
408	(CH$_2$)$_2$OCH$_2$Ph	Me	Me	H	H	−1.80

Note: [a]The −OMe group is in the α-position in this compound.

Table 15. Miscellaneous Artemisinin Derivatives

Compd	Structure	LogRA	Compd	Structure	LogRA
409	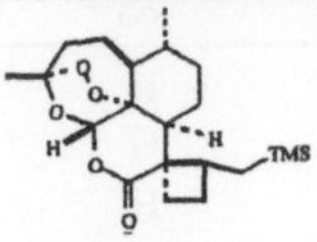	−1.24	411	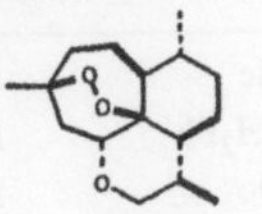	−0.24
292	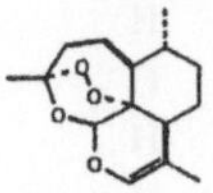	0.78	258	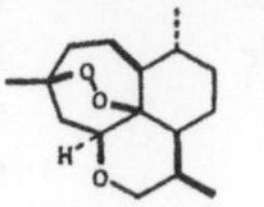	−2.59
10	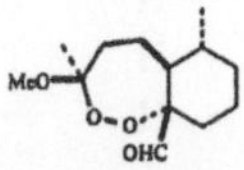	−4.00	240	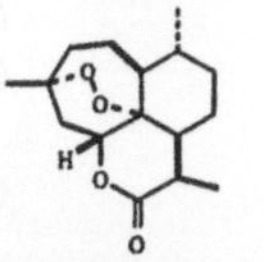	−0.96
149	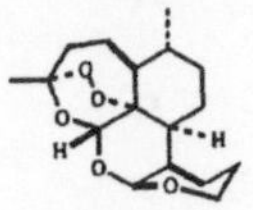	0.23	260	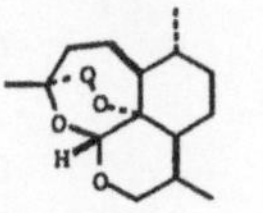	−0.79
410	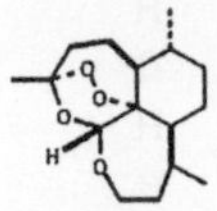	−1.20	412	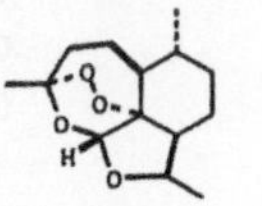	−0.64
177	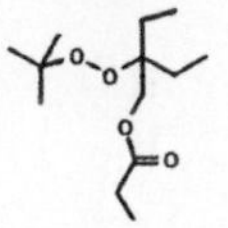	−4.00	413	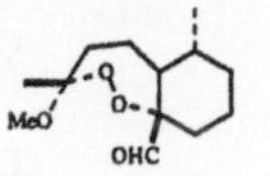	−3.30
178	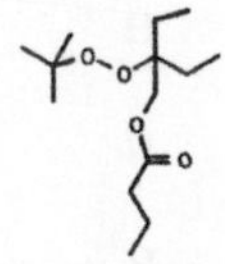	−4.00	414	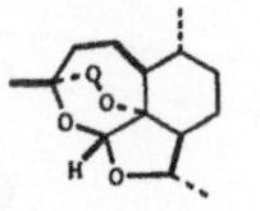	0.353
156	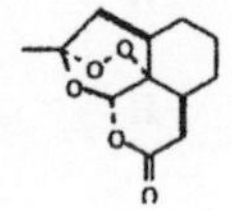	−1.79			

Molecular Modeling

The 3-D artemisinin analogue database was created with Sybyl 6.2 (Tripos Assoc., St. Louis, MO)[154] using the reported crystal structure for artemisinin as a template. The structures were minimized using the Tripos force field in Sybyl 6.2, producing structures close to that of artemisinin. To best describe the aromatic side chains of some analogues, as well as the aliphatic nature of the artemisinin backbone, Gasteiger–Hückel charges were calculated for each of the compounds.

In the development of our 3-D QSAR models, choice of analogue conformation played an important role in providing a realistic pharmacophore model. The

Table 16. Bicyclo Artemisinin Derivatives

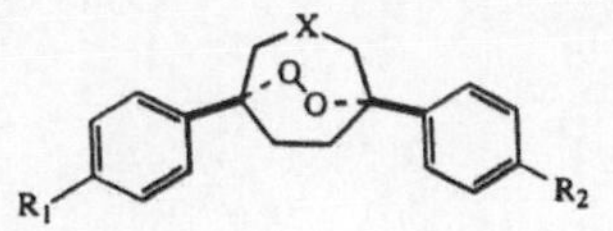

Structure	X	R_1	R_2	LogRA
415	CH_2	H	H	−0.91
416	CH_2	OMe	OMe	−0.67
417	*i*-Pr	H	H	−2.17
418	SO_2	H	H	−2.29
419	N-S(O)$_2$Ph	H	H	−2.26
420	O	H	H	−2.26

conformation of each analogue, assumed to be that of the active species, might be approximated if a crystal structure of artemisinin interacting with hemin were available. Since this data is not likely to become available (hemin reacts rapidly with artemisinin), different approaches to select the "active" conformation were used.

First, if the interaction of a peroxidic drug with hemin is a relatively rapid and structurally insensitive step in a potentially complex sequence of events, then perhaps its contribution to defining the potency of this class of molecules could be ignored. In this event, it may be possible to use the E_{min} structures to provide a realistic 3-D QSAR.

Alternatively, by modeling the manner in which artemisinin interacts with hemin, a conformation can be predicted by which this interaction is minimized, though the absolute energy of the peroxidic ligand alone might be slightly higher in energy than the lowest energy conformation determined by molecular mechanics methods.

Standard Alignment

All 202 minimized analogues were aligned with the assumption that the core ring structure and peroxide bridge are most meaningful with respect to biological activity. Though a number of flexible and static alignments were tried, the best model was obtained by fitting atoms C-1a, C-3, C-5a, and C-8a. Not every analogue in the database contained the peroxide bridge thought to be needed for activity; specifically, 1-deoxyartemisinin analogues (Table 9) were included in the dataset. The best fit was thus accomplished using the atoms which held the peroxide or oxide bridge in close alignment, atoms C-1a and C-3. Overlap of the 202 molecules can be seen in Figure 9.

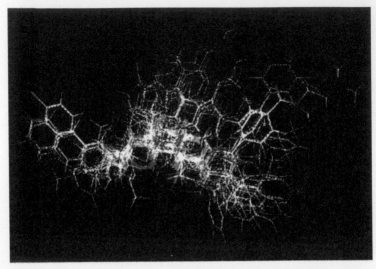

Figure 9. Standard alignment of 202 artemisinin analogues used in the CoMFA model development.

Dock Alignment

Current studies point toward the interaction of the peroxy moiety with the hemin iron(II) core as a major component of the antimalarial mode of action. This artemisinin–hemin association has been modeled, and it has been proposed that activation of artemisinin proceeds through a radical intermediate occurring at a point near the minimum intermolecular energy of interaction.[132] In an attempt to create a model that might more accurately reflect the active conformation of each analogue, the Sybyl Dock protocol was used to approximate analogue–hemin binding. Dock calculates the fields of nonbonded electrostatic and Leonard–Jones steric interaction between points on a lattice, by default 0.25 Å spacing, and the docking site. The ligand–site interaction is approximated by associating each atom of the ligand with the nearest lattice point, and summing the fields over all ligand atoms.

Initially, an energetically favorable interaction was achieved by manual manipulation of the ligand with respect to hemin. Consideration of a large number of artemisinin analogues required an assumption that the peroxide docked to the hemin iron in a manner identified by Shukla et al.[132] Docking was complete when the minimal combined electrostatic and steric docking energies were found.

To take into consideration the effect the metalloporphyrin molecule plays in the active conformation of the ligand, the ligand structural conformation was minimized with respect to hemin using the Tripos Force Field. In the minimization process, a certain small amount of translation occurs, the ligand arriving at a

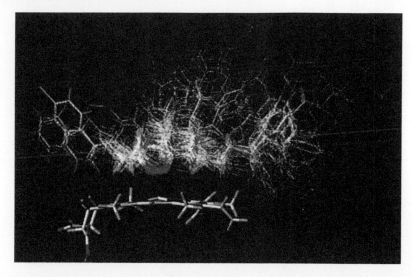

Figure 10. Alignment of 202 artemisinin analogues based on proposed docking (to hemin) conformation.

minimal conformation when a selected (0.05 kcal/mol) energy difference is found between conformations. Ligands within the database were realigned on the basis that a common orientation of the peroxy group with the Fe(II) core of hemin is likely a mechanistic bottleneck, and would therefore be more important than minor difference in gas–phase docking energies. The overlap of the compounds can be seen in Figure 10.

In some cases, substituents at C-9 and C-3 had significant impact on the docking interaction to an extent where the critically important contribution of the peroxide bridge interacting with the hemin iron atom was diminished. To reduce the significance of non-peroxide bridge interaction with hemin, analogue substituents at C-3 or C-9 were rotated to a "local minima" energy conformation (e.g. angle C10–C9–C16–C17 = 58°, Figure 11) so that the substituent did not play the primary role in the docking interaction. It is important to note that these local minima are within energy differences reasonably attainable under the test conditions. Although some analogues are not in their lowest energy conformation as calculated in gas-phase nondocking mode, it is thought that this process leads to useful binding information, potentially mimicking the actual active conformation.

3-D QSAR Analysis

Comparative molecular field analysis (CoMFA) is comprised of relating measurements of the electrostatic and steric fields around a template molecule to the molecule's biological activity. The CoMFA methodology is based on the assump-

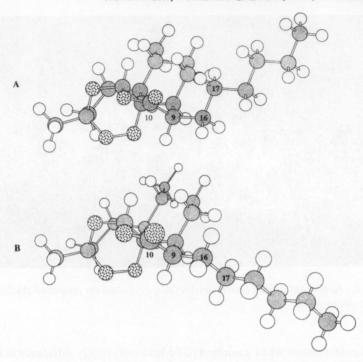

Figure 11. Torsional angle of **B** (dihedral C10–C9–C16–C17 = 174°, global minimum) modified in docking process to give **A** (dihedral C10–C9–C16–C17 = 58°).

tion that the drug–receptor interactions are noncovalent, and the changes in the biological activities or binding affinities of sample compounds correlate with changes in the steric and electrostatic fields of these molecules. An aligned database is placed in a lattice, and these measurements are taken at regular intervals throughout the lattice using a probe atom of designated size and charge. In this study, lattice spacing was 2 Å, and the probe atom was either carbon or hydrogen, using +1 charge. These interaction data were analyzed by multivariate statistical analysis using partial least squares (pls) and cross-validation. A cross-validated r^2 (q^2) obtained as a result of this analysis served as a quantitative measure of the predictive ability of the final CoMFA model. The q^2 value is a statistical indication of how well a model can *predict* the activity of members left out of the model formation. This is different from that of the conventional r^2, which is simply a reflection of how well the fit equation reproduces input values.

The standard and docked (n = 202) aligned databases were analyzed using CoMFA/pls. The results indicated three compounds having especially high residual values (differences between predicted and actual activities). These "outliers" were removed, and the pls analysis repeated (n = 199).

A number of the analogues in the database were racemates, though the database itself was constructed using the enantiomer having the absolute configuration corresponding to (+)-artemisinin. Analysis using CoMFA–pls of the non-racemic compounds ($n = 162$), as well as removal of two high residual chiral compounds and reanalysis (pls, $n = 160$), was done in an effort to find a model that might be able to predict the activities of each individual enantiomer, taking into consideration that only the total activity of the mixture is experimentally available.

VII. RESULTS AND DISCUSSION

A. 2-D QSAR

Individually or combined, the various predicted properties showed little correlation with activity. The final group of six data columns were the best correlated of the entire prediction, thought still statistically unimpressive with $r^2 = 0.518$, $q^2 = 0.485$, $s = 0.738$, $F = 28.6$. This would lead to the speculation that there is no simple linear correlation between any one of the physical properties and the activity of artemisinin analogues. Smaller, less structurally diverse datasets were examined in hopes of revealing correlations by reducing the overall number of variables. As discussed above, sets of closely related 10-deoxo analogues of artemisinin seem to show a modest correlation of molecular length (B_1 parameter) and lipophilicity with relative activity, but with a poor correlation coefficient: $LogRA = 0.65(CLogP) - 0.16B_1 - 0.14$; $r^2 = 0.366$.

The use of molecular orbital parameters in this regard remain to be done, but may be of utility in drug design efforts.

B. 3-D QSAR (CoMFA)

As shown in Tables 17 and 18, the pls analysis with respect to all 202 artemisinin analogues indicated a modest r^2 (0.79) value and reasonably predictive q^2 (0.64). High residual values ($|\log| > 2$) were associated with three compounds in particular (**127**, **344**, **172**). The poor prediction of these compounds might be attributed to differences in metabolic behavior or cell absorption. For instance, lipophobicity of an acid moiety (**127**) could very well adversely affect drug accumulation in the parasite, and subsequently lead to lower measured antimalarial activity. Analogue **127** exhibited a large negative residual value indicating an overprediction. Another carboxylic acid (**396**), where the acid is located at the end of a longer alkyl chain, is more lipophilic and less affected by potential cell penetration problems. The high residual values identified with the other outliers (**344** and **172**) might be attributed to structural features greatly affecting conformation; for example, the C-11 methyl in analogue **172** (Table 7) imposes different conformational preferences than analogue **170** which lacks this methyl group. These three compounds were removed from the model, leading to a subsequent pls analysis using an $n = 199$ dataset. A

Table 17. Statistical Results of Crossvalidated PLS Analyses

Number of Compounds	Probe[a]	q^2	Optimum Number of Components
202[b]	2 Å/C.3[a]	0.64	5
202	2 Å/H	0.66	5
202 (Dock)	2 Å/C.3	0.62 (0.61)	6 (5)
199[c]	2 Å/C.3	0.69	5
199	2 Å/H	0.71 (0.68)	10 (5)
199 (Dock)	2 Å/C.3	0.65 (0.64)	6 (5)
162[d]	2 Å/C.3	0.69	5
160[e]	2 Å/C.3	0.74	5
160 (Dock)	2 Å/C.3	0.74 (0.71)	7 (5)

Notes: [a]Lattice spacing (in angstroms), probe atom type (sp^3-C or proton).
[b]The entire database.
[c]The same database without compounds **127, 344** and **172**.
[d]The 202 database without 40 racemates.
[e]The 162 database without compounds **127** and **344**.

significant increase in the r^2 (0.929) and q^2 (0.69) values further justified the deletion of these outliers.

One particularly interesting facet of the model is the observation of predicted vs. actual values displayed for the 15 most active compounds in the dataset. These compounds were invariably underpredicted by the model, though each was correctly predicted to be more active than artemisinin. Residual values for these compounds were reasonably low as a whole, with only two of the residuals

Table 18. Statistical Results of Conventional PLS Analyses

Number of Compounds	Probe	r^2	s	F
202	2 Å/C.3	0.79	0.71	149.62
202	2 Å/H	0.92	0.47	204.92
202 (Dock)	2 Å/C.3	0.81	0.70	151.54
199	2 Å/C.3	0.83	0.65	184.97
199	2 Å/H	0.93	0.43	231.50
199 (Dock)	2 Å/C.3	0.93	0.42	244.61
162	2 Å/C.3	0.87	0.62	201.96
160	2 Å/C.3	0.89	0.55	251.11
160 (Dock)	2 Å/C.3	0.96	0.35	317.56

exceeding +1.0. This might imply that the increased activity for these most active analogues is due to a secondary effect not adequately sampled by this model.

With respect to the entire dataset ($n = 199$), the two compounds with the highest residual values (worst predictions) included an analogue with a known lack of stability (**127**, residual value = -1.77) and a highly inactive compound that was still predicted to be highly inactive (calculated activity relative to artemisinin, 5.3×10^{-3}; residual value = -1.76).

In the literature many racemates (40) have been tested for in vitro activity. The artemisinin-like enantiomer of each pair was included in the $n = 199$ database under the assumption that this was the active isomer. In an attempt to better understand the contribution of each enantiomer, these compounds were removed from the dataset, and a third model ($n = 202 - 2$ high residual compounds $- 40$ racemates = 160) was developed. The notable predictivity of the model was indicated by a high q^2 (0.89) value. Judging by the standard error, r^2, and F test values (0.55, 0.89, and 251.11, respectively), the model reproduced the input data somewhat better than the $n = 199$ precursor.

Prediction of the in vitro activity for the racemic pair of compounds leads to a less clear interpretation of the data. Is only one enantiomer active? If so, should the biological activity value be "doubled," assuming only half of the assayed material is active? If the enantiomers are unequally active, how should the data be treated? This final model was used to predict the activities of each enantiomeric pair, providing some insight into the solution to these questions.

Here again alignment of the enantiomeric pair resulted in apparent differences in models. A new alignment rule was used (see Figure 12), and activities were predicted for each enantiomer. For a given analogue and its enantiomer, the predicted activities were compared with the actual biological data in an attempt to find a mathematical correlation. Though in fact most of the enantiomers having the natural configuration (that of (+)-artemisinin) showed considerably more activity than the corresponding enantiomer with the unnatural configuration, in a number of cases the unnatural enantiomer had significant predicted activity. When the predicted activity of each enantiomeric pair was combined and then averaged, those values did not correspond well to the actual test data for the racemates. The model did not adequately predict the activity of the racemates under the assumption that drug binding occurs in a chiral environment.

To aid in visualization, Figure 13 shows the electrostatic and steric maps for the pls analysis with $n = 199$, 2 Å/C.3 (Table 17). In the steric map, yellow contours correspond to regions in space where steric bulk would be predicted to decrease antimalarial activity. The yellow regions appear around the endoperoxide and other areas on the α-face of the molecule, suggesting that increased steric bulk on the underside of the molecule would be detrimental to the activity. Conversely, green contours represent areas around the template molecule where an increase in antimalarial activity due to increased steric bulk is anticipated. In Figure 13, green polyhedra surround the regions near the C-9 methyl, C-3 methyl and the 7β-H of

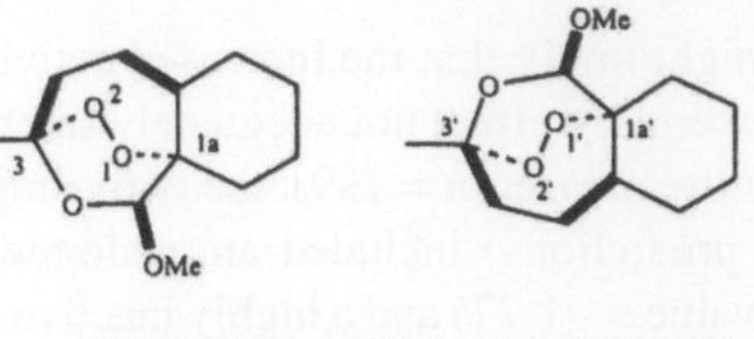

Alignment (analog 137): 1a-1a', 1-1', 2-2', 3-3'

Figure 12. Alignment rule used in the prediction of antimalarial activity of racemic artemisinin analogues.

artemisinin, suggesting that greater steric bulk in these areas would increase activity. It is interesting to note how an active analogue (9-butyl) fits within the contours compared to artemisinin or relatively inactive analogues.

Within the CoMFA electrostatic map, red contours are displayed in areas where negative charge is associated with increased activity of the database analogues. Red contours are visible near the peroxide bridge, supporting the important role the peroxide plays in activity; and near O-11 (or N-11). There are also red contours in

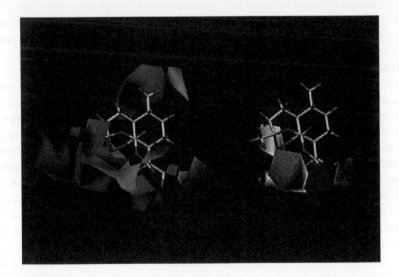

Figure 13. CoMFA contour maps about (+)-artemisinin for the standard alignment database (n = 199, 2 Å/C.3). In the steric contour map to the left, green contours indicate areas where steric bulk is predicted to increase antimalarial activity, while red contours indicate regions where steric bulk is predicted to decrease activity. The electrostatic contour map on the right displays yellow polyhedra where partial negative charge is correlated with antimalarial activity; the blue polyhedra indicating a relationship between partial positive charge and activity.

the vicinity of the C-9 methyl, a point supported by the increase in activity from arteether (**145**) to the more negative-charged difluoro analogue **360**. Blue contours indicate areas within the lattice where electropositive properties of a molecule are predicted to increase activity. These regions include the partial positive charge associated with hydrogen atoms bound to carbon, and can be correlated with lipophilic interaction. A broad band of blue extends from the C-9 methyl toward the carbonyl at C-10. This indicates that alkyl groups added to these areas may increase activity due to their inherent electropositive charge. An increase in lipophilicity by removal of the carbonyl at C-10 can be favorable—as seen in Figure 13, the contour map surrounding **112**: this analogue is simply the reduced product of the slightly less active **44**.

Dock Aligned Database

The CoMFA protocol was repeated for the $n = 202$ dock-aligned database, and pls analysis indicated that the model was comparable to that found with the standard alignment. Overall, the "dock conformation" and alignment led to a model with higher r^2 (0.93) and F-test (244.61), and reduced standard error (0.62), but the cross-validated analysis produced a predictive model with $q^2 = 0.62$, slightly less than that of the standard-aligned model. The optimal number of components increased, implying a statistically "less robust" model. Similarities between the models can be noticed in the high residual value of compounds **299** (a polar dihydroxy compound), as well as overprediction of inactive (LogRA < 1.0) compounds. In addition to statistical improvements relative to the standard alignment, the magnitude of this over/under prediction was greatly decreased, resulting in noticeably smaller residual values for active analogues. All of the 15 most active compounds were predicted to have greater activity than the parent artemisinin, though not all were overpredicted. Pruning the database to 199 as before provided a statistical improvement with $s = 0.42$. Finally, deletion of the same analogues as before to 160 compounds improved the statistical measures producing a model with a q^2 (0.74), indicating significant predictive capability. Supporting the soundness of the pharmacophore is the improved standard error ($s = 0.35$), the lowest among the models.

Contour maps for the $n = 199$ 2 Å/C.3 Dock Model (Table 17), representing electrostatic and steric volumes deemed important to antimalarial activity, can be seen in Figure 14. Some major visual differences are apparent between the dock and nondock models. For example, in the dock alignment rule models the contribution of the red contours is diminished relative to the nondock alignment rule model (Figure 13). Perhaps this difference can be explained by the possibility that groups which interfere with hemin binding are negatively correlated with potency. In the dock models, where the side chains have been artificially moved out of the region occupied by hemin, the red contours are much smaller. The green contours in the dock model are much stronger than those in the nondock models, and are

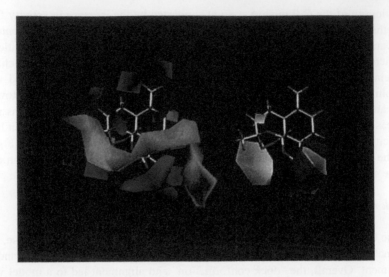

Figure 14. CoMFA contour maps about (+)-artemisinin for the dock aligned database (*n* = 199, 2 Å/C.3). In the steric contour map to the left, green contours indicate areas where steric bulk is predicted to increase antimalarial activity, while red contours indicate regions where steric bulk is predicted to decrease activity. The electrostatic contour map on the right displays yellow polyhedra where partial negative charge is correlated with antimalarial activity; the blue polyhedra indicating a relationship between partial positive charge and activity.

incidentally along the periphery of the artemisinin skeleton corresponding to the C-9, N-11, and C-3 substituent space (see Figure 10). Finally in the electrostatic CoMFA, yellow contours surround the oxygen atoms of the trioxane and pyran rings. There is variation from dock to nondock models where in the dock model, the yellow contour just below the peroxide assumes much greater magnitude than the nondock model. Further, the exact location of the contours about the peroxy group is somewhat different in the nondock model with the contours tracking the location of the lone pair electrons of the peroxide (Figure 14 versus Figure 13).

Validation of the 3D-QSAR Model

The q^2 value of a CoMFA model, together with other statistical information from the pls analysis, provides information on the predictive capability of the model. In this study we have generated CoMFA models that describe the pharmacophore either with or without the involvement of hemin, both of which provide good q^2 values. Selection of the model that most accurately depicts reality is not trivial since many variables are inherent in the cell-culture bioassay results. However, it may be possible to distinguish between models by appropriate choice of test molecules. If one model is more consistent in predicting the activities of novel compounds, then it is surely more reflective of reality and should be preferred over the other model in drug design efforts.

We chose to design and then synthesize a class of compounds heretofore unknown in the artemisinin area: 8,8-disubstituted-D-norartemisinins. While the parent molecule **421** was not predicted in the CoMFA models to be highly active, homologues of **421** were.[155]

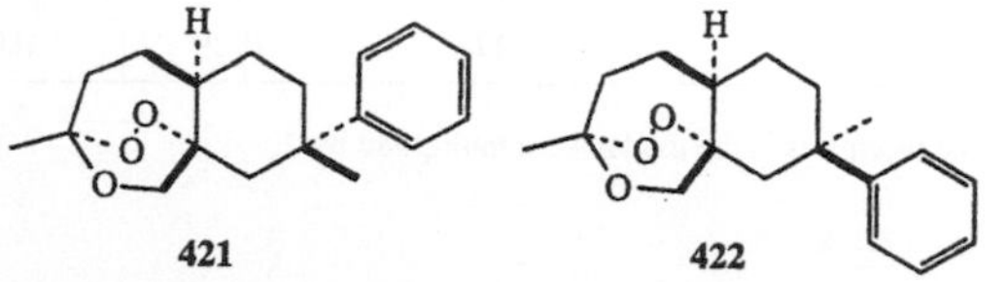

(32)

421   **422**

Further, there does not exist sufficient data regarding the biological activity for each enantiomer in a racemic mixture of artemisinin analogues. To help define the antimalarial contribution made by each enantiomer, a model was developed in which 40 racemates were placed as single (+)-enantiomers in the original database of 202 compounds. By doing this it was implied that the (−)-enantiomers are as active as their counterpart, and an optically pure compound is not necessary for antimalarial activity. These "achiral" models have been discussed as 199 (2 Å/C.3) and 199 Dock. A database without the racemates, or "chiral" models 160 (2 Å/C.3) and 160 Dock, were also included in our study. Finally, adjustment to an "achiral" model was attempted in which the activities of the racemates in the 199 (2 Å/C.3) model were multiplied by 2, or model 199* (2 Å/C.3). In each case, data was subjected to the statistical pls analysis (Table 18). Relative antimalarial activities were predicted for two racemic compounds **421** and **422**,[155] using these datasets as shown in Tables 19 and 20.

The predictions from the achiral CoMFA model 199 (2 Å/C.3), which includes the racemates, best reflects experimentally derived antimalarial activities. The predicted value for the natural enantiomer of **421** was 110% of artemisinin and 12%

Table 19. Relative In Vitro Antimalarial Activity of ABC-ring Artemisinin Analogues Against *P. falciparum*

Structure	IC_{50} (ng/mL)	Relative Activity[a]
1 (Artemisinin)	4.2	1
421	12.86	0.33
422	132.03	0.03

Note: [a]Relative activity = [IC_{50} artemisinin (control value/IC_{50} analogue] MW analogue/MW artemisinin.

Table 20. Prediction Values for **421** and **422** in Different CoMFA Models

			Predicted Relative Activity			
	Achiral Model			*Chiral Model*		
Natural Configuration	*199 Dock*	*199 (2 Å/C.3)*	*Unnatural Configuration*	*199* (2Å/C.3)* [a]	*160 (2 Å/C.3)*	*160 Dock*
421	0.20	1.10		2.16	4.73	0.70
			421	0.37	0.82	2.25
422	0.71	0.12		0.02	0.05	2.07
			422	0.29	0.02	0.81

Note: [a] The LogRA for the racemates in the model were multiplied by 2.

for **422**, compared with 33% and 3%, respectively, for the in vitro results. While both figures were off by a factor of 4, the log difference is closer to the *s* value for this model of about 0.65 (Eq. 33). The 199 Dock model closely predicted the activity of **421** as 20% of artemisinin (found, 33%), but was inaccurate in relation to prediction of **422** at 70% of artemisinin (found, 3%).

In an attempt to clarify the chirality issue, the original dataset with 202 compounds was modified. Using the assumption that only the (+)-enantiomers are active, the LogRA for the racemates were multiplied by 2 with the idea that experimental IC_{50} values actually are 0.5 that measured. The new dataset 199* was statistically analyzed by pls, and the results (Eq. 34) follow the same pattern of the previous model, though we now find the model is poorer at predicting activities than if we assume the enantiomers show equivalent activity.

$$r^2 = 0.83, \ F = 184.97, \ s = 0.65 \qquad (33)$$

$$r^2 = 0.83, \ F = 188.74, \ s = 0.65 \qquad (34)$$

In light of these results, chirality does not seem to be necessary for antimalarial activity in this class of drugs. This was further evidenced by the over prediction of **421** in the 160 (2 Å/C.3) model, where no racemates are present.

As can be seen in the contour map generated from the "dock-aligned" database, interaction with the hemin iron is a predominant component of the dock-minimized model. It follows that the predictivity of the 160 Dock model might favor compounds in which this interaction is influenced by the dock-minimization process.

Analogue **421** is a reasonable example of the class of compounds most affected by the docking strategy. Here, the special positioning of the phenyl ring on the same face of the molecule as the peroxide bridge is noticeably modified in the dock minimization process. This repositioning results in a bending upwards of the phenyl group. The predicted activity of **421** in the 160 Dock model (70% of artemisinin) is close to the activity measured experimentally (33% of artemisinin) and is certainly within the standard error of the model (Table 20).

Analogue **422** is much less affected by the dock minimization and coincidentally is much less well predicted by the 160 Dock model. Though only 3% as active as artemisinin, this compound is predicted to be much more active than analogue **421**. This added activity might be attributed to the focus by the dock model on the peroxide face interaction with hemin; compounds with substituents on the β-face are not as thoroughly represented by the model as a whole, and are less likely to be predicted well.

To better understand the importance of optical activity towards antimalarial activity and to further improve this CoMFA study, the synthesis of (–)-artemisinin and various other analogues having unnatural configuration are underway. In addition, further analogues needed to complete this study are β-face substituents in the B and C rings of artemisinin, or primarily the axial positions at C-5, C-7, and C-8. Analogues such as **422** pave the way for the latter of these cases, C-8β substituted analogues.

VIII. CONCLUSIONS

Synthetic methodology has been developed for introducing substituents at C-9, O-11 (as N-R), and C-3 via a total synthetic intermediate; for substitution of carbon for oxygen at O-13 by a separate route; and for a variety of simplified, tricyclic versions of artemisinin. Analogues accrued from these studies have been bioassayed and their SAR discussed. However, in terms of rational design, these relationships are complex and the bioassay data more easily utilized by computer aided pharmacophore modeling. Accordingly, with a dataset of over 200 artemisinin analogues, various CoMFA pharmacophore models have been developed based on: (a) the conformational hypotheses that the active conformation of the analogues is the globally minimized structure (b) the conformational hypotheses that the active conformation of the analogues is based on their interaction with the proposed molecular target in the parasite, hemin; (c) the assumption that enantiomers have identical activity; and (d) the assumption that enantiomers have differing activities. For example, in some models racemates have been included or excluded from the analysis, while in other studies the analogues are docked to hemin prior to being placed in the molecular database or the analogues are simply minimized and templated to artemisinin. The results initially indicate that the achiral docked model may best represent reality, lending support to Meshnick's original hypothesis regarding parasitic hemin being the mechanistic trigger for the action of the drug, and to the assertion that chirality is not important for antimalarial activity. However, this is a hypothesis best tested by synthesis of unnaturally configured analogues of artemisinin, and the unnatural enantiomer of the natural product, (–)-artemisinin. Once armed with this data, other modeling possibilities can be excluded, thus simplifying the overall approach to development of 3D-QSAR intended for drug design efforts.

In terms of validation of such a model, a measure of internal consistency is available in the form of the q^2 value. However, the ultimate test of a model is its ability to predict activities for newly reported compounds. While the model can do well at predicting variation within its own dataspace, when confronted with unquantified regions not represented in the original dataset, such as those encountered with bulky 8β-substituted analogues, the model is unreliable. In order to provide full predictive capability, studies in progress address the synthesis of additional top-face or β-oriented analogues that will be bioassayed and added to the model having the correct conformational and chirality hypotheses.

Finally, it should be emphasized that extensive synthetic studies, in some cases utilizing complex pathways, were a necessary prelude to these modeling efforts. The ultimate goal of employing computer-aided drug design to define a simple, yet potent series of analogues for economical treatment of malaria is ongoing. In addition to structural simplification as an approach to affordable antimalarials, we have recently reported a three step semisynthesis of some of our best analogues (e.g. **114**) from artemisinic acid, a natural product that occurs along with artemisinin in *A. annua* (2.6% dry weight yield).[156] Encouraging results of in vivo bioassay and neurotoxicity testing will be reported shortly for these analogues.

ACKNOWLEDGMENTS

We would like to thank Drs. Robert Engle, Jean Karle, Dennis Kyle, Wil Milhous and Robert Miller of the Walter Reed Army Institute of Research for their advice and collaborations in preceding years, members of the Medicinal Chemistry Department (University of Mississippi), and Dr. James D. McChesney (NaPro Biotherapeutics). Support of this work was funded by the Department of Defense USAAMRDC (WRAIR), UNDP/World Bank/WHO Special Programme for Research and Training in Tropical Diseases (TDR), NIAID, and the University of Mississippi Research Institute of Pharmaceutical Sciences.

REFERENCES

1. White, N. J.; Miller, K. D.; Churchill, F. C.; Berry, C.; Brown, J.; Williams, S. B.; Greenwood, B. M. *N. Engl. J. Med.* **1988**, *319*, 1493–1500.
2. Miller, L. H. *Science* **1992**, *257*, 36.
3. Marshall, E. *Science* **1990**, *247*, 399–402.
4. Miller, K. D.; Greenberg, A. E.; Campbell, C. C. *N. Engl. J. Med.* **1989**, *321*, 65–70.
5. Krogstad, D. J.; Herwaldt, B. L. *N. Engl. J. Med.* **1988**, *319*, 1538–1540.
6. Editorial. *Lancet* **1992**, *339*, 649–651.
7. Qinghaosu Research Group. *Scientia Sinica* **1980**, *23*, 380–385.
8. Shen, C.; Zhuang, L. *Med. Res. Rev.* **1984**, *4*, 47–86.
9. Klayman, D. L. *Science* **1985**, *228*, 1049–1055.
10. Luo, X.-D.; Shen, C.-C. *Med. Res. Rev.* **1987**, *7*, 29–52.
11. Clark, G. R.; Nikaido, M. M.; Fair, C. K.; Lin, J. *J. Org. Chem.* **1985**, *50*, 1994–1996.

12. Jung, M.; ElSohly, H. N.; Croom, E. M.; McPhail, A. T.; McPhail, D. R. *J. Org. Chem.* **1986**, *51*, 5417–5419.
13. Binns, F.; Wallace, T. W. *Tetrahedron Lett.* **1989**, *30*, 1125–1128.
14. Ravindranathan, T.; Kumar, M. A.; Menon, R. B.; Hiremath, S. V. *Tetrahedron Lett.* **1990**, *31*, 755–758.
15. Schmid, G.; Hofheinz, W. *J. Am. Chem. Soc.* **1983**, *105*, 624–625.
16. Xu, X. X.; Zhu, J.; Huang, D.Z.; Zhou, W. S. *Tetrahedron* **1986**, *42*, 819–828.
17. Avery, M. A.; Chong, W. K. M.; Jennings-White, C. *J. Am. Chem. Soc.* **1992**, *114*, 974–979.
18. Vennerstrom, J. L.; Acton, N.; Lin, A. J.; Klayman, D. L. *Drug Des. and Deliv.* **1989**, *4*, 45–54.
19. Brossi, A.; Venugopalan, B.; Gerpe, L. D.; Yeh, H. J. L.; Flippen-Anderson, J. L.; Buchs, P.; Luo, X. D.; Milhous, W.; Peters, W. *J. Med. Chem.* **1988**, *31*, 645–650.
20. Lin, A. J.; Lee, M.; Klayman, D. L. *J. Med. Chem.* **1989**, *32*, 1249–1252.
21. Lin, A. J.; Li, L.; Aayman, D. L.; George, G. F.; Flippen-Anderson, J. L. *J. Med. Chem.* **1990**, *33*, 2610–2614.
22. Lin, A.; Li, L.; Andersen, S. L.; Klayman, D. L. *J. Med. Chem.* **1992**, *35*, 1639–1642.
23. Roth, R.; Acton, N. *J. Nat. Prod.* **1989**, *52*, 1183–1185.
24. Acton, N.; Roth, R. J. *J. Org. Chem.* **1992**, *57*, 3610–3614.
25. Lansbury, P. T.; Nowak, D. M. *Tetrahedron Lett.* **1992**, *33*, 1029–1032.
26. Haynes, R. K.; Vonwiller, S. C. *Trans. Royal Soc. Trop. Med. Hyg.* **1994**, *88*, 23–26.
27. Avery, M. A.; Jennings-White, C.; Chong, W. K. M. *J. Org. Chem.* **1989**, *54*, 1792–1795.
28. Schwaebe, M.; Little, R. D. *J. Org. Chem.* **1996**, *61*, 3240–3244.
29. Haynes, R. K.; King, G. R.; Vonwiller, S. C. *J. Org. Chem.* **1994**, *59*, 4743–4748.
30. Jung, M.; Li, X.; Bustos, D.; ElSohly, H.; McChesney, J.; Milhous, W. K. *Tetrahedron Lett.* **1989**, *30*, 5973–5976.
31. Jung, M.; Li, X.; Bustos, D.; ElSohly, H.; McChesney, J.; Milhous, W. K. *J. Med.Chem.* **1990**, *33*, 1516–1518.
32. Jung, M.; Yu, D.; Bustos, D.; ElSohly, H. N.; McChesney, J. D. *Bioorg. Med. Chem. Lett.* **1991**, *1*, 741–744.
33. Avery, M. A.; Mehrotra, S.; Johnson, T.; Miller, R. *J. Med. Chem.* **1996**, *39*, 4149–4155.
34. Haynes, R. K.; Vonwiller, S. C. *SynLett.* **1992**, 481–483.
35. Haynes, R. K.; Vonwiller, S. C. *Acc. Chem. Res.* **1997**, *30*, 73–79.
36. Jefford, C. W.; Boukouvalas, J.; Kohomoto, S.; Bernardinelli, G. *Tetrahedron* **1985**, *41*, 2081–2088.
37. Jefford, C. W.; Favarger, F.; Ferro, S.; Chambaz, D.; Bringhen, A.; Bernardinelli, G.; Boukouvalas, J. *Helv. Chim. Acta* **1986**, *69*, 1778–1786.
38. Jefford, C. W.; McGoran, E.; Boukouvalas, J.; Richardson, G.; Robinson, B.; Peters, W. *Helv. Chim. Acta* **1988**, *71*, 1805–1812.
39. Chang, H. R.; Jefford, C. W.; Pechère, J.-C. *Antimicrob. Agents Chemother.* **1989**, *33*, 1748–1752.
40. Kepler, J. A.; Philip, A.; Lee, Y. W.; Morey, M. C.; Carroll, F. I. *J. Med.Chem.* **1988**, *31*, 713–716.
41. Singh, C. *Tetrahedron Lett.* **1990**, *31*, 6901–6902.
42. Bunnelle, W. H.; Isbell, T. A.; Barnes, C. L.; Qualls, S. *J. Am. Chem. Soc.* **1991**, *113*, 8168–8169.
43. Vennerstrom, J. L.; Fu, H.-N. E. W. Y.; Ager, A. L., Jr.; Wood, J. K.; Anderson, S. L.; Gerena, L.; Milhous, W. K. *J. Med. Chem.* **1992**, *35*, 3023–3027.
44. Avery, M. A.; Jennings-White, C.; Chong, W. K. M. *J. Org. Chem.* **1989**, *54*, 1789–1792.
45. Avery, M. A.; Chong, W. K. M.; Detre, G. *Tetrahedron Lett.* **1990**, *31*, 1799–1802.
46. Posner, G. H.; Oh, C. H.; Gerena, L.; Milhous, W. K. *J. Med. Chem.* **1992**, *35*, 2459–2467.
47. Posner, G. H.; Oh, C. H.; Wang, D.; Gerena, L.; Milhous, W. K.; Meshnick, S. R.; Asawamahasadka, W. *J. Med. Chem.* **1994**, *37*, 1256–1258.
48. Posner, G. H.; Oh, C. H.; Milhous, W. K. *Tetrahedron Lett.* **1991**, *32*, 4235–4238.
49. Avery, M. A.; Chong, W. K. M.; Bupp, J. E. *J. Chem. Soc., Chem. Commun.* **1990**, *21*, 1487–1489.

50. Imakura, Y.; Yokoi, T.; Yamagishi, T.; Koyama, J.; Hu, H.; McPhail, D. R.; McPhail, A. T.; Lee, K. H. *J. Chem. Soc., Chem. Commun.* **1988**, 372–374.
51. Imakura, Y.; Hachiya, K.; Ikemoto, T.; Kobayashi, S.; Yamashita, S. *Heterocycles* **1990**, *31*, 2125–2129.
52. Zhou, W.-S.; Xu, X.-X. *Chem. Res.* **1994**, *27*, 211–216.
53. Liu, H. J.; Yeh, W. L.; Chew, S. Y. *Tetrahedron Lett.* **1993**, *34*, 4435–4438.
54. Büchi, G.; Wüest, H. *J. Am. Chem. Soc.* **1978**, *100*, 294–295.
55. Gröbel, B.-T.; Seebach, D. *Chem. Ber.* **1977**, *110*, 852–866.
56. Sekiguchi, A.; Ando, W. *J. Org. Chem.* **1979**, *44*, 413–415.
57. Bates, T. F.; Thomas, R. D. *J. Org. Chem.* **1989**, *54*, 1784–1785.
58. Still, W. C. *Synthesis* **1976**, 453–454.
59. Katsuhara, J. *J. Org. Chem.* **1967**, *32*, 797–799.
60. Caine, D.; Procter, K.; Cassell, R. A. *J. Org. Chem.* **1984**, *49*, 2647–2648.
61. Oppolzer, W.; Petrzilka, M. *Helv. Chem. Acta* **1978**, *61*, 2755–2762.
62. Roush, W. R.; Walts, A. E. *Tetrahedron* **1985**, *41*, 3463–3478.
63. Stowell, J.; Keith, D.; King, B. *Org. Synth.* **1984**, *62*, 140–148.
64. Still, W. C.; Macdonald, T. *J. Am. Chem. Soc.* **1974**, *96*, 5561–5563.
65. Ireland, R. E.; Varney, M. D. *J. Am. Chem. Soc.* **1984**, *106*, 3668–3670.
66. Ireland, R. E.; Mueller, R. H.; Willard, A. K. *J. Am. Chem. Soc.* **1976**, *98*, 2868–2877.
67. Hill, R. K. In *Stereodifferentiating Addition Reactions Part B*; Morrison, J. D., Eds; Academic Press: San Diego, 1984, Vol. 3, pp. 521–526.
68. Evans, D. A. In *Stereodifferentiating Addition Reactions, Part B;* Morrison, J. D., Eds; Academic Press: San Diego, 1984, pp 14–21.
69. Avery, M. A.; Chong, W. K. M.; Jennings-White, C. *Tetrahedron Lett.* **1987**, *28*, 4629–4632.
70. Huet, F.; Lechevallier, A.; Pellet, M.; Conia, J. M. *Synthesis* **1978**, 63–65.
71. Vitorelli, P.; Winkler, T.; Hansen, H. J.; Schmid, H. *Helv. Chim. Acta* **1968**, *51*, 1457–1461.
72. Rhoads, S. J.; Raulins, N. R. *Org. React.* **1975**, *22*, 1–252.
73. Avery, M. A.; Bonk, J. D.; Bupp, J. *J. Label. Cpds. Radiopharm.* **1996**, *38*, 263–267.
74. Avery, M. A.; Bonk, J. D.; Mehrotra, S. *J. Label. Cpds. Radiopharm.* **1996**, *38*, 249–254.
75. Meshnick, S. R.; Thomas, A.; Ranz, A.; Xu, C. M.; Pan, H. *Molec. Biochem. Parasitol.* **1991**, *49*, 181–190.
76. Leskovac, V.; Theoharides, A. D.; Peggins, J. *Comp. Biochem. Physiol: C: Comp. Pharmacol. Toxicol.* **1991**, *99C*, 383–390.
77. Leskovac, V.; Theoharides, A. D.; Peggins, J. *Comp. Biochem. Physiol., C: Comp. Pharmacol. Toxicol.* **1991**, *99C*, 391–396.
78. Acton, N.; Karle, J. M.; Miller, R. E. *J. Med. Chem.* **1993**, *36*, 2552–2557.
79. Paitayatat, S.; Tarnchompoo, B.; Thebtaranonth, Y.; Yuthavong, Y. *J. Med. Chem.* **1997**, *40*, 633–638.
80. Posner, G. H.; Park, S. B.; Gonzalez, L.; Wang, D.; Cummings, J. N.; Klinedinst, D.; Shapiro, T. A.; Bachi, M. D. *J. Am. Chem. Soc.* **1996**, *118*, 3537–3538.
81. Bergbreiter, D. E.; Newcomb, M. *Tetrahedron Lett.* **1979**, *43*, 4145–4148.
82. Yamashita, M.; Matsumiya, K.; Tanabe, M.; Suemitsu, R. *Bull. Chem. Soc. Jpn.* **1985**, *58*, 407–408.
83. Avery, M. A.; Gao, F.; Chong, W. K.; Mehrotra, S.; Milhous, W. K. *J. Med. Chem.* **1993**, *36*, 4264–4275.
84. Avery, M. A.; Gao, F.; Mehrotra, S.; Chong, W. K.; Jennings-White, C. *The Organic and Medicinal Chemistry of Artemsinin and Analogs*. Trends in Organic Chemistry; Research Trends: Trivandrum, India, 1993, pp 413–468.
85. Still, W. C. *Macromodel*; New York, 1994, Vol. 4.5.
86. Kraus, G. A.; Frazier, K. A.; Roth, B. D.; Taschner, M. J.; Neuenschwander, K. *J. Org. Chem.* **1981**, *46*, 2417–2419.

87. Torok, D. S.; Ziffer, H. *Tetrahedron Lett.* **1995**, *36*, 829–832.
88. Lin, A. J.; Miller, R. *J. Med. Chem.* **1995**, *38*, 764–770.
89. Rong, Y.-J.; Wu, Y.-L. *J. Chem. Soc., Perkin Trans.* 1 **1993**, 2149.
90. Claus, R. E.; Schreiber, S. L. *Org. Syn.* **1985**, *64*, 150–156.
91. Earnshaw, C.; Wallis, C. J.; Warren, S. *J. Chem. Soc., Perkin Trans. 1* **1979**, 3099–3106.
92. Francotte, E.; Lohmann, D. *Helv. Chim. Acta* **1987**, *70*, 1569–1582.
93. Lee, R. A.; McAndrews, C.; Patel, K. M.; Reusch, W. *Tetrahedron Lett.* **1973**, *14*, 965–968.
94. Rösch, L.; Altnau, G.; Otto, W. *Angew. Chem., Int. Ed. Eng.* **1981**, *20*, 581–583.
95. Meshnick, S. R.; Jefford, C. W.; Posner, G. H.; Avery, M. A.; Peters, W. *Parasitol. Today* **1996**, *12*, 79–82.
96. Meshnick, S. R.; Taylor, T. E.; Kamchonwongpaisan, S. *Microbiol. Rev.* **1996**, *60*, 301–315.
97. Meshnick, S. R.; Jefford, C. W.; Posner, G. H.; Avery, M. A.; Peters, W. *Parasitol. Today* **1996**, *12*, 79–82.
98. Meshnick, S. R. *Ann. Trop. Med. Parasitol.* **1996**, *90*, 367–372.
99. Meshnick, S. R.; Yang, Y.; Lima, V.; Kuypers, F.; Kamchonwongpaisan, S.; Yuthavong, Y. *Antimicrob Agents Chemother.* **1993**, *37*, 1108–1114.
100. Gund, T. M. S., K. L.; Meshnick, S. R. *J. Mol. Graphics* **1995**, *13*, 215–222.
101. Cramer III, R. D.; Patterson, D. E.; Bunce, J. D. *J. Am. Chem. Soc.* **1988**, *110*, 5959–5967.
102. Cramer III, R. D.; Bunce, J. D.; Patterson, D. E.; Frank, I. E. *Quant. Struct. Act. Relat. Pharm. Chem. Biol.* **1988**, *7*, 18–25.
103. Posner, G. H.; Wang, D.; Cumming, J. N.; Oh, C. O.; French, A. N.; Bodley, A. L.; Shapiro, T. A. *J. Med. Chem.* **1995**, *38*, 2273–2275.
104. Posner, G. H.; Cummings, J. N.; Polypradith, P.; Oh, C. O. *J. Am. Chem. Soc.* **1995**, *117*, 5885–5886.
105. Burkert, U.; Allinger, N. L. *Molecular Mechanics*; American Chemical Society: Washington DC, 1982.
106. Ye, B.; Wu, Y.-L. *Tetrahedron* **1989**, *45*, 7287–7290.
107. Ye, B.; Wu, Y. L.; Li, G. F.; Jiao, X. Q. *Yaoxue Xuebao* **1991**, *26*, 228–230.
108. Shulte-Elte, K.; Ohloff, G. *Helv. Chim. Acta* **1967**, *50*, 153.
109. Cumming, J. N.; Ploypradith, P.; Posner, G. H. *Adv. Pharmacol.* **1997**, *37*, 253–297.
110. Desjardins, R. E.; Canfield, C. J.; Haynes, D. E.; Chulay, J. D. *Antimicrob. Agents Chemother.* **1979**, *16*, 710–718.
111. Milhous, W. K.; Weatherley, N. F.; Bowdre, J. H.; Desjardins, R. E. *Antimicrob. Agents Chemother.* **1985**, *27*, 525–530.
112. Lin, A. J.; Klayman, D. L.; Milhous, W. K. *J. Med. Chem.* **1987**, *30*, 2147–2150.
113. Lin, A. J.; Zikry, A. B.; Kyle, D. E. *J. Med. Chem.* **1997**, *40*, 1396–1400.
114. Wesche, D. L.; DeCoster, M. A.; Tortella, F. C.; Brewer, T. G. *Antimicrob. Agents Chemother.* **1994**, *38*, 1813–1819.
115. Smith, S. L.; Fishwick, J.; McLean, W. G.; Edwards, G.; Ward, S. A. *Biochem. Pharmacol.* **1997**, *53*, 5–10.
116. Kamchonwongpaisan, S.; McKeever, P.; Hossler, P.; Ziffer, H.; Meshnick, S. R. *Am. J. Trop. Med. Hyg.* **1997**, *56*, 7–12.
117. Fishwick, J.; Edwards, G. *Chemico-Biological Interact.* **1995**, *96*, 263–271.
118. Thomas, C. G.; Ward, S. A.; Edwards, G. *J. Chromatogr.* **1992**, *583*, 131–136.
119. Brewer, T. G.; Peggins, J. O.; Grate, S. J.; Petras, J. M.; Levine, B. S.; Weina, P. J.; Swearengen, J.; Heiffer, M. H.; Shuster, B. G. *Trans. Royal Soc. Trop. Med. Hyg.* **1994**, *88*, 33–36.
120. Zaman, S. S.; Sharma, R. P. *Heterocycles* **1991**, *32*, 1593–1638.
121. Butler, A. R.; Wu, Y. L. *Chem. Soc. Rev.* **1992**, *21*, 85–90.
122. Casteel, D. A. In *Burger's Medicinal Chemistry and Drug Discovery*; Wolff, M. E., Eds; John Wiley & Sons: New York, 1997, Vol. 5, pp 3–91.

123. Gerpe, L. D.; Yeh, H. J. C.; Yu, Q.-S.; Brossi, A.; Flippen-Anderson, J. L. *Heterocycles* **1988**, *27*, 897–901.
124. Vennerstrom, J. L.; Eaton, J. W. *J. Med. Chem.* **1988**, *31*, 1269–1277.
125. Jefford, C. W.; Velarde, J.; Bernardinelli, G. *Tetrahedron Lett.* **1989**, *30*, 4485–4488.
126. Jefford, C. W. *Bicyclic Artemisinin Substitutes: Looking for the Pharmacophore*; 43rd Annual Meeting of the American Society of Tropical Medicine and Hygiene; Cincinnati, OH, 1994.
127. Posner, G. H.; Oh, C. H.; Webster, K.; Ager, A. L.; Rossan, R. N. *Am. J. Trop. Med. Hyg.* **1994**, *50*, 522–526.
128. Posner, G. H.; Gonzalez, L.; Cumming, J. N.; Klinedinst, D.; Shapiro, T. A. *Tetrahedron* **1997**, *53*, 37–50.
129. de Almeida Barbosa, L.-C.; Cutler, D.; Mann, J.; Crabbe, M. J.; Kirby, G.; Warhurst, D. C. *J. Chem. Soc., Perkin Trans 1* **1996**, 1101–1105.
130. Wu, J.; Ji, R. *Acta Pharmacologica Sinica* **1982**, *3*, 55–60.
131. Mekonnen, B.; Ziffer, H. *Tetrahedron Lett.* **1997**, *38*, 731–734.
132. Shukla, K.; Gund, T.; Meshnick, S. R. *J. Mol. Graphics* **1995**, *13*, 215–222.
133. CLogP, PCModels, version 4.42, Irvine, CA, 1996.
134. Avery, M. A.; Bonk, J.; Mehrotra, S.; Chong, W. K. M.; Miller, R.; Milhous, W.; Goins, K. D.; Venkatesan, S.; Wyandt, C.; Khan, I.; Avery, B. A. *J. Med. Chem.* **1995**, *38*, 5038–5044.
135. Tsar, v. 2.41, 1996: Cache Scientific, Beaverton, OR.
136. Avery, M. A.; Fan, P.-C.; Karle, J.; Miller, B.; Goins, K. *Tetrahedron Lett.* **1995**, *36*, 3965–3968.
137. Avery, M. A.; Fan, P.-C.; Bonk, J.; Karle, J.; Miller, B.; Goins, K. *J. Med. Chem.* **1996**, *39*, 1885–1897.
138. Gu, H. M.; Warhurst, D. C.; Peters, W. *Trans. Roy. Soc. Trop. Med. Hyg.* **1984**, *78*, 265–270.
139. Avery, M. A.; Mehrotra, S.; Bonk, J. D.; Vroman, J. A.; Goins, K.; Miller, R. *J. Med. Chem.* **1996**, *39*, 2900–2906.
140. Leban, I.; Golic, L.; Japelj, M. *Acta Pharm. Jugosl.* **1988**, *38*, 71–77.
141. Avery, M. A.; Gao, F.; Chong, W. K.; Hendrickson, T. F.; Inman, W. D.; Crews, P. *Tetrahedron* **1994**, *50*, 957–972.
142. Haynes, R. K.; Vonwiller, S. C. *Tetrahedron Lett.* **1996**, *37*, 257–260.
143. Augustijns, P.; D'Hust, A.; Van Daele, J.; Kinget, R. *J. Pharm. Sci.* **1996**, *85*, 577–579.
144. Cabantchik, Z. I. *Blood Cells* **1990**, *16*, 421–432.
145. Pu, Y. M.; Torok, D. S.; Ziffer, H.; Pan, X.-Q.; Meshnick, S. R. *J. Med. Chem.* **1995**, *38*, 4120–4124.
146. Jung, M.; Bustos, D. A.; El Sohly, H. N.; McChesney, J. D. *Synlett.* **1990**, 743–744.
147. Vroman, J. A.; Mehrotra, S.; Avery, M. A.; Miller, R. *Three Synthetic Routes to 9-Butyl Artemisinin.* 22nd Annual MALTO Medicinal Chemistry-Pharmacognosy Meeting, Xavier University, New Orleans, LA, 1995.
148. Posner, G. H.; Wang, D.; Gonzalez, L.; Tao, X.; Cumming, J. N.; Klinedinst, D.; Shapiro, T. A. *Tetrahedron Lett.* **1996**, *37*, 815–818.
149. Bustos, D. A.; Jung, M.; ElSohly, H. N.; McChesney, J. D. *Heterocycles* **1989**, *29*, 2273–2277.
150. Jefford, C. W.; Velarde, J. A.; Bernadinelli, G.; Bray, D. H.; Warhusrt, D. C.; Milhous, W. K. *Helv. Chim. Acta* **1993**, *76*, 2775–2788.
151. Khalifa, S. I.; Baker, J. K.; Jung, M.; McChesney, J. D.; Hufford, C. D. *Pharm. Res.* **1995**, *12*, 1493–1498.
152. Vroman, J. A.; Avery, M. A.; Mehrotra, S.; Bonk, J. D.; Srivastaba, R.; Srivastava, A.; Ager, A. L. J.; Miller, R. *Continuous Studies on the Synthesis and Activity of Some 9-Substituted-10-Deoxo-Artemisinin Analogs;* 23rd Annual MALTO Medicinal Chemistry-Pharmacognosy Meeting; University of Mississippi, 1996.
153. Posner, G. H.; McGarvey, D. J.; Oh, C. H.; Kumar, N.; Meshnick, S. T.; Asawamahasadka, W. *J. Med. Chem.* **1995**, *38*, 607–612.
154. Sybyl, v. 6.2, St. Louis, MO, 1992.

155. Haraldson, C. A.; Karle, J. M.; Freeman, S. G.; Duvadie, R. K.; Avery, M. A. *Bioorg. Med. Chem. Lett.* **1997**, *7*, 2357–2362.
156. Vroman, J. A.; Khan I. A.; Avery, M. A. *Tetrahedron Lett.* **1997**, *38*, 6173–6176.

DESIGN OF COMPOUND LIBRARIES FOR DETECTING AND PURSUING NOVEL SMALL MOLECULE LEADS

Allan M. Ferguson and Richard D. Cramer

Advances in Medicinal Chemistry
Volume 4, pages 219–244.
ISBN: 0-7623-0064-7

ABSTRACT

The selection of compounds for synthesis in high-throughput synthesis experiments and for use in high-throughput biological assays is of prime importance in the detection of exploitable biological properties in novel small molecule leads. We discuss factors associated with the design of compound libraries for use in such high-throughput methods, and how these impact both on finding new leads and on following up on them. Specifically four key design elements are discussed, viz. the relevance of the compounds to lead discovery, the collective diversity of the library, the "uniqueness" of individual compounds, and their synthetic accessibility. The design of the Optiverse library is discussed as an example and a comparison is made of the structural similarities between Optiverse, a database of recent therapeutic patents, and a database of commercial reagents.

I. INTRODUCTION

The introduction of high-throughput methods, principally high-throughput screening (HTS) and combinatorial chemistry, in the pharmaceutical industry was born out of the continued pressure to shorten the cycle for lead discovery and development. The sometimes dazzling successes resulting from the application of these new experimental methods are already being presented in the scientific community for peer review. It is clear however that the likelihood of detecting pharmaceutical and other bio-relevant agents with exploitable properties is a key factor in the selection of compounds for use in high-throughput biological assays. In the search for new or better bioactivity, it is essential to ensure that a molecular design process takes into account such bio-relevant properties.

For most companies, accelerating the process of lead discovery and lead development is a key motivation for having access to HTS technology and a high-throughput chemistry capability. However by themselves, these tools have little value without input from, among other disciplines, medicinal chemistry, biochemistry, molecular biology, and molecular design—to guide synthesis and testing to new lead compounds in specific disease areas. Placing aside key biological considerations, such as genomics, assay development, and HTS, the likelihood of finding active compounds depends critically on the appropriateness and range of properties present in the compounds tested. Simply screening large numbers of compounds is not the answer. One needs an assurance that collections of compounds assembled for screening contain sufficient chemical variation—variation that has relevance to the target(s) of interest—to have the best chances of detecting useful pharmaceutical and other leads.

Finding activity, however, is only half the battle. Following up on a lead quickly can be just as difficult and often more critical to the drug discovery process. Once a qualified lead has been detected, it is most likely that analogues will be needed to enhance the activity or optimize the properties in some desirable way. For many, this is a major hurdle that must be overcome. Experience with peptides, for example, has shown that it can often be very difficult to translate the information from say a flexible hexapeptide to a "small molecule" lead. Even where the lead is already a small molecule, synthetic accessibility to this compound or its analogues can be fraught with difficulty more often if the lead has come from natural sources.

In a sense, the advent of high-throughput chemistries has subtly altered the role and importance of each individual compound. In one respect compounds have become more important because when synthesized they are viewed less as simple by-products of optimizing activity against a particular target and more as a means to ensure that future leads will be obtained. In another respect, the high-throughput nature of synthesis combined with a consequential reduction in chemical analysis effort has lessened the individual worth and distinctiveness of compounds. Their intrinsic value is therefore greater as part of a collection or library which implies somewhat paradoxically that the sum of the parts is worth less than the whole.

This chapter discusses, in general, how perceptions of drug discovery have been altered by application of the new high-throughput technologies and, in particular, those factors that influence the value of a library of compounds. The design of such libraries of compounds for use principally for lead discovery programs is also discussed with particular reference to the processes associated with the Optiverse library. Finally, the chapter also considers how a careful design can make it easier to progress rapidly from leads to potential clinical candidates.

II. LEAD FINDING AND LEAD FOLLOWING PARADIGMS

In the search for novel bioactive compounds the twofold challenge, stated succinctly, remains the same, viz.: to detect new and novel leads with exploitable properties and to rapidly follow up on these leads to optimize their properties for use in biological systems.

The coupling of combinatorial chemistry and HTS capabilities represents one of the most important steps forward in addressing these challenges in recent years. In a sense, paradigms associated with lead discovery have changed with the advent of these new tools. Of course, increased synthesis throughput comes at a cost, mainly the complexity of the compounds obtained is often lower and analytical assurance that the compound is present and of a purity (in the case of nonmixtures) that is appropriate for biological testing is often ambiguous.

There are other factors which also need to be taken into account. The larger numbers of compounds that arise from these chemistries typically pale into insignificance when set against all potential compounds that could arise even from available reagents within a single reaction. ("Reagents" are taken to represent building blocks used to synthesize reaction products.) For example, a reaction involving individual couplings to a single diamino scaffold, which is used in the Optiverse library, can give rise to almost a billion potential reaction products just using reagents listed as being currently available.[1] This is equivalent to more than 50 times the total number of CAS-registered compounds, currently listed at 18 million.[2] Even to attempt to test all of these compounds from this one reaction would require more than 50 years of HTS screening at 50,000 compounds per day. In essence, a comprehensive diamine-scaffold-based library is still well beyond the practical reach of today's pharmaceutical and agrochemical industries and will remain so for the foreseeable future.

Coupling to diamines is but a single group of reactions that is amenable to the methods of combinatorial chemistry or other high-throughput synthesis methods. The list of high-throughput reactions is expanding but it still does not represent all organic chemistry, nor will it ever. This is an important and often overlooked factor when considering not only the types of chemicals and chemistries desired in a lead-finding experiment but also the very limits of high-throughput synthesis.

Another key limitation of high-throughput synthesis lies in the reagents themselves. The number of reagents listed as being commercially available at any given time is typically more than 180,000.[1] Those reagents that are listed as being available for synthesis in various databases are more often biased by commercial and economic pressures rather than property variations. Their suitability must therefore be scrutinized even at the outset. In our experience, for example, synthesis of libraries based simply on available reagents often lead to high lipophilicities. Also, as is the norm with synthesis chemistry, there is a tendency to find "clusters" of reagents that share a common intermediate. This can translate into a bias in the synthesized library towards the target of interest and often results in chemical redundancy.

Another consideration is that reagent "availability" status has different meanings to different chemical suppliers. The variability in delivery times for compounds can cause delays of the order of months to high-throughput synthesis experiments. In such circumstances, it is often easier to simply truncate the experiment by eliminating the candidate compounds corresponding to the missing reagents. By truncation one makes the sometimes dangerous assumption that the eliminated compounds do not provide any useful information when in fact they might or, worse still, one runs the risk of overlooking leads. To this end, there are numerous anecdotal examples of company X finding a drug that was overlooked by company Y perhaps for synthesis reasons. However, practical issues dictate that the line has to be drawn somewhere since experiments cannot wait indefinitely for the "available" reagents to be resynthesized by tardy suppliers.

Since it is not possible to make all compounds, one must always be pragmatic. The goal is to make the best use of what is available, using accessible chemistries that can generate compounds to the degree of quality that the testing regiments demand plus access to sources of readily available reagents.

One more obvious factor must not be overlooked is the availability of relevant biological target information where it exists. If good target-related information has been collated (perhaps in the form of a pharmacophore or structure–activity trend), a high-throughput chemistry experiment, particularly using molecular design methodology, can be designed to utilize this information perhaps more effectively than with traditional synthetic methods. Researchers at Eli-Lilly recently reported a combinatorial chemistry experiment where a CNS lead structure had led to an optimized compound that was progressed to clinical trials within approximately 12 months.[3] In essence this lead follow-up process involved "exploding" and testing new analogues around a given lead, followed if necessary by some more lead refinement studies.

In concert with target-related information often comes the need to make compounds for which constituent reagents are simply not available. In this lead-following scenario one may be required to consider synthesis of appropriate intermediates or chemical components in order to generate relevant starting materials. More often than not, such preceding syntheses are not amenable to high-throughput synthesis and can be costly and time-consuming. One must therefore have sufficient confidence in the target information to justify the additional costs.

Where there is little prior information relating to the target or to the pharmacophore—the more predominant scenario in most drug discovery programs—it is essential to specify some criteria for the compounds being synthesized in the high-throughput chemistry experiment. This is necessary to justify the synthesis of a few thousand compounds, instead of a billion as is the case for a diamino scaffold library. Thus, it is important to cut down the number of potential compounds to be synthesized to a realistic number using rational selection criteria that preserve, as far as possible, all the chemical information in the full set yet simultaneously filter out irrelevant functionalities or properties.

For convenience, we suggest that these high-throughput lead-finding and lead following experiments fall largely within three high-throughput chemistry design paradigms. These are:

1. *Lead discovery* where libraries of compounds are designed to maximize the chances of detecting new leads either within the defined bounds of a given therapeutic area or spanning as wide a range of bio-diversity as possible.
2. *Lead explosion* where a lead or leads are followed up by an analoguing experiment designed to probe the neighboring (bioisosteric) property space with the intention of finding active analogues and detecting structure–activity trends.
3. *Lead refinement* where sufficient information has been accumulated to use (Q)SAR-based design of compounds to achieve higher (optimal) efficacy and/or better property profiles.

Of course, the activities described in these "paradigms" are not new to the pharmaceutical and agrochemical industries. However, in the past the distinction between the three has often been less clear owing to the stepwise nature of medicinal chemistry, where new leads either came from unrelated analogue synthesis programs or from natural product broths. HTS and combinatorial chemistry enforces more rigorous boundaries between these paradigms by the nature of their scale and array-based form. In essence, these new technologies typically demand separate experiments which are formally designed to address lead discovery, lead explosion, or lead refinement requirements.

The design of compounds within each of these paradigms is discussed in the following sections in the context of the Optiverse library,[4,5] utilizing a design process developed at Tripos Inc.[6] and subsequent synthesis done at MDS Panlabs.[7] The process is also similar to that employed in Tripos' LeadQuest library and one may infer some similarities to other synthesized libraries involving preselection of reaction products.

III. LEAD DISCOVERY: DESIGNING A LIBRARY FOR FINDING NEW LEADS

The design of lead discovery libraries is discussed here largely in the context of work carried out at Tripos in association with Panlabs though reference is also made, where appropriate, to other groups.

One collection of chemicals that has been tailored for the discovery of new and novel leads is the Optiverse library. The design process does not attempt to be specific to biological targets or therapies but instead seeks to select reaction products from high-throughput chemistries that have properties appropriate for exploitable biological activity. At the time of this writing, 100,000 reaction products (one per well) have been designed, synthesized, and added to the Optiverse library.

A more comprehensive description of the Optiverse library design process,[4,5] discussion of some of the descriptors of chemical structure and properties,[8] and the validation methods used to ensure how well the descriptors represented molecular diversity[9] appear elsewhere. However, for clarity, some of the relevant factors associated with these are discussed in this section. Note also that explicit discussions of Panlabs' experimental chemistry are described elsewhere.[10]

To make the assertion that libraries of compounds are designed for lead discovery assumes, as far as possible, that important rules and empiricisms of medicinal chemistry are being used to guide the synthesis of compounds. There are four key guidelines to the design process. These are:

1. *Relevance*, an assurance that the compounds have properties that make them sufficiently "drug-like" to have, in the event of activity being found, a low probability of unacceptable pharmacokinetics, toxicity, or other undesired trait.
2. *Diversity*, a measure of the ranges of relevant chemical properties (often referred to as property space) that are spanned by the library as a whole.

3. *Nonredundancy*, an assurance that individual compounds in the library are sufficiently different from other compounds in the library to avoid sometimes costly duplication of synthetic and screening efforts.
4. *Synthetic accessibility*, an assurance that once a lead has been detected there is sufficient chemical information available to enable a lead follow-up program to proceed.

All of these criteria are important to designing a library for lead discovery, and all will be discussed in the following paragraphs and sections in the context of the work carried out to develop the Optiverse and LeadQuest libraries.

A conceptual image that underlies the design process used for the Optiverse library is given in Figure 1. In this figure, all chemical properties that are of relevant to biological activity are reduced idealistically to two dimensions, corresponding to the x and y directions of the plot. Thus any compound can be represented as a point in this x-y space. In Figure 1a, a hypothetical corporate library shows significant clustering of compounds (represented as hexagons) with the conse-

Figures

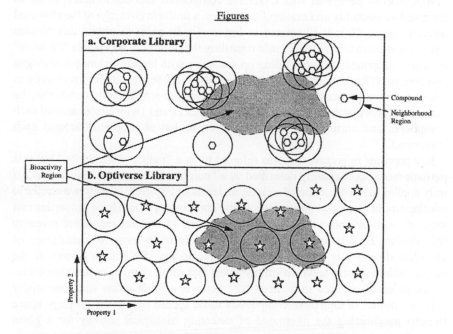

Figure 1. A simplified plot of the distribution of molecules (symbols) in an idealized two-dimensional chemical property space. The upper plot (**a**) represents a database consisting of discrete sets of close analogues. The lower plot (**b**) represents the Optiverse library. (Adapted from ref. 4. Copyright 1996 The Society for Biomolecular Screening, Inc.).

quence that there are large areas of property space that are unsampled. This is a view that many companies take or have taken of their own corporate databases. This situation is likely to arise where exhaustive analoguing has taken place in the pursuit of historically important targets. While these databases have undoubted value particularly in areas associated with the historically important targets, the barren regions of property space make them far from ideal for discovering new leads.

An awareness of apparent large gaps in their databases, often measured by conducting a "diversity audit" study, have led companies to inject new, often acquired, compounds with significantly different chemical properties into their corporate databases. This retrospective correction can give better coverage of property space and lessens the risk of missing areas where biological activity could be detected (indicated in Figure 1 by the dashed line). As an aside, it should be noted that not all of the thousands of dimensions of property space are relevant to biological activity. Martin,[11] for example, suggests that 16 properties are required to describe the side chains of peptoids. The relevance (and validity) of the descriptors of properties used in the Optiverse design are discussed later.

Rather than retrospectively correcting an existing library of compounds, the design process associated with Optiverse compounds sets out to make, as far as chemical accessibility and availability dictates, a uniform coverage of the chemical property space. This is illustrated in Figure 1b. In order to obtain this blanket coverage, decisions have to be made regarding the minimum acceptable "distance" between compounds to avoid filling up the space with largely redundant analogue compounds. ("Redundant" is used in a lead discovery sense, but not necessarily in a lead follow-up process, such as analoguing, where similar compounds may be desirable.) This distance is represented in Figures 1a and 1b by circles around each compound, and corresponds to an exclusion region of properties around each compound.

In a previous reference, this was referred to as a "radius of similarity," but it is perhaps more appropriately described as a "neighborhood region," since a radius only applies to the depiction of a circle. In terms of Figure 1, for any two circles to overlap would mean that the corresponding compounds would have properties that were too similar and would thus be considered to contain redundant property information. Exactly what these distances relate to in terms of descriptors of chemical diversity is discussed later. Suffice to say, for Figure 1, use of the neighborhood principle, which simply states that small changes in structure correspond to small changes in bioactivity,[12] permits a finite rather than essentially infinite number of Optiverse compounds to be spread out across property space thereby maximizing the likelihood of detecting biological activity for a given expenditure of resources.

All Optiverse compounds are designed and synthesized using reactions that have been pre-validated for high-throughput chemistry and in terms of available reagents. High-throughput synthesis protocols are used to combine reagents in single- and (predominantly) multistep reactions, utilizing protection/deprotection chemistry where necessary. As is common in high-throughput chemistry experiments, "scaffolds" or "cores" are used to generate classes of compounds which are considered to be biologically important and which give rise to sufficient numbers of diverse structures. An ethylene diamine scaffold, for example, is relatively common in drugs; the NCCN fragment appears in more than 20% of drugs. While some of the reaction protocols remain proprietary and thus cannot be discussed here, several "simple" reactions such as amidation, esterification, epoxidation, and reductive amination[13] are also utilized. Experimental details pertaining to the high-throughput synthesis and quality control testing of Optiverse compounds is beyond the scope of this review, but are described in detail elsewhere.[10]

The assertion that the use of such straightforward chemical reactions gives rise only to simple and obvious compounds is not at all the case, in the experience of the authors. Despite depending largely on the chemical suppliers for reagents, the complexity of these starting materials alone can allow access to some surprisingly complicated structures, even via single-step reactions. Furthermore, one has only to compare Optiverse molecules with databases of drugs to see that it is possible to generate close analogues or, more rarely, the identical structures from commercially available starting materials and high-throughput chemistries. (At the time of writing there are around half a dozen "accidental drug" structures in Optiverse.)

A point made earlier is worth restating here: that is, our dependence on the availability and price of commercially available reagents is more pronounced in the realms of high-throughput chemistry, and will have a more profound influence on the regions of property space that are explored. In the search for new leads, we must at least have an awareness that we are looking largely in regions of property space that are shaped and bounded by what is available and what is affordable. In terms of Figure 1b, this translates into gaps in an otherwise more uniformly covered property space.

Compounds are added to the Optiverse library on a reaction-by-reaction basis. A scheme illustrating the overall design process for a reaction involving the coupling of two groups of reagents to a series of scaffolds (e.g. diamines) is given in Figure 2. This process differs slightly to that reported previously[4] in that all relevant reaction products are now utilized in the design, rather than carrying out a formalized preselection of reagents first. This is possible because a new and proprietary tool called ChemSpace[14] which is able to build and process libraries of molecules in many orders of magnitude faster than conventional database software. In essence, from a knowledge of the chemistry in each reaction and the starting materials, ChemSpace allows the definition of a virtual library of candidate molecules that are all synthetically accessible.

Our design process calls for ab initio specification of all reaction products for direct use in subsequent molecular selection (diversity) processes. Other approaches base the selection of compounds for synthesis solely on reagent properties. Thus the diversity of the library is based on reagent properties rather than those of the final products. Although this is inherently easier to carry out it can detrimentally

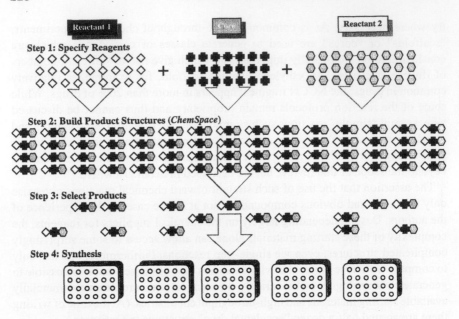

Figure 2. A scheme depicting the Optiverse library design for a two-step reaction, where symbols denote generic reagent classification.

affect the quality of the library. Reagent-based versus product-based selection is discussed further later in the chapter.

The first step in the Optiverse design process is to collect all the available reagents within each class (symbols represent reagent classes in Figure 2). Although the numbers of available reagents amassed can be in the tens of thousands, there are fewer after unsuitable and unavailable materials have been filtered out. Unsuitability is judged mainly in terms of structure (incompatible groups, undesired chemical reactivity, inappropriate groups) and molecular properties (excessive lipophilicity and molecular weight), though the options to restrict other properties such as reagent price and availability are also possible.

In step two, the reagents are used by ChemSpace to build all possible reaction products within the bounds of the defined chemical equation. This defines virtual libraries containing hundreds of thousands to hundreds of millions of synthetically accessible molecules. In the third step, a number of operations are carried out. First, ChemSpace ensures that the molecular weight and calculated log P values (e.g. CLOGP[15]) are within acceptable thresholds, viz. < 750 AMU and $-2 <$ CLOGP $<$ 7.5, respectively. Studies have shown that most published drugs and drug-like compounds lie within these bounds.[4] Subsequently, molecular diversity calculations are undertaken to select a representative diverse subset from the virtual library using both 2-D and 3-D descriptors. The subset is then used as a basis for the construction of the synthesis arrays, denoted as step four.

Clearly a critical factor in the design process is the type of compound selection process (molecular diversity) being carried out and indeed the selection of molecular descriptors. The 2-D descriptors adopted are Unity 2-D molecular fingerprints which are part of UNITY software.[16] These encode the presence of all molecular fragments of length 2–6 atoms contained within each structure in the form of "bits." In total, there are 988 bits that may be set by molecular fragments, and the distribution of these bits gives a fingerprint encoding of all structures. The 3-D descriptor used, called "topomeric field," in essence measures the shape of the molecules in a fashion similar to that of CoMFA.[8, 17–19] Both descriptors are utilized simultaneously via a normalizing factor. Discussions of neighborhood properties in this chapter will be restricted mainly to the 2-D fingerprints, though equivalent analyses for the topomeric fields was also carried out.

Our approach to selecting a diverse subset is based on utilizing a minimum similarity between each molecule and all other molecules in the virtual library. For the 2-D fingerprints, the similarity is measured by a Tanimoto coefficient[20] which measures similarity on a pair-wise basis. A Tanimoto coefficient for any pair of molecular structures lies in the range of zero (dissimilar) to one (similar). It is defined as the ratio of the number of common bits (in this case molecular fragments) set in two molecules divided by the number of bits set in either.

It is on the Tanimoto scale that we choose to estimate our neighborhood regions corresponding to the circles in Figure 1. Studies of 20 datasets drawn from recent references (listed in ref. 9) were used in conjunction with an array of molecular descriptors to test the principle of neighborhood behavior, whether small changes in biological activity between pairs of compounds produce only small changes in molecular descriptor.[12] The aim of these studies was to validate the choice of descriptor used to measure molecular diversity and uniqueness.

The studies found that 2-D fingerprints ranked among the best at reproducing this neighborhood behavior. This is illustrated in Figure 3 for one of these datasets[21] where pair-wise activity differences are plotted against 2-D fingerprints "differences," Tanimoto coefficients (left plot) and against random number differences (right plot). While the right plot exhibits no apparent relationship between the descriptor and the activity, neighborhood behavior is clearly evident for the fingerprints. The left plot of Figure 3 clearly shows that a lower right triangular subdivision of the graph is heavily populated while the upper left triangular region is almost empty.

Figure 3 indicates that large changes in bioactivity comparing two compounds with dissimilar activities are produced only by larger differences in structure (numerically smaller Tanimoto coefficients) only. (Note that the direction of the x-axis on this graph is reversed compared with others in this paper.) However small changes in bioactivity (comparing two compounds with similar activities) can result from smaller or larger changes in structure (larger or smaller Tanimoto coefficients,

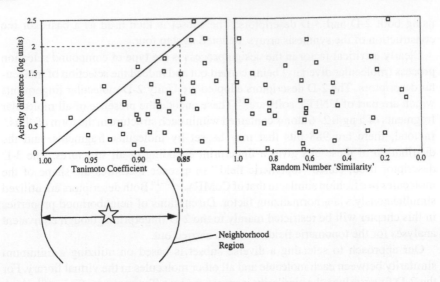

Figure 3. Two plots to illustrate adherence to the neighborhood principle. The left hand plot compares pair-wise activity differences with the corresponding Tanimoto coefficients. The right hand plot compares pair-wise activity differences with differences between random numbers assigned to each molecule.

respectively). Large structural differences that apparently induce minimal differences in activity are consistent with the fact that some structural alterations may not be part of or affect the pharmacophore.

Calculation of the slope in the left hand Figure 3 plot is described in ref. 5. In essence, it corresponds to a slope which maximizes the density of points in the lower right triangle (strictly a trapezoid[5]). This plot also indicates that for this dataset an activity difference of ca. 2 log units corresponds to a Tanimoto coefficient of 0.85. While it was found that 2 log unit differences in activity from different biological tests in the literature datasets gave rise to significantly different activity and descriptor plots, it was clear overall that an activity difference of ca. 2 log units corresponded roughly to a Tanimoto coefficient of 0.85. This corresponds to our neighborhood region of similarity, depicted by the circles in Figures 1 and 3. Hence this is our measure of structural uniqueness: any two compounds with a Tanimoto coefficient of >0.85 are considered nonunique, thus one must be discarded.

The selection of a Tanimoto coefficient corresponding to a 2 log unit difference in activity between a compounds in the test studies effectively equates to a resolution of our bioactivity space. We consider a 2 log unit "resolution" to this space to be ideal for searching for lead structures. Note that two independent research groups arrived at similar Tanimoto values in slightly different studies.

Brown,[22] following earlier work by Willett,[23] published results that indicated that 85% of compounds having a Tanimoto coefficient of 0.85 to any active compounds are themselves active. Taylor[24] in a simulation study adopted a Tanimoto coefficient of 0.80 as a threshold to distinguish similar from dissimilar compounds and was more recently adopted in related studies by Delaney.[25]

To have chosen a bioactivity resolution of say 1 log unit in our studies would have assigned a Tanimoto coefficient of 0.90–0.95 between molecules which would in essence have halved the radii of our circles. An obvious consequence of this would have been to increase the number of possible "unique" compounds that could have been synthesized and added to Optiverse by a factor of 10 or more. However, visual inspection of a such a set of molecules suggests that a 1 log unit threshold is too permissive in a lead searching program: a pair of structures with a neighborhood region of >0.90 typically differ at a single point of substitution. Such small variations are of course more important in the context of a lead explosion program (discussed later) where we may wish to ensure that we stay within the structural neighborhood of an identified lead. In addition, a smaller sampling distance between compounds might decrease the chances of missing activity as a result of false negatives in screening. However, many times more compounds would have to be synthesized and tested to search the same property space.

Alternatively, a coarser pair-wise activity difference of say 3 log units could have been adopted. This would have given rise to a neighborhood distance of 0.75–0.80 (larger radii of the circles), which would have reduced the number of compounds in Optiverse by a factor of 10 or more. Here the distance between compounds would have been unacceptably large, and the risk of missing entire regions of bioactivity between compounds would be much higher. Thus the selection of a neighborhood region of 0.85 for Optiverse design is a compromise in the sense that it dictates the number of compounds that can be considered unique in Optiverse. However, it also places the boundary to similarity between Optiverse structures typically beyond minor substituent changes, at a point where activity differences are unlikely to be as small as 2 log units. It is noted that in analogous neighborhood principle studies, for the topomeric field descriptor we estimate a neighborhood region value of 91 kcal/mol (corresponding to a Tanimoto coefficient of 0.85 for fingerprints).

An alternative metric to describe 3-D properties of molecules is discussed by Ashton et al.[26] In their approach, a pharmacophore fingerprint is used in conjunction with conformational searching to determine possible 3-D shapes that molecules can adopt. Tools from Tripos and CDL are available to carry out this type of analysis. However as with other methods there are limitations, the completeness of conformational searching being one. Perhaps the most important limitation of the approach is that it has a tendency to pick the most flexible molecules (that set the most pharmacophore bits). In a lead discovery experiment, following up on flexible molecules can be a long and sometimes fruitless process.

It is perhaps also appropriate to mention at this juncture Pearlman's DiverseSolutions[27,28] approach to measuring diversity. Among other capabilities it defines a method, referred to as the BCUT approach, which utilize atomic charge, polarizability, hydrogen bond donors, and acceptors plus several others to describe molecules. The best of these descriptors are then selected from an analysis of the full set of compounds and used subsequently to make a diverse subset selection.

Returning to the Optiverse design process, with the 0.85 definition of a neighborhood region for each product structure, it now becomes possible to ensure that there are no two structures that are too similar, thereby eliminating all products that are essentially redundant. With this assurance of diversity and uniqueness it is now possible to complete the last operation in step four of the scheme depicted in Figure 2. Thus for each set of theoretical reaction products resulting from each modeled reaction, a blanket coverage of structures spread out across structural property space as illustrated in Figure 1 is possible. The similarity/dissimilarity threshold of 0.85 is also used to determine if any structures from the theoretical reaction products are similar to compounds that have already been synthesized and added to the Optiverse library. Typically a small percentage of structures from each designed reaction is eliminated for this reason.

It is noted that several of the filtering steps have the added consequence of removing the purely combinatorial-based form to the library. One has to balance two competing priorities: synthesis efficiency and screening value. The former makes it easy to generate large libraries, while the latter takes a more accurate measure of molecular properties by eliminating inappropriate product molecules and thus facilitates the specification of a better library.

On the other hand, high-throughput experiments optimized for synthesis efficiency are compatible with reagent-based selection and diversity. However, this approach was not favored for Optiverse or LeadQuest design mainly because preliminary tests along with independent studies by Gillet[29] showed that using only reagent properties could sometimes make a poor selection of the compounds for synthesis. An example of this is given in Figure 4. In this figure markedly different commercially available aldehydes and amines can react in a reductive amination experiment to give identical products. It is therefore possible to couple pairs of dissimilar reagents to give products that are similar or even identical. Thus reagent-based, while ensuring the appropriateness of reagents, cannot guarantee selecting the best reaction products.

Product-based selection can also have difficulties. Notably, to design an idealized library might require synthesis of large numbers of "singleton" compounds where the advantages of high-throughput array-based synthesis are lost. To overcome this, reagents or intermediates have to be arranged into smaller groups that facilitate the definition of synthetic subarrays whereby, for example, a subset of aldehydes react with a subset of amines. Synthetic efficiency is determined by the smallest array size that one considers practical. For single step chemistries, this is all the way down to 1-by-1 combinations. For multistep chemistries, where time-consuming synthesis of intermediates may be necessary, larger arrays such as 1×8 or greater are often more practical.

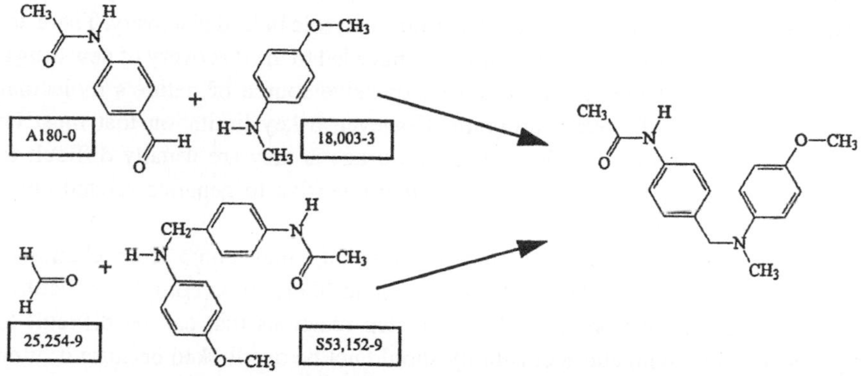

Figure 4. An example of the use of diverse reagents resulting in the synthesis of the same product (supplier IDs are given in the boxes).

Another key element in the design of the Optiverse library is that the selection of molecules for synthesis is based on a similarity hypothesis. This is in essence the antithesis of diversity-based selection which is in part why we have used the term "diversity" sparingly in this chapter. Most diversity approaches need to define or at least estimate how large the associated chemical universe is prior to selecting compounds. Those selected first are typically "chemical oddball" compounds since they are taken from the extremes with later selections representing more "reasonable" compounds. This can complicate the library design process since atypical compounds may be more difficult to work with or may result in undesired biological activity. Clark suggested a means to bias the selection of compounds towards the middle of cloud of molecules through a method called OptiSim.[30] The method makes selection of compounds on the basis of randomized subsets which are less likely to choose from the extremes than normal diversity methods and thereby gives rise to better libraries.

The scope for expansion of the Optiverse library is not based on defining the boundaries to a chemical structure space. Rather, application of the neighborhood principle ensures that compounds that are not too similar to each other are chosen. Conceptually, the Optiverse and LeadQuest libraries grow outwards at the dictate of synthetic feasibility and reagent availability factors rather than grow inwardly by filling interstitial spaces between compounds as is implicit in many diversity based methods. We believe this to be a more solid foundation for synthesizing compounds libraries. Medicinal chemistry, after all, is founded on describing small similarities between analogues rather than large dissimilarities between unrelated compounds.

Synthesized libraries of compounds such as Optiverse are not the only source of new leads. Natural products also play an important role in lead discovery. There are numerous examples where natural products have led to the discovery of new drugs. It is also clear that they represent an inexpensive source of nature's molecular diversity in a lead discovery program. However, a key limitation that must be overcome if leads are to be developed into drugs is they are usually difficult to synthesize, thereby making it more resource-intensive to generate related compounds with better properties.

Another important source of chemical diversity may come from chemistry research programs in academia. These are again likely to be complex structures more often resulting from extended multistep reactions that can be difficult to reproduce. Thus, synthetic accessibility should not be overlooked because it often determines how easily a lead can be followed up. The importance of the design of libraries of compounds to the lead follow-up process is discussed in more detail in the following sections.

The overall aim of a lead discovery experiment is to acquire new leads by screening an appropriate set of compounds. Thus, we want to get into the "neighborhood" of property space where detectable activity is located. A design process must therefore increase the probabilities of finding a lead by taking into account what is already known about those properties when constructing a library of compounds. Alternatively, where no specific property information is available, the design process needs to cover relevant regions of property space as efficiently as possible. Generally, it is improbable that initial leads found in a lead discovery program will have properties directly suitable for drug development, but this won't be known until we have examined more fully the appropriate region of property space. Usually this is done via synthesis of compounds containing functionalities that are related to the lead. Such explosion of compounds around a bioactive starting point is discussed in the next section.

IV. LEAD EXPLOSION: FOLLOWING UP ON A LEAD COMPOUND

For many organizations actively involved in lead discovery and development programs, the rate-limiting step is not always finding hits in HTS (or other testing protocols) or converting these to qualified leads. Informal discussions indicate that pharmaceutical and biotechnology companies have numerous leads that have proven difficult to follow up. One of the main obstacles is being able to convert the lead compound into a set of (bioisosteric) analogues—being able to develop a suitable synthetic route to the lead compound and related structures, of course, without even the guarantee that bioactivity in these analogues will be preserved or enhanced.

Since Optiverse compounds are all synthesized rather than simply acquired from various sources, synthetic information from at least one synthetic route relating to each structure is always available and becomes a valuable asset in following up on leads. In such circumstances, one may be in a position to begin synthesis of lead analogues almost immediately, particularly if available reagents are well suited to the analoging process.

Perhaps more important from a lead follow up perspective is that Optiverse compounds represent the proverbial "tip of the iceberg" in that only a small proportion of compounds that could have been synthesized are actually synthesized. A large proportion of potential compounds are not considered for synthesis due to their being too similar to other synthesis candidates or compounds already in the library. Even within a given reaction set there can be upwards of a thousand alternative molecules for every Optiverse compound, generally involving minor substituent variations. There is a good chance therefore that some of these may be much more active than the Optiverse compounds first found to exhibit the desired bioactivity.

To look for these alternative molecules one has to, in a sense, reverse the neighborhood region principle. Instead of identifying only those structures that have large enough dissimilarities from one another, one searches for structures that are within some limit of similarity. In terms of 2-D fingerprints, structures could be identified that have Tanimoto similarities greater than 0.85 to known active compounds. In terms of the depiction of the Optiverse process in Figure 1b, alternative compounds would be sought inside the circles associated with Optiverse compounds rather than restricting ourselves to the more distant compounds outside these circles.

This explosion of compounds around a given lead or leads is depicted in Figure 5. In this figure, we have essentially zoomed in to the "interesting" region of property space where bioactivity has been detected. From a knowledge of the active compounds, we are thus in a position to generate large numbers of alternative structures for each active compound. Experience has also shown us that widening the neighborhood region (to Tanimoto values of say 0.70) around each active lead is sometimes prudent since it allows the inclusion of a wider variety of chemical structure, particularly for patent filings.

Keeping track of all possible compounds that could have been synthesized from more than just a single reaction is not necessarily an easy task. Some sort of repository is required to process potentially billions of molecules. We do this in terms of a virtual database defined using ChemSpace which currently contains structural and property information relating to more than a trillion of such reaction products. Coupled to this, ChemSpace provides an ultrafast search engine that allows us to carry out similarity searching in realistic time periods, thus enabling us rapidly to identify potential synthesis targets for analoguing programs.

The lead explosion process works by comparing the structures of the lead compounds with those in ChemSpace followed by compiling a list of ChemSpace

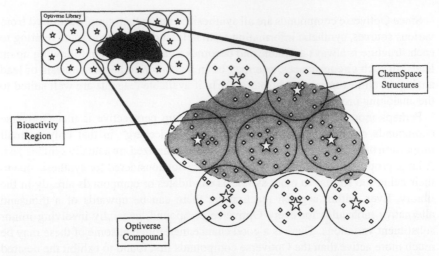

Figure 5. A depiction of lead explosion where Optiverse compounds (stars) are supplemented by structurally similar ChemSpace structures (diamonds).

structures that are similar to the leads. Two-dimensional fingerprints and topomeric fields, which have been shown to obey the neighborhood behavior rule,[11] are especially valuable in this operation. It is also possible and sometimes desirable to bias results towards the biological targets from which the leads were detected. Biasing may include searching only for structures that contain specific functionality, applying intergroup geometric distance constraints, and restricting the range of CLOGPs. Additionally, different similarity studies can be carried out in terms of the same compounds but against different targets, especially where selectivity is important. Even after applying such properties to further restrict the virtual library compounds being considered, the number of similar molecules within the neighborhood regions of actives may still number in the thousands or millions. It may then be necessary to use molecular selection calculations once again to choose one or more representative subsets.

Frequently, it is desirable to carry out lead explosion studies in a series of successive steps whereby the information generated from earlier steps is incorporated into the design and selection of subsequent subsets. One major obstacle associated with this feedback approach to lead analoguing is the dependence on reagent ordering and delivery, since typically all reagents within a synthesis array are required before synthesis can commence. Realistically, the waiting time of for all reagents to be delivered is 4 to 6 weeks, but can be even longer for some suppliers. Longer waiting times are also realized where starting materials have to be synthesized in-house.

Since a lead explosion program involves a much more intensive examination of a localized area of property space of established relevance, it gives one a better chance of obtaining a compound with optimal properties to progress to development as was the case for researchers at Eli-Lilly.[3] At the very least, with the assumption that the properties and descriptors used to select these compounds are valid, HTS or perhaps other more quantitative testing may reveal activity variations that permits the definition of QSAR and more complete pharmacophore maps.

A lead explosion experiment is a knowledge-generating experiment that utilizes high-throughput chemistry methods rapidly to generate lead-related compounds. However, the fact that commercially available reagents are used in conjunction with reaction protocols amenable to high-throughput synthesis must not be overlooked. This does limit the areas of property space in a lead explosion experiment to those whose starting materials are most immediately accessible. While this may permit us to find good compounds, there is still a realistic chance that even better compounds lie somewhere in property space that is is not immediately accessible to a lead explosion or other high-throughput chemistry experiment. To remain in the domain of high-throughput chemistry, to access these areas of property space, we must consider synthesizing the appropriate starting materials or defining reaction protocols for new chemistries. The latter may require that we adopt more a more stepwise synthesis and testing procedure that rests in the domain of conventional medicinal chemistry. This takes us to a third design paradigm–lead refinement.

V. LEAD REFINEMENT: HONING THE PROPERTIES

This section need not be long since it in essence represents a paradigm that has driven pharmaceutical and related research for many decades—the premise that small changes in structure induce proportional changes in bioactivity in a logical and hence exploitable way.[12]

Figure 6, which propagates the same theoretical design as in Figures 1 and 5, illustrates how a lead refinement process would relate to leads generated from a lead discovery program and compounds synthesized in a subsequent lead explosion experiment. The contouring of bioactivity is used to illustrate the point that more structure–activity information has been generated from the lead explosion experiment and indeed that better compounds can be obtained.

In a lead refinement paradigm, one relies on empiricisms that are sometimes formalized in models such as QSARs and pharmacophores, or sometimes "encoded" in the mind of an experienced medicinal chemist. The compounds made in a lead refinement experiment tend to be even more focused on a given region of property space, and this may require different chemistries in order to access. The types of structural modifications suggested to be beneficial may not however be amenable to high-throughput chemistry, nor may the starting materials be available.

Of course, there may be factors other than efficacy that become increasingly important at this stage, including selectivity and bioavailability. Also, the leap from

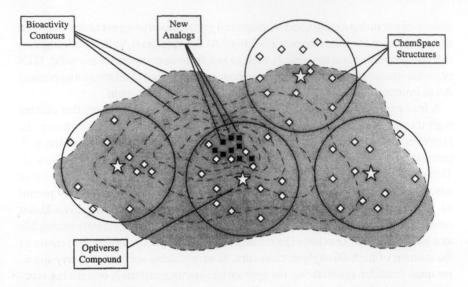

Figure 6. An illustration of lead refinement where new analogue compounds (squares) are specifically made from a knowledge of structure–activity to obtain the best properties. Optiverse compounds are represented by stars and ChemSpace structures by diamonds.

in vitro to in vivo systems often occurs here, which in itself can have a profound effect on the types of compounds designed and indeed the directions of the research project. Thus, since the numbers of compounds will typically be smaller at this step and may involve a number of distinct synthesis routes, lead refinement may well proceed via more conventional synthetic chemistry and not by means of high-throughput synthesis methods.

We do not therefore foresee combinatorial chemistry or other variants of high-throughput chemistry as a method universally to displace more conventional "round-bottom flasked" chemical synthesis. Both have a role to play in the discovery of new drugs. As far as molecular design is concerned, in a lead refinement program emphasis generally moves away from the construction of libraries to the modeling of individual molecules or molecular interactions.

VI. DESIGNED LIBRARY COMPARISON WITH CURRENT DRUGS

We have asserted that explicit design of compound libraries will enhance one's ability to discover new leads of value in pharmaceutical and related research. However, categorical proof of whether or not this is the case is difficult to present.

Simply finding leads from HTS or other assays is insufficient evidence since it is conceivable that almost any reasonably sized collection of compounds can yield active compounds. Moreover, leads from a designed library can improve the probability of finding bioactivity but will not necessarily give rise to compounds of higher efficacy or in larger numbers than those from other libraries.

While proprietary leads with <10 nm activities in Optiverse have been detected for client companies by Panlabs,[7] the deliberate lack of bias in Optiverse towards any specific target or therapeutic area makes it unlikely that the highest levels of activity will be found.

Stated briefly, the Optiverse library has been designed utilizing available reagents and reactions amenable to high-throughput synthesis, to produce structures that are similar but not too similar (for patent reasons) to known drugs. Figure 7 shows two distributions corresponding to comparisons between Optiverse and the Patent Fast Current Drugs[31] (shaded) and between Maybridge compounds[32] and Current Drugs (dotted line). Each distribution represents the Tanimoto similarities of the nearest-neighbor compound in Optiverse or Maybridge to Current Drugs. It is evident that the Optiverse/Current Drugs distribution is centered around 0.65 (indicating that on average the Tanimoto similarity for an Optiverse compound to a Current Drug compound is 0.65), while the Maybridge/Current Drugs distribution lies at ca. 0.50.

Since many of the compounds in Optiverse are based on Maybridge reagents which are used extensively in many discovery research programs and not just as

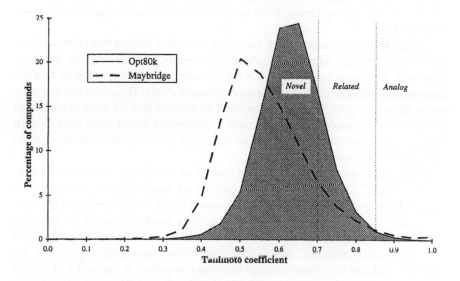

Figure 7. Similarity plot comparing the distribution of the nearest Optiverse compounds to Current Drugs[31] (shaded) and Maybridge[32] compounds to Current Drugs.

starting materials, it is evident from this plot that employing a rational design, such as is used for Optiverse can result in libraries of compounds that are inherently more drug-like in nature. We suggest, therefore, that such a library would have a higher probability of finding new leads, not just in the therapeutic areas to which existing drugs belong, but also in new areas. Also, we suggest that purely random libraries of compounds, if such libraries exist, would likely yield fewer or perhaps no good leads in screening. This assertion is based on the assumption that too many of the compounds would contain molecular structures and have properties that would be irrelevant to drug discovery.

The percentage of Optiverse compounds above the 0.85 neighborhood threshold is around 0.5% (~400 compounds) which is a relatively small part of the library. Arguably, one would not want to have too many compounds this similar to known drugs for patent reasons, as stated earlier. This makes the Tanimoto range below 0.85 more interesting for the discovery of new leads since it suggests that Optiverse compounds are sufficiently different to have exploitable properties. By inspection of pairs of nearest neighbors in the range of 0.70–0.85 (~4000 compounds), we find that structural differences fall roughly into two categories: molecules with the same backbone structure but with different pendant groups attached, or molecules have differing backbones but with the same pendant groups.

We consider that Optiverse compounds for which Tanimotos are <0.7 to the Current Drug structures will have structural differences that are larger and more significant. However, it should be noted that the structural building blocks that define these Optiverse compounds are no different from many drug structures. These differences arise largely from significantly different arrangement of functional groups.

A final comparison study pulled four drug structures chosen from thousands of possibilities from the database for which there were Optiverse neighboring structures with Tanimotos >0.7 using an arbitrary search query that specified that the drug structures should contain two or more nitrogen atoms. These four comparisons are given as examples in Figure 8. It should be stressed that the pairs of structures in Figure 8 do not represent the most similar Optiverse/Current Drug pairs, though they are representative of the types of structural similarity that corresponds to Tanimotos in this range.

Clearly the structural similarity to a class III anti-arrhythmia compound developed by Searle[33] is very high, differing only by a p-methoxy group, reflecting the highest Tanimoto in Figure 8. It should be noted that the corresponding Optiverse compound was made by a single-step amidation reaction, one of the simplest Optiverse reactions. A subsequent scan of ChemSpace indicated that the actual anti-arrhythmic compound was indeed part of the virtual library defined for Optiverse and that the corresponding Optiverse compound in Figure 8 was chosen as representative of the neighborhood. This implies that a follow-up lead explosion study would have identified the Current Drugs-quoted structure. Discussions with Searle chemists[34] who developed these compounds indicated, not surprisingly, that

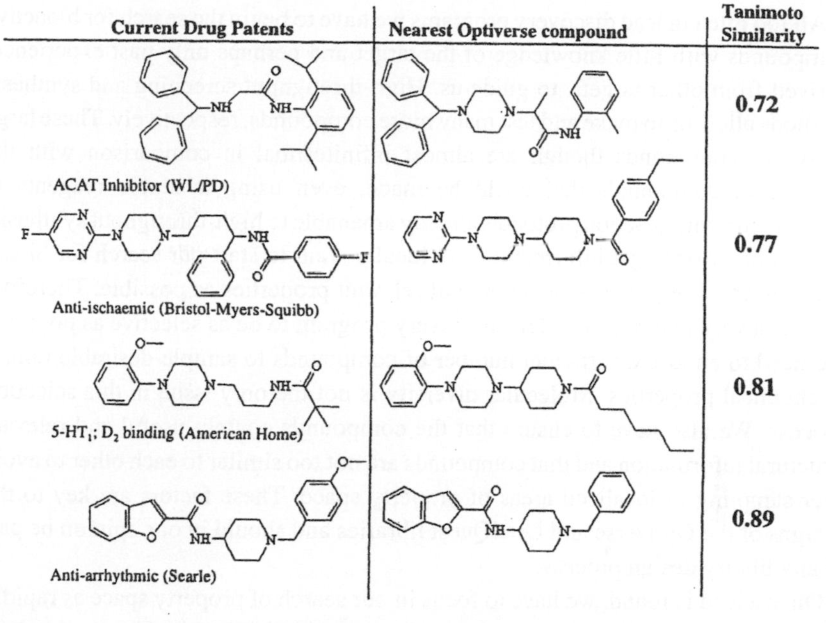

Current Drug Patents	Nearest Optiverse compound	Tanimoto Similarity
ACAT Inhibitor (WL/PD)		0.72
Anti-ischaemic (Bristol-Myers-Squibb)		0.77
5-HT$_1$; D$_2$ binding (American Home)		0.81
Anti-arrhythmic (Searle)		0.89

Figure 8. Examples of four "typical" drug/Optiverse comparisons for Tanimoto coefficients in the range 0.70 to 0.90.

they had indeed made the Optiverse compound as part of their anti-arrhythmic research and that it too was active, which indicates that the neighborhood principle using 2-D fingerprints is valid for at least some of these anti-arrhythmia compounds. This was also evident from their issued patent.[33]

VII. CONCLUSIONS

A library design process used in association with high-throughput chemistry and HTS sets out to make knowledge- and experience-based decisions regarding the selection of compounds. In each of the lead discovery, explosion, and refinement paradigms, an efficient and effective design, using good estimates of important properties, allows us to look at a far greater number of potential molecules than it is possible to synthesize. In essence, a good design process should set out to bias the selection of compounds towards those that are appropriate for screening, taking into account synthetic limitations. We assert therefore that compound library design is an essential component that must not be overlooked in any high-throughput experiment which seeks to facilitate the progression from lead discovery to clinical candidates.

All too often in lead discovery programs we have to begin the search for bioactive compounds with little knowledge of the target and perhaps only past experience, derived from other targets, to guide us. High-throughput screening and synthesis methods allow us to make and test many more compounds, respectively. These large arrays of compounds though are almost infinitesimal in comparison with the number of compounds that could be made, even using available reagents in conjunction with reaction protocols already amenable to high-throughput synthesis.

With no prior target information, we ideally want to start our search for bioactivity by considering as wide a range of relevant properties as possible. Therefore there is a very real need in a lead discovery program to be as selective as possible. We need to choose a sufficient number of compounds to sample desirable ranges of chemical properties. Molecular diversity is not the only issue in this selection process. We also have to ensure that the compounds contain useful and relevant structural information and that compounds are not too similar to each other to avoid over-sampling in localized areas of property space. These factors are key to the designs of the Optiverse and LeadQuest libraries and should in our opinion be part of any library design process.

Once a lead is found, we have to focus in our search of property space as rapidly as possible to understand structure–activity relationships and perhaps build pharmacophore maps. To do this, we need to design more compounds to explore ranges of properties centered around our lead. This explosion of compounds is a knowledge-gathering process which, when the compounds are tested, should increase the understanding of bioactivity. By application of the neighborhood principle to eliminate similar molecules, any molecules selected for synthesis will likely have several hundred or a thousand similar structures that are stored in a virtual library. These often represent an ideal starting point for lead follow-up.

Lead explosion and other directed synthesis methods may uncover better compounds, but these may not be the best. Further lead refinement may be necessary, more likely using chemistries that are less amenable to high-throughput methods, to design compounds which will enhance the properties still further. This activity still lies firmly in the domain of conventional medicinal chemistry.

ACKNOWLEDGMENTS

The authors would like to acknowledge the invaluable help of R. D. Clark, R. C. Glen, M. S. Lawless, D. E. Patterson, and P. Hecht of Tripos Inc. for useful comments. The authors would like to especially thank Prof. P. Willett for helpful suggestions throughout the course of this work and for numerous valuable comments. AMF also wishes to thank his wife for her endless patience during the writing of this chapter.

REFERENCES

1. Available Chemicals Directory (ACD) Documentation, Version 95.2, MDL Information Systems, Inc., 1995.
2. Number of CAS registered structures listed as of March 1997, Chemical Abstracts Service, American Chemical Society NW Washington DC.
3. Kaldor, S. W.; Fritz, J. E.; Tang, J.; McKinney, E. R. *Bioorg. Med. Chem. Lett.*, **1996**, *6*, 3041–3044.
4. Ferguson, A. M.; Patterson, D. E.; Garr, C. D.; Underiner, T. L. *J. Biomolec. Screening* **1996**, *1*, 65–73.
5. Patterson, D. E.; Ferguson, A. M.; Cramer, R. D.; Garr, C. D.; Underiner, T. L. In *High-throughput Screening: The Discovery of Bioactive Substances*; Devlin; Wallace, Eds.; Marcel Dekker: New York. In press.
6. Accelerated Discovery Services, Tripos Inc., St Louis, MO.
7. Panlabs Inc., Bothell, WA.
8. Cramer, R. D.; Clark, R. D.; Patterson, D. E.; Ferguson, A. M. *J. Med. Chem.* **1996**, *39*, 3060–3069.
9. Patterson, D. E.; Cramer, R. D.; Ferguson, A. M.; Clark, R. D.; Weinberger, L. E. *J. Med. Chem.* **1996**, *39*, 3049–3059.
10. Garr, C. D.; Peterson, J. R.; Schultz, L; Underiner, T. L.; Cramer, R. D.; Ferguson, A. M.; Lawless, M. S.; Patterson, D. E. *J. Biomolec. Screening* **1996**, *1*, 179–186.
11. Martin, E. J.; Blaney, J. M.; Siani, M. A.; Spellmeyer, D. C.; Wong, A. K.; Moos, W. H. *J. Med. Chem.* **1995**, *38*, 1431–1436.
12. *Concepts and Applications of Molecular Similarity;* Johnson, M. A.; Maggiora, G. M., Eds.; Wiley: New York, 1990.
13. *Advanced Organic Chemistry: Reactions, Mechanisms and Structures, 4th ed.;* March, J., Ed.; Wiley: New York, 1992.
14. ChemSpace, Tripos, Inc., St. Louis, MO.
15. *Substituent Constants for Correlation Analysis in Chemistry and Biology;* Hansch, C.; Leo, A., Eds.; Wiley-Interscience: New York, 1979.
16. UNITY Chemical Information Documentation, Version 2.3, Tripos, Inc., St. Louis, MO.
17. Cramer, R. D.; Patterson, D. E.; Bunce, J. D. *J. Am. Chem. Soc.* **1988**, *110*, 5959–5967.
18. Clark, M.; Cramer, R. D.; Jones, D. M.; Patterson, D. E.; Simeroth, P. E. *Tetrahedron Comp. Meth.* **1990**, *3*, 47–59.
19. SYBYL Molecular Modeling Documentation, Version 6.1, Tripos Inc., St. Louis, MO.
20. Willett, P.; Winterman, V. *Quant. Struct. Activ. Relat.* **1986**, *5*, 18–25.
21. Uehling, D. E.; Nanthakamur, S. S.; Croom, D.; Emerson, D. L.; Leitner, P. P.; Luzzio, M. J. *J. Med. Chem.* **1995**, *38*, 1106–1118.
22. Brown, R. D.; Martin, Y. C. *J. Chem. Inf. Comput. Sci.* **1996**, *36*, 572–584.
23. Willett, P. *Similarity and Clustering in Chemical Information Systems;* Research Studies Press: Letchworth, UK, 1987.
24. Taylor, R. *J. Chem. Inf. Comput. Sci.* **1995**, *35*, 59–67.
25. Delaney, J. S. *J. Molec. Diversity* **1996**, *1*, 217–222.
26. Ashton, M. J., Jaye, M. C., Mason, J. S. *DDT*, **1996**, *1*, 71–78.
27. Pearlman, R. S.; Smith, K. M.; In *3D QSAR and Drug Design: Recent Advances*, Eds. Kubinyi; Martin; Folkers,Eds.; ESCOM. In press.
28. DiverseSolutions User Guide v2.0.1 Tripos Inc., St Louis, MO.
29. Gillet, V. J.; Willett, P.; Bradshaw, J. *J. Chem. Inf. Comput. Sci.* **1997**, *37*, 731–740.
30. Clark, R. D. *J. Chem. Inf. Comput. Sci.* Submitted.
31. Patent Fast-Alert Current Drugs, February 1995.

32. Maybridge Electronic Catalogue, July 1995.
33. U.S. Patent Number 5,098,915, March 24, 1992.
34. Russell, M. Searle & Co. Private communication.

A THEORETICAL MODEL OF THE HUMAN THROMBIN RECEPTOR (PAR-1), THE FIRST KNOWN PROTEASE-ACTIVATED G-PROTEIN-COUPLED RECEPTOR

Mary Pat Beavers, Bruce E. Maryanoff,
Dat Nguyen, and Brian D. Blackhart

Advances in Medicinal Chemistry
Volume 4, pages 245–271.
Copyright © 1999 by JAI Press Inc.
All rights of reproduction in any form reserved.
ISBN: 0-7623-0064-7

ABSTRACT

A three-dimensional molecular model of the human thrombin receptor (TR; PAR-1), a member of the seven-transmembrane G-protein-coupled receptor (GPCR) superfamily of glycoproteins and the first GPCR to have a tethered-peptide ligand (ref. 7), was constructed by using de novo computer-based modeling techniques. Because of numerous shortcomings associated with the use of the bacteriorhodopsin template for building GPCR models, we employed an alternative modeling strategy involving GPCR alignment and topological properties, as first described by Moereels (ref. 10). To initiate assembly of the seven-helix bundle for PAR-1, we relied on specific, conserved residues in the transmembrane (TM) domains, which in this instance are Asn-120 in TM1, Asp-148 in TM2, and Asp-367 in TM7. The side chains of these residues then comprise a hydrogen-bonding network that orients the three helices so that they are properly positioned to interact with each other. The energy-minimized three-helix bundle (TM1-TM2-TM7) served as a cornerstone for assembly of the complete seven-helix bundle, through systematic addition of TM3 through TM6 involving suitable orientation followed by limited translation, rotation, and tilting. The three extracellular (EC) loops were added to the energy-minimized, seven-helix construct via loop-search routines, and the disulfide bridge between two GPCR-conserved cysteines on loops EC1 and EC2 was formed using the protein manipulation software program SCULPT. Finally, part of the extracellular N-terminus, from Tyr-95 (at the extracellular end of TM1) to Asn-75, was attached to the receptor. The known structure–activity relationship surrounding the thrombin receptor agonist peptide motif SFLLRN, coupled with the results from our receptor-based, site-directed mutagenesis, were used to define and enhance our computer-generated, energy-minimized models of the ligand-bound thrombin receptor. We examined the acidic residues in the three extracellular loops (seven total: D167 and E173 in EC1; E241, D256, E260, and E264 in EC2; and E347 in EC3) to identify electrostatic interactions that could account for the ammonium N-terminus and/or the Arg residue of SFLLRN. When both EC1 acidic residues were mutated to Ala, there was no change in receptor functional responses. However, when all four acid residues in EC2 were mutated to Ala, there was a significant reduction in receptor activation in the presence of thrombin or agonist peptide. When the single acidic residue in EC3 was mutated to Ala, reduced functional activity was also observed. Contrary to published observations (refs. 20 and 24), we were unable to find an important role for Glu-260 (E260A mutant TR), whereas replacement of Asp-256 (D256A mutant TR) had a significant effect on receptor function, especially in response to the TR-agonist peptide SFLLRNP-NH$_2$ (TRAP-7). Our mutagenesis work also confirmed the importance of the peptide triad Asn-120/Asp-148/Asp-367 (viz. refs. 10 and 18) and the C-175/C-254 disulfide bridge (viz. ref. 25) in receptor function. On the basis of our studies, we suggest that the ammonium group at the N-terminus of the peptide ligand interacts with the carboxylic acid side chain of Glu-347, at the extracellular surface of TM7, and that the aromatic ring of the ligand's critical Phe residue binds within a hydrophobic pocket on the receptor defined by the side chains of Phe-182, Leu-340, Tyr-337, and Phe-339.

I. INTRODUCTION

The G-protein-coupled receptors (GPCRs) represent a large superfamily of cell-surface glycoproteins that are activated by diverse small and large molecules, including biogenic amines, peptides, and proteins; moreover, such receptors are responsible for the action of a wide variety of drug molecules.[1] Although the three-dimensional (3-D) structure of a GPCR in the presence of its endogenous ligand could be a valuable tool for drug design, experimental structures for these membrane-bound proteins are virtually unknown, for obvious reasons. Given the current inadequacies of X-ray crystallographic and NMR techniques for the structural analysis of GPCRs, computational methods have emerged as the principal alternative.

Molecular modeling of a GPCR comes under the aegis of a protein-folding problem, where the final structure happens to be membrane-bound. Such protein structures are characterized by a recurrent helix–loop–helix motif with seven transmembrane-spanning regions, presumably assembled in a seven-helix bundle topology. This view has been supported by the only available three-dimensional structures for seven-helix transmembrane proteins, those of bacteriorhodopsin (1BRD),[2] the atomic coordinates of which define seven helices without loops, and of halo-rhodopsin.[3a,b] Both of these receptors are retinal-binding, photoresponsive proton pumps whose structures were solved with substantial difficulty by electron cryomicroscopy (resolution of ca. 7–9 Å). In a recently refined structure of bacteriorhodopsin (2BRD),[4] having a resolution of 3.5 Å, one helix (helix D) was moved toward the cytoplasm by nearly 4 Å compared to the earlier set of coordinates (1BRD). Although this newer model provides an improved template for studying how retinal docks into the protein, it departs significantly from the earlier complex.

For several years, many types of GPCRs have been computationally modeled on the basis of the low-resolution 1BRD structure, despite *fairly poor protein sequence homology (under 20%) between bacteriorhodopsin and diverse GPCRs within the superfamily*. Most of these cases have involved the construction of biogenic amine receptors,[5] the largest and most well-studied subfamily of GPCRs. From site-directed mutagenesis, chimeric mutagenesis, and fluorescence labeling, the small organic ligands for these receptors have been found to bind within the transmembrane (TM) regions and, for the most part, not to interact directly with the extracellular (EC) surfaces of the receptors. In fact, most published 3-D models of biogenic amine receptors do not possess the connecting loops between the helices nor the N-terminal extension that precedes transmembrane helix one (TM1). On the other hand, GPCRs that have peptides as their natural ligands are more likely to involve extracellular surfaces, at least in their initial binding modes. In fact, a key role has been demonstrated for the extracellular domains of GPCRs, the N-terminal extension, and the three loops in binding and activation with endogenous peptide ligands.[6]

Others have used both the electron microscopy structural data for rhodopsin[3a] and models of human rhodopsin[3d,e] as templates for constructing GPCR models. In 1995, Herzyk and Hubbard[3e] constructed a template for rhodopsin that they claimed could be used as a starting point for the development of other GPCR models. This rhodopsin template was generated by using a set of geometrical and structural restraints derived from both experimental and theoretical data. The experimental data, which included 2-D information from the electron microscopy and neutron diffraction studies on rhodopsin, were used to identify the locations and tilts of the helices in the seven-helix bundle. The theoretical data, which included multiple sequence alignments and calculations of periodicity in hydropathy, were used to determine the orientations of the helices with respect to the facing of side chains either towards the outside lipid-exposed face or towards the inside buried face. Additional experimental and theoretical constraints were used as input for the generation of their initial model. This model was then refined by using a combined simulated annealing/Monte Carlo optimization technique to create a family of configurations of helices. This family was then analyzed further, helices were assigned to their 2-D images from the projection maps, and the different groups of restraints were applied to the model. They applied their methodology to the bacteriorhodopsin sequence, and found that the rms deviation between the resulting model and the EM structure for bR was 1.8 Å.

In summary, the rule-based techniques of Herzyk and Hubbard[3e] allowed for the construction of a model or structural template that could serve for the modeling of other GPCRs. This template, however, is not currently available through public databases, so it has proved difficult to assess the quality of a thrombin receptor model built according to this structural model. Furthermore, as the authors suggest, the quality of the final GPCR model built with their methodology is highly dependent on the availability of experimental and theoretical restraints. The authors allude to the mutagenesis data available for bR and rhodopsin, where the mutations directly affect the spectral characteristics of the ligand retinal. By contrast, mutations in most other GPCR systems have only indirect effects on the respective ligands. Nonetheless, Du et al.[3d] used the rhodopsin model of Herzyk and Hubbard to construct a model of the human neuropeptide Y1 receptor, the coordinates of the rhodopsin template being personally provided to Du by Herzyk.

Baldwin and Schertler[3c] also constructed a template for the transmembrane helices in the rhodopsin family of GPCRs. Their model was derived from alignment information on some 500 sequences in the rhodopsin family. Their helix packing rules were derived from the 3-D density maps of frog rhodopsin determined by electron cryomicroscopy. Since neither the Herzyk/Hubbard[3e] nor the Baldwin/Schertler models are publically available for comparison, it is not clear what the differences are between them. One notable point of controversy between the two studies lies in the helix packing between TM regions 1, 2, and 7. There are three conserved amino acid residues, one from each of these regions, the side chains of which are proposed to be in close proximity to each other. A direct interaction between two of these residues was experimentally supported for the GnRH receptor.[18] Du et al., who utilized the Herzyk-Hubbard[3e] template in their study of the neuropeptide Y1 receptor, support these interhelical interactions and claim that Asp-86 of TM2 and Asn-316 of TM7 lie adjacent to each other in their model. Baldwin and Schertler, however, do not observe this analogous interaction in their rhodopsin-family template. They noted that the α-carbon positions for the conserved Asp in TM2 and Asn in TM7 are too far apart (10.4 Å) for there to be any hydrogen bonding between the side chains of these residues. They also question the experimental findings for the GnRH receptor that support the direct interaction between the Asp in TM2 and the Asn in TM7, and suggest that the mutational findings reported by Zhou et al.[18] must be attributable not to hydrogen bonding between the side chains, but to a more complex network of other interacting side chains. Given this controversy between the two rhodopsin templates, it is difficult to say whether or not they would serve as a worthwhile basis for the construction of new GPCR models. Only when the experimental coordinates for the high-resolution structure of rhodopsin become available will this controversy be settled.

We have been interested in the thrombin receptor (TR, PAR-1), a novel type of GPCR that mediates the cellular actions of α-thrombin, a trypsin-like serine protease important in blood coagulation and hemostasis. The thrombin receptor is particularly notable because of its activation by *proteolytic cleavage* of an unusually long extracellular N-terminal peptide chain, between Arg-41 and Ser-42, *to expose a new N-terminus that bears the agonist motif SFLLRN* (TR 42-47; human sequence) (Figure 1).[7,8] Thus, the thrombin receptor is *the first GPCR identified that is activated by a protease* (protease-activated receptor 1, or PAR-1) and *that has an intramolecular mechanism of ligand activation*, via a "tethered-peptide" epitope. There are now two other members of this class of GPCRs: protease-activated receptor 2 (PAR-2),[9a] which can be activated by trypsin and possesses the N-terminal recognition sequence SLIGKV (human; SLIGRL in mice) and protease-activated receptor 3 (PAR-3), which is a second thrombin receptor and possesses the N-terminal sequence TFRGAP (human).[9b] Because of the significant role expected for the extracellular loops and the N-terminal extension in binding of the peptide ligand, incorporation of these domains is crucial for constructing a ligand-receptor

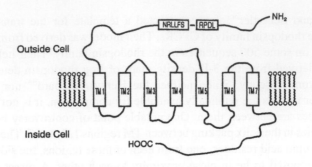

Figure 1. Diagram of the human thrombin receptor, viewed edge-on, showing the sequence around the proteolytic cleavage site (LDPR-SFLLRN) in the long extracellular amino-terminus and the seven transmembrane domains (TM1-TM7).

3-D model for the first-known thrombin receptor (PAR-1). To accommodate the positioning of the helices toward or away from the cytoplasmic side of the receptor, which would directly affect the positioning of the extracellular loops with respect to each other, we needed to develop a receptor model that addressed the packing and tilting of helices in an initial seven-helix bundle assembly. Subsequent elaboration resulted in a model containing the extracellular loops with a key disulfide bridge, a significant part of the N-terminus, and an agonist hexapeptide ligand. This paper presents our computational and mutagenesis studies directed toward establishing a useful 3-D working model of the ligand-bound thrombin receptor (PAR-1). As a consequence, we were able to identify some intermolecular interactions that may contribute to activation by the tethered ligand domain, ultimately leading to thrombin receptor-mediated signal transduction.

II. CONSTRUCTION OF A WORKING MODEL OF PAR-1

A. Model-Building Strategy and Assembly of the Seven-Helix Bundle

Initially, we attempted to build a 3-D homology model of the transmembrane regions of the human thrombin receptor from the structure of 1BRD, which is available from the Brookhaven Protein Database. Unfortunately, the suboptimal placement of several amino acid side chains resulted in severe deviations from structural standards for membrane-bound receptors with a seven-helix bundle topology (Figure 2). For example, carboxylate and ammonium groups on amino acid side chains at a mid-helix location were directed into the membrane, rather than toward the inside of the helix bundle, and the hydrophobic packing between some helices was either poor or nonexistent.

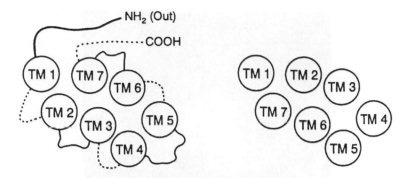

Figure 2. End-on view of the GPCR seven-helix bundle topology, showing the counterclockwise (left) and clockwise topologies, with the helices assembled in sequential order.

Therefore, we opted to employ a de novo GPCR modeling strategy, via a methodology originally developed by Moereels[10] and applied to the construction of a very reasonable model of the 5-HT2 (serotonin-2) receptor. This approach, which diverges from the bacteriorhodopsin-based homology modeling used by many other researchers, relies on specific, conserved residues in the transmembrane (TM) domains, referred to as "receptophores".[10] Three key residues, drawn from the set Asp, Glu, Asn, and Gln are present in TM1, TM2, and TM7, and are aligned such that the carboxylate and amide side chains comprise a hydrogen-bond network for holding the three helices together. For the thrombin receptor, these residues are Asn-120 in TM1, Asp-148 in TM2, and Asp-367 in TM7 (Figure 3). The energy-minimized three-helix bundle (TM1–TM2–TM7) forms a cornerstone for constructing the ultimate seven-helix bundle of the GPCR. After the TM1–TM2–TM7 triad is assembled, the seven-helix bundle is arrived at by systematically adding TM3 through TM6 with minor reorientation and energy minimization. When viewed from the extracellular surface, the helices that comprise the seven-helix bundle were added sequentially to the TM1–TM2–TM7 triad to generate a counterclockwise arrangement. Although it is not known for sure whether this counterclockwise disposition of helices exists in GPCRs, it has been widely accepted as the most probable. In this vein, Kontoyianni et al.[11] constructed models of the β_2 adrenergic receptor and proposed two general bundle classes—a clockwise and a counterclockwise model. They claimed that both models were consistent with most available experimental data, but that their clockwise model better explained the stereoselective ligand binding data. However, Schwartz and coworkers[12] pointed out that, although the cryomicroscopy structures of bacteriorhodopsin and rhodopsin do not reveal definitively the clockwise versus counterclockwise nature of the seven helices in a given bundle, much evidence favors the counterclockwise

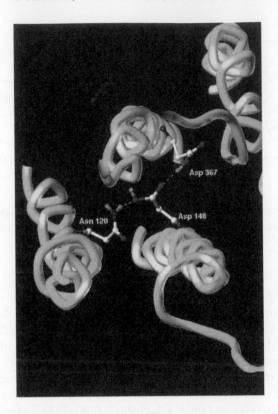

Figure 3. Key hydrogen-bonding network between side chains of the GPCR-conserved amino acids: Asn-120 (TM1), Asp-148 (TM2), and Asp-367 (TM7) for human thrombin receptor.

arrangement. They cited the work of several research groups, primarily predicated on reciprocal mutational studies, in support of both the spatial proximity of GPCR-conserved residues in transmembrane regions that are presumed to face each other and the counterclockwise orientation of helices one through seven. In our studies on PAR-1, we adopted this counterclockwise arrangement of the helices.

Protein sequence alignments of the transmembrane regions within a GPCR subfamily containing the thrombin receptor from human and other species (hamster,[13] rat,[14] *X. laevis*,[15] and mouse[16]) revealed the conservation of key amino acids in TM1, TM2, and TM7: Asn in TM1, Asp in TM2, and Asp or Asn in TM7, allowing for assembly of the cornerstone (Figures 4–6, alignments). Recently, Moereels et al.[17] identified 10 conserved key residues distributed throughout the transmembrane regions and loops in a GPCR subfamily of 113 neurotransmitter and opioid receptors. Since these amino acids were conserved throughout a large subfamily of the GPCRs, they were deemed important for architectural and structural reasons. Two of the amino acids included in this classification are Asn in TM1 and Asp in TM2. The TM7-conserved Asp (Asn in the thrombin receptor family), although not included in the Moereels' classification scheme, is present in many other peptide–ligand GPCRs. Weinstein and coworkers[18] have experimentally validated the spatial proximity of such GPCR-conserved residues in TM2 and TM7 in reciprocal mutation studies. The conserved Asn and Asp residues in TM2 and TM7 of the GnRH receptor were found to interact with each other, presumably with their side chains arranged in a hydrogen-bonding network that holds the two helices together.

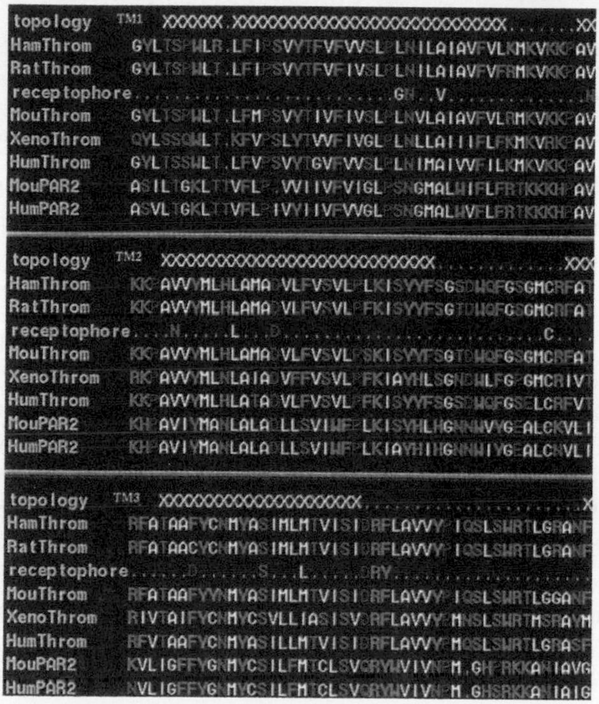

Figure 4. Amino acid sequence alignment of thrombin receptors and PAR-2s from different species along TM1, intracellular loop 1 (IC1), TM2, extracellular loop 1 (EC1), TM3, and IC2, with the highly conserved "recepto-phore" residues. Legend for amino acid color-code: blue = basic, including histidine; red = acidic; orange = hydroxyl-bearing; glutamine, and asparagine; white = aliphatic hydrophobic, glycine, and methionine; green = phenylalanine, tyrosine, and tryptophan; yellow = cysteine; magenta = proline.

```
topology    TM4   XXXXXXXXXXXXXXXXXX.....................
HamThrom      NF CLVI VMA IMGVV LLLK QT RV GLN I TCH VLN TLL G
RatThrom      NF CVVI VMA IMGVV LLLK QT QV GLN I TCH VLN TLLHG
receptophore........H.............P..................
MouThrom      NF CVVI VMA IMGVV LLLK QT RV GLN I TCH LLS LM G
XenoThrom     VMAC F I LI IA T I LLV EQTQK I RL  I TCH VLD LK LK
HumThrom      SF CLAI ALA IAGVV LVLK QT IQV GLN I TCH VLN TLL G
MouPAR2       VGVSLA LLI FLV  I LYVM QT IYI ALN I TCH VL E VLVG
HumPAR2       IGISLA LLILLV  I LYVW QT IFI ALN I TCH VL QLLVG

topology    TM5   XXXXXXXXXXXXXXXXXXXXXXXXXXXX...............X
HamThrom      GFYSYYF AF AVFFLV LI I T ICVM I IRCL SSS VA RS  KKSRALFL A
RatThrom      GFYSYYF AF AIFFLV LI I  VCY S I IRCL SSS AVA R  KKSRALFL A
receptophore............F..P..Y....................
MouThrom      GFYSYYF AF AIFFLV LIVS VCY S I IRCL SSS AVA R  KKSRALFL A
XenoThrom     DFYIVYF  FCLL FFF V FI I T ICYIGI IRSL SSS I NSC  KK RALFLAV
HumThrom      GYYAYYF AF AV FF V LI I  VCYVS I IRCL SSS AVA R  KKSRALFL A
MouPAR2       G MFNYFL LA IGVFLF ALL A AYVLM K LRSS AM HS KKR A IRLI I
HumPAR2       G MFNYFL LA IGVFLF AFL A AYVLM RML RSS AM NSE KKRKRA IKLIV
```

Figure 5. Amino acid sequence alignment of thrombin receptors and PAR-2s from different species along TM4, EC2, and TM5. (See caption to Figure 4 for more details.)

Although the GnRH receptor contains a conserved Asn in TM1, this TM1 residue was not included in Weinstein's study.

Our mutational studies with this helix triad of the thrombin receptor focused on key amino acids in TM1, TM2, and TM7 to explore the possibility for spatial proximity of Asn-120 in TM1, Asp-148 in TM2, and Asp-367 in TM7 (Figure 7 and Table 1). When Asp-148 in TM2 was mutated to Ala, a loss in receptor function occurred for both treatment with thrombin or the agonist peptide SFLLRNP-NH$_2$ (TRAP-7), possibly due to disruption of a hydrogen-bond network involving Asp-148, Asn-120, and Asp-367. However, when Asp-148 was mutated to Asn, wild-type receptor activity was observed with thrombin or TRAP-7. In this case, the amide-containing side chain of Asn could compensate for the carboxylic acid side chain of Asp, while participating in a hydrogen-bonding network. The Ala mutation for the Asn-120 residue in TM1 caused changes in receptor activity that paralleled those observed for the TM2-conserved residue in that there was a loss in receptor function in the presence of both thrombin and TRAP-7. The effect of the N120D mutant was not nearly as substantial as that of the N120A mutation, causing only a minor decrease in thrombin receptor activation in the presence of both agonists. Similar mutations in the TM7 area included D367A, which caused the same reduction in receptor activity that was noted for the TM2 mutant, and D367N,

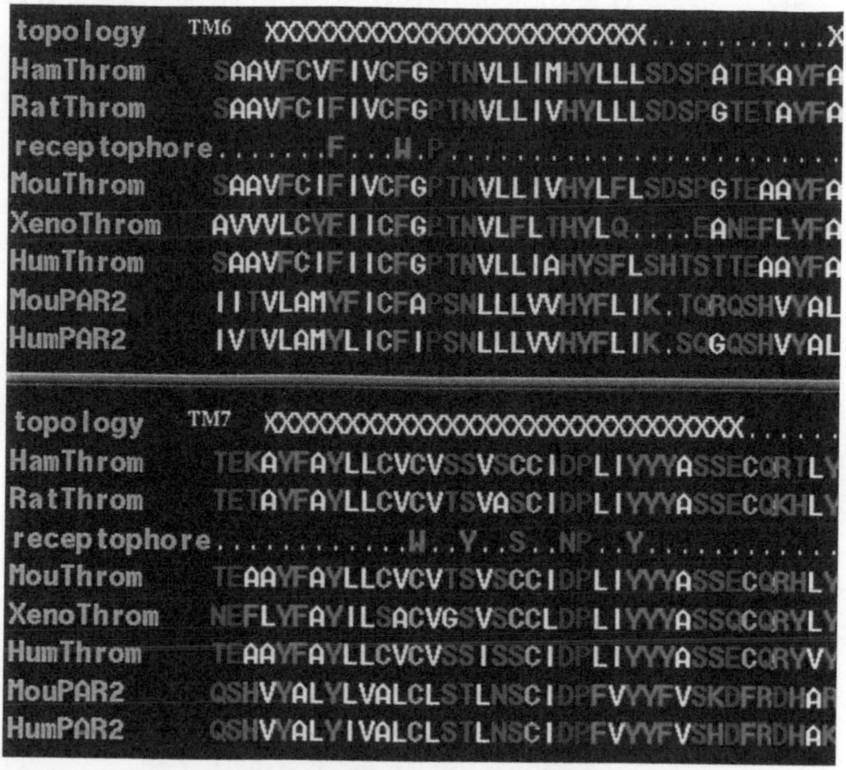

```
topology    TM6  XXXXXXXXXXXXXXXXXXXXXXXXX...........X
HamThrom    SAAVFCVFIVCFGPTNVLLIMHYLLLSDSPATEKAYFA
RatThrom    SAAVFCIFIVCFGPTNVLLIVHYLLLSDSPGTETAYFA
receptophore.......F...H.P...................
MouThrom    SAAVFCIFIVCFGPTNVLLIVHYLFLSDSPGTEAAYFA
XenoThrom   AVVVLCYFIICFGPTNVLFLTHYLQ....EANEFLYFA
HumThrom    SAAVFCIFIICFGPTNVLLIAHYSFLSHTSTTEAAYFA
MouPAR2     IITVLAMYFICFAPSNLLLVVHYFLIK.TQRQSHVYAL
HumPAR2     IVTVLAMYLICFIPSNLLLVVHYFLIK.SQGQSHVYAL

topology    TM7  XXXXXXXXXXXXXXXXXXXXXXXXXXXXXXXX......
HamThrom    TEKAYFAYLLCVCVSSVSCCIDPLIYYYASSECQRTLV
RatThrom    TETAYFAYLLCVCVTSVASCIDPLIYYYASSECQKHLV
receptophore...........H..Y..S..NP..Y.........
MouThrom    TEAAYFAYLLCVCVTSVSCCIDPLIYYYASSECQRHLV
XenoThrom   NEFLYFAYILSACVGSVSCCLDPLIYYYASSQCQRYLV
HumThrom    TEAAYFAYLLCVCVSSISSCIDPLIYYYASSECQRYVV
MouPAR2     QSHVYALVLVALCLSTLNSCIDPFVYYFVSKDFRDHAR
HumPAR2     QSHVYALYIVALCLSTLNSCIDPFVYYFVSHDFRDHAK
```

Figure 6. Alignment of thrombin receptors and PAR-2's along TM6, EC3, and TM7. (See caption to Figure 4 for more details.)

which retained wild-type activity. The double mutant D148N and D367N also caused a loss of receptor function. These results support a requirement for at least one acidic residue in the hydrogen-bonding network, similar to the findings of Weinstein.[18] Disruption of this network by individual Ala replacements also caused a loss of receptor function. Our results, coupled with those of Weinstein,[18] support the spatial proximity of Asn-120, Asp-148, and Asp-367, probably amidst a hydrogen-bonding network, and establish a foundation for de novo thrombin receptor model building, à la Moereels.[10,17]

From a procedural standpoint, the start and stop residues of each of the seven helices was first identified by hydropathy analysis by using Kyte–Doolittle[19] hydropathy indices. Each of the seven helices was then built individually, essential hydrogens were added, and the peptides were minimized with the Kollman United Atoms force field within the Sybyl software suite (Tripos Associates, Inc.). Typically, steepest descent was used initially as the minimization method (25–50

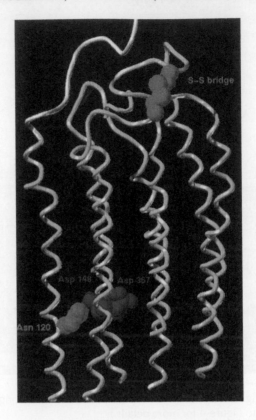

Figure 7. Spatial arrangement of the GPCR-conserved amino acids Asn-120 (TM1), Asp-148 (TM2), and Asp-367 (TM7).

Table 1. Receptor Mutagenesis Data (WT = Wild-Type Activity)

Mutants	Calcium Efflux EC_{50} (Thrombin, nM)	Calcium Efflux EC_{50} (SFLLRNP-NH_2, μM)
Wild-Type Receptor	0.01–0.1	0.1
N-Terminus		
K76A	WT	WT
P79A	WT	WT
K82A	WT	WT
P85A	WT	WT
F87A	WT	WT
I88A	>10	>4
S89A	WT	>40
S93A	WT	WT

(*continued*)

Table 1. (Continued)

Mutants	Calcium Efflux EC_{50} (Thrombin, nM)	Calcium Efflux EC_{50} (SFLLRNP-NH$_2$, μM)
L96A	WT	1–10
Y95A	WT	WT
T97A	WT	WT
D91N	WT	WT
EC1		
D167A/E173A	WT	0.1–1.0
EC2		
E241A/D256A/E260A/E264A	>1.0	
E241A	0.1–1.0	0.1–2.0
D256A	0.1–1.0	>40
EC2		
N259A	>1.0	>2.0
N259P	0.1–1.0	>2.0
E260A	WT	0.1–2.0
E264A	WT	WT
EC3		
E347A	0.1–1.0	>50
E347S	0.1–1.0	>40
E347Q	>6	>40
TM2		
F182A	WT	WT
TM6		
L340A	WT	WT
Y337A	WT	>2
F339A	WT	WT
TM2/TM6		
F182A/Y337A	>10	>40
Disulfide Bridge		
C175S/C254S	>10	>40
C175S	1.0	>40
C254S	>10	>40
H-Bond Network		
TM1 N120D	WT	WT
TM1 N120A	0.2–0.4	>40
TM2 D148N	WT	WT
TM2 D148A	>10	>40
TM7 D367N	WT	>40
TM7 D367A	>10	>40
TM2/TM7 D148N/D367N	1.0	2.0

iterations), followed by the conjugate gradient minimization method, until the minimum energy change between iterations was 0.001 kcal/mol. Helices one, two, and seven of the thrombin receptor were placed in three separate Sybyl work areas and a model of the serotonin-2 (5HT2) receptor (atomic coordinates supplied by H. Moereels) was placed in a fourth work area. Then, helices one, two and seven of the 5HT2 receptor were overlaid with the same helices of thrombin receptor, so that the three conserved amino acids of thrombin receptor, Asn-120, Asp-148, and Asp-367, were fit to the corresponding conserved amino acids of the 5HT2 receptor, Asn-92, Asp-120, and Asn-376. This alignment routine allowed for proper orientation of these architecturally important building blocks within the protein, such that the side chains of these complementary residues could interact with each other in a hydrogen-bonding array. Before merging helices one and two, the remaining groups along the helical axes were inspected in 3-D, and the helices were moved and rotated as necessary to relieve unfavorable interhelical steric interactions, while retaining the hydrogen-bonding network between Asn-120, Asp-148, and Asp-367. After helices one and two were merged and energy-minimized, helix seven was merged to the helix 1,2 ensemble to provide the starting TM 1,2,7 cluster, which was again energy-minimized. The remaining helices were then attached, one at a time to the nucleus of TM 1,2,7 triad (in the order: TM3, TM6, TM5, TM4) with limited manipulation (translation, rotation, tilting) to achieve optimal interhelical contacts. Each helix was manipulated to attain maximal hydrophobic packing of the aromatic and aliphatic side chains of the helices and to avoid strongly unfavorable steric situations. Polar groups were made to point inwards toward the core of the helix bundle, unless they were situated at the ends of the helices, where they could interact, at least hypothetically, with phospholipid head-groups at the membrane surface. The predominant forces that held the helices together were hydrophobic packing between aromatic and aliphatic side chains. This furnished a seven-helix bundle topology related to that presented in Figure 2.

B. Addition of the Extracellular Loops

With the seven-helix bundle construct in hand, we turned our attention to incorporation of the three extracellular (EC) loops. The cytoplasmic loops were disregarded because they would not be involved in molecular recognition between the receptor and the SFLLRN ligand. Extracellular loop 3, the smallest of the three EC loops, was added first via the loop-search routine in the Biopolymer mode of Sybyl. The loop backbone choices found in the Brookhaven PDB were examined in 3-D and selected on the basis of their fit to the overall protein structure. After the side chains were added to EC3, some of them had to be rotated to avoid unfavorable steric interactions with other parts of the protein. Then, the entire protein was energy minimized. Extracellular loop 1 was then added, followed by EC2, and each time the loop selection was made after analyzing the protein in 3-D. Side chains of amino acids in the loops that were less than 2 Å away from other atoms in the protein were rotated as needed, and the entire protein was energy minimized.

C. Formation of the Disulfide Linkage between EC1 and EC2

A disulfide bridge between extracellular loops one and two had to be incorporated into our model. This disulfide bridge involved the two GPCR family-conserved Cys residues that occur at the border of EC1 at the top of TM3 and in the middle of EC2. In the thrombin receptor, these correspond to C175 (EC1) and C254 (EC2). The SCULPT molecular modeling software package[20] was used to manipulate these two loops so that the Cys side chains could be brought into close enough proximity to each other to form a disulfide bond. The starting distance between the Cα carbons of the two Cys residues was around 21 Å, quite far from the optimal Cα–Cα distance of 5.5–6.0 Å. Prior to formation of the disulfide bridge, the Sybyl-formatted protein structure was written out as a PDB (protein database) file for transfer to the SCULPT environment. Within SCULPT, the entire protein was "frozen", then the residues that comprise EC loops 1 and 2 were "thawed", thus allowing for manipulation of each loop as a unit without breaking any covalent bonds. As the loops were moved toward each other, the C175/C254 Cα–Cα distance became gradually shorter, approaching the optimal 6 Å distance. While the loops were being manipulated, the energy of the entire protein was continually updated to the screen by the SCULPT force field, and high-energy shells were displayed on those regions in which bad contacts were being generated, thus guiding the user through specific movements of the loops. When near-optimal Cα–Cα (5.6 Å) and S–S (2.3 Å) distances were achieved in SCULPT, and the energy was –909 kcal/mol, the final SCULPT structure was written out to a PDB file and read into SYBYL. The disulfide bond was then formed in the Biopolymer mode within Sybyl, since covalent bonds cannot be formed within the SCULPT environment; the resulting structure was energy-minimized. The final structure from SCULPT was already low in energy as measured by the Tripos force field (–1400 kcal/mol), although it was further minimized using the Kollman United Atoms force field. The protein required only 500 iterations of conjugate gradient minimization using the Kollman United Atoms force field to achieve a minimum energy change of 0.001 kcal/mol between iterations. The final energy in the Sybyl environment was –2520 kcal/mol. Although the overall energy for the complex was relatively low, there were still some violations of phi–psi space in regions around the conserved cysteine residues. The complex required further relaxation with molecular dynamics to relieve these constrained regions, which could have arisen from the loop pulling during the SCULPT manipulation.

D. Mutational Analysis and Addition of the N-Terminal Extension

The extracellular N-terminus of the thrombin receptor has been demonstrated to be important for the binding of the endogenous, tethered peptide–ligand,

SFLLRN.[15] The cleaved human thrombin receptor has an extracellular N-terminus of approximately 55 amino acids (before the beginning of TM1 and up to the SFLLRN motif). Given also the key role of the extracellular epitopes in other peptide–ligand GPCRs, we were interested in examining the role of the N-terminus of the thrombin receptor, in addition to its extracellular loops.

The analysis of chimeras of the human and *Xenopus laevis* thrombin receptors demonstrated the significance of the extracellular domains.[15] The tethered-ligand agonist motif for *Xenopus* receptor, TFRIFD, is quite different from the motif of the human thrombin receptor, SFLLRN. This difference is responsible for the specificities of the agonist peptides on their respective receptors, and was exploited to gain valuable information about the localization of the binding site for SFLLRN. In these studies,[15] epitopes from the amino terminus and all three extracellular loops of the *Xenopus* receptor were substituted with the corresponding sequences from the human thrombin receptor. This chimera, in which the *Xenopus* thrombin receptor's entire extracellular surface (all three EC loops and the N-terminus), were replaced with the human thrombin receptor's sequence, caused a 3 log potency gain for SFLLRN and a 3 log loss for TFRIFD. Additional chimeric mutagenesis pointed more specifically to the N-terminus and the second extracellular loop (EC2). Since these changes contributed significantly to the changes in specificity observed for the ligands with their respective receptors, we focused our 3-D-docking experiments on the interface between the loop regions and the N-terminus, with some involvement from the border regions between the loops and transmembrane domains.

Given the functional importance of at least part of the N-terminal extension, that closest to TM1, we decided to append the amino acid residues from Tyr-95 to Asn-75. The Protein Build command in the Biopolymer mode of Sybyl was used to attach this peptide epitope onto the protein in an extended conformation (with the Biopolymer Conformation Set command). Individual bonds were rotated to bring the N-terminal extension above the plane of the extracellular loops, thus allowing enough space for the agonist peptide SFLLRN to dock within the extracellular region between the loops and the N-terminus of the receptor.

After the model was assembled, SFLLRN was built in a "turn-like" geometry (Biopolymer Conformation Set command; 3_{10} helix), again using the Protein Build command within the Biopolymer mode of Sybyl. Both extended and turn geometries were investigated for the docking of SFLLRN to the receptor, but the turn geometry appeared to be superior, fundamentally on inspection and on the basis of the correlation of site-directed mutagenesis results to our 3-D model.

At this point, we had assembled a preliminary model of the human thrombin receptor that contained all three extracellular loops linked to the seven TM domains, as well as the N-terminus out to residue 75. Docking studies with agonist peptide SFLLRN, and additional mutagenesis experiments, were carried out in parallel. Measurement of the effects of site-specific mutations in human PAR-1 identified several key amino acids of the receptor as being important for activation (Table 1), and aided our understanding of the binding mode for the SFLLRN/thrombin receptor complex. In addition, published studies describing the effects of chimeric constructs between extracellular epitopes of human PAR-1 and human PAR-2[9a,21,22] were utilized to refine our molecular modeling.

PAR-2 is also activated by cleavage of its N-terminal domain and its agonist peptide sequence, SLIGKV, is quite different from the human agonist peptide sequence. Although SFLLRN can activate PAR-2, SLIGKV has little activity at the human thrombin receptor.[22] By analogy to the *Xenopus*/human thrombin receptor chimera studies, these specificity differences can be exploited to learn more about possible docking interactions of SFLLRN to the human thrombin receptor. The work of Gerszten et al.[15] with the *Xenopus* receptor, and subsequently of Lerner et al.[22] with the PAR-2 receptor, demonstrated that the entire extracellular face of this GPCR subfamily of protease-activated receptors is responsible for activity and, more specifically, that EC2 is a major determinant of agonist specificity. Earlier studies with antibodies[23] also showed that the extracellular domains are important for receptor function.

E. Point Mutation of the Acidic Residues in the Extracellular Loops

The N-terminal ammonium group of SFLLRN that is revealed on proteolytic cleavage is important for receptor activation. Systematic structure–activity studies with thrombin receptor-activating peptides have pinpointed the critical components of the side chains that furnish key information for modeling purposes.[24] The minimum sequence length is five amino acids (SFLLR): the Phe residue at position 2 is crucial for activity, while an N-terminal free amino group, a basic or aromatic residue at position 5, and a bulky aliphatic residue at position 4 are moderately important; and a *p*-fluorophenylalanine at position 2 yields agonist peptides with a significant increase in potency.[25] Since chimeric mutagenesis has established that the extracellular region of thrombin receptor is sufficient for agonist activity, we examined the contribution of the acidic residues in the three extracellular loops of the human thrombin receptor. We hoped to identify electrostatic interactions that could account for both the ammonium N-terminus and the Arg residue of SFLLRN. Seven acidic residues are distributed throughout the extracellular loops in human PAR-1: two in EC1 (D167 and E173), four in EC2 (E241, D256, E260, and E264), and one in EC3 (E347). When both EC1 acidic residues were mutated to Ala, there was no change in receptor functional responses to either human α-thrombin or SFLLRNP-NH$_2$ (Table 1). However, when all four acid residues in EC2 were mutated to Ala, there was a significant reduction in receptor activation in the presence of thrombin or TRAP-7. When the single acidic residue in EC3 was mutated to Ala, reduced functional activity of the receptor in the presence of each agonist was also observed. Results from these Ala replacements pointed to loops EC2 and EC3 as potential agonist peptide binding loci.

Further studies involving individual mutations of each of the acidic amino acids in EC2 to Ala, showed that only Asp-256 had a major effect on the functional response to agonist peptide (TRAP-7), with a modest response to thrombin itself (Table 1). The E264A mutant behaved like the wild-type receptor in response to both TRAP-7 and thrombin, E260A behaved like the wild-type receptor in response to thrombin but was modestly affected in response to TRAP-7, and E241A was modestly affected with both thrombin and TRAP-7. From these data, the most notable amino acid is Asp-256, which is conserved in human, rat, mouse, and *X. laevis* thrombin receptors, as well as in mouse and human PAR-2. Asp-256 is also only two positions away from the GPCR family-conserved EC2 Cys residue, which is presumably involved in a key disulfide bridge with a Cys in EC1 (Cys-175); the Asp and Cys occur in the PAR family-conserved sequence ITTCHD. Alignments of this subfamily of thrombin receptors against other peptide–ligand and biogenic amine–ligand GPCRs revealed that, although this Cys is present in all receptor protein sequences, the Asp is absent at this corresponding position in EC2 for all GPCRs except thrombin receptors. This is encouraging in that conservation among a small subfamily of GPCRs suggests a functional and/or binding significance for a receptor epitope. Conservation among the superfamily of receptors, on the other hand, is thought to confer architectural significance to the protein, like the conserved cysteines comprising the disulfide bridge between EC1 and EC2.

The choice of negatively charged amino acids on the extracellular face of the receptor that are potentially involved in an interaction with the Ser-42 ammonium group was now narrowed down to either Asp-256 or Glu-347. This warranted further single-point mutagenesis studies to select from these two acidic residues, as well as an examination of alternative binding conformations for SFLLRN in our model. *When Glu-347 in EC3 was mutated to Gln, a dramatic reduction in receptor function occurred for both thrombin and the agonist peptide.* The E347A and E347S mutations also caused a significant reduction in function. These data provide strong support for an electrostatic interaction between the carboxylate side chain of Glu-347, the only acidic residue in EC3 (at the end of TM7), and the Ser N-terminal ammonium group of the ligand. We ruled out having just a hydrogen-bonding interaction between the ammonium group and the carboxylate of Glu-347 because Ser or Gln at position 347 were not adequate for receptor function.

Inspection of our 3-D model of the thrombin receptor suggested that the Glu-347/Ser-42–NH_3^+ interaction is superior to one involving Ser-42–NH_3+ and Asp-256 in EC2. This was partly because the disulfide bridge involving Cys-254 created a constraint that anchored Asp-256 away from the junction between EC1 and EC2, and away from the transmembrane interface (Figure 8). With SFLLRN in a turn-type conformation, the Asp-256 side chain was pointed towards the N-terminal extension in the receptor model. This important acidic residue might interact directly with a basic side chain within a segment of the receptor N-terminal peptide. However, since alanine mutations for Lys-76 or Lys-82 yielded receptors with wild-type behavior, these amino acids apparently do not interact with Asp-256.

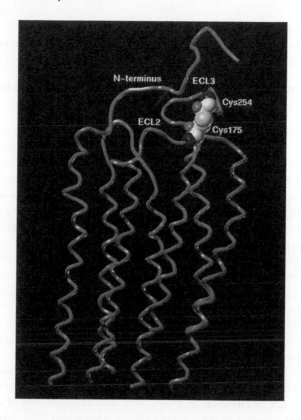

Figure 8. Disulfide bridge between EC1 and EC2 involving Cys-175 and Cys-254.

Further Ala scanning of the basic amino acids in the receptor N-terminal portion (e.g. Arg-70) may help to address this issue.

It is noteworthy that the E260A mutant thrombin receptor behaved like wild-type receptor in the presence of thrombin, and nearly so with the agonist peptide. Thus, this residue was discarded as a meaningful interacting group with the agonist peptide. By contrast, Nanevicz et al.[26] identified Glu-260 as an important amino acid and proposed that the Glu side chain could interact as a counterion with the terminal ammonium group of Arg-46 of the agonist peptide. Their studies also revealed that SFLLEN did not interact with the wild-type human thrombin receptor. In contrast, when the E260R mutant receptor was tested in the presence of SFLLEN, receptor function was restored. This complementarity supports an interaction between Arg-46 and Glu-260; however, a possible role for the N-terminal ammonium group of the agonist peptide was unfortunately not addressed.[26] Since the Arg residue of SFLLRN can be effectively replaced by a nonbasic Phe residue (such as with the *X. laevis* peptide TFRIFD or SFLLF[25b]), a strong charged interaction is apparently not required at this end of the agonist peptide sequence. Thus, the importance of Glu-260 to agonist–receptor interaction is currently unclear.

F. Mutational Analysis of the Extracellular Disulfide Bridge between EC1 and EC2

In nearly all GPCRs, there are two conserved Cys residues, one preceding transmembrane region three, and one preceding transmembrane region five; occurrence in these locations would place the Cys residues within EC loops 1 and 2. It has been postulated, and even experimentally validated for certain GPCRs,[27] that these two conserved cysteines are covalently bonded through an intramolecular disulfide bridge. The incorporation of a disulfide bridge between these two EC loops would conformationally constrain this region of the receptor and provide a more appropriate 3-D model for ligand docking studies.

We sought to validate this proposed disulfide bridge for the thrombin receptor by replacing Cys-175 and Cys-254 with serine residues to give the double mutant C175S/C254S and the single mutants C175S and C254S. Functional activity, as measured by calcium efflux, was altered in the case of the double mutant C175S/C254S for thrombin or SFLLRNP-NH$_2$ as agonists (Table 1), suggesting some disruption of the loop geometries leading to a reduction in activation in the absence of the bridge constraint. In addition, the calcium efflux activity was reduced for the single mutant C254S with each agonist. However, in the case of the single mutant C175S, the reduction in calcium efflux activity was observed only with TRAP-7 and not with thrombin itself. This wild-type activity observed for the C175S mutant receptor in the presence of thrombin is puzzling. Nonetheless, the other supporting data for the importance of the disulfide bridge between loops EC1 and EC2 are very encouraging, and recommend incorporating this feature into the 3-D model.

G. Site-Directed Mutagenesis Procedures

A cassette-replacement approach was used to facilitate the introduction of amino acid mutations at various sites of the thrombin receptor. First, unique endonuclease restriction enzyme sites were generated at several positions within the thrombin receptor cDNA by mutating the nucleotide sequences. Second, the polymerase chain reaction (PCR) with primers encoding for the desired mutations was used to generate the cDNA cassette with the appropriate endonuclease restriction enzyme sites for replacement of the wild-type sequence. The locations for the introduction of the sites were chosen based on two requirements. They needed to be at or near regions of the cDNA sequence that codes for amino acids at junctions of transmembrane domains and extracellular loops. Also, introduction of the sites did not alter the amino acid sequence of the protein. The site-directed mutagenesis method of Kunkel et al.[28] was used to introduce the mutations required for generating the endonuclease restriction sites. The presence of the mutation and the fidelity of the substituted region were confirmed by sequence analysis of both strands of the cDNA.

The effects of the mutations on functional responses were determined on thrombin receptors that were expressed in *Xenopus* oocytes. Agonist-stimulated calcium efflux with human α-thrombin (10 nM) and SFLLRNP-NH$_2$ (40 μM), as described previously,[29] was employed to determine the functional responsiveness of the receptor.

Expression levels of the mutant thrombin receptors were determined in transiently transfected COS cells by assaying the level of surface expressed thrombin receptors by FACS analysis, usually for mutant receptors that showed a significant diminution of function. COS cells were transfected by calcium phosphate–DNA co-precipitation[30] with 10 ng of DNA. The transfected cells were analyzed by FACS 48 h after transfection. A monoclonal antibody specific for the amino terminal region was used as the primary antibody to detect thrombin receptors present on the surface of the cells.[31] A FITC conjugated goat anti-mouse IgG was used as a secondary antibody. Expression levels of the mutant receptors were determined relative to expression levels of wild-type controls. Expression-level data (% of control) for some of the mutant receptors are listed in Table 2 (wild-type thrombin

Table 2. Relative Expression Levels of Mutants as Measured in Transiently Transfected COS Cells

Mutants	Expression (%)
Wild-type receptor	100
Vector control	3
P85A	80
I88A	100
S89A	100
L96A	100
N120A	50
D148A	<10
C175S	80
F182A	92
C254S	48
D256A	100
Y337A	100
E347A	100
D367A	<10
C175S/C254S	72
D148N/D367N	80
F182A/Y337A	100

receptor = 100%; vector control = 3%). Attempts to quantify thrombin receptor expression levels in oocytes were unsuccessful. Therefore, relative expression levels of the mutant thrombin receptors were measured in transiently transfected COS cells.

III. LIGAND–RECEPTOR INTERACTIONS

A. Hydrophobic/Aromatic Pocket

Coincidentally, by aiming the peptide N-terminal amino group in our 3-D ligand–receptor complex toward Glu-347, rather than Asp-256,[32] we identified a hydrophobic pocket as a possible component for interaction with Phe-43 in the agonist peptide motif. This hydrophobic cluster involves residues Phe-182, Leu-340, Tyr-337, and Phe-339 (Figure 9). Assignment of the stop and start residues for

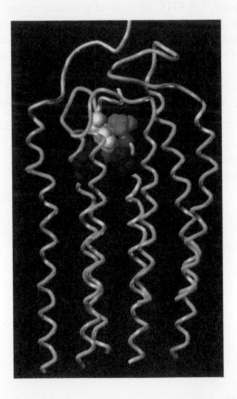

Figure 9. Hydrophobic cluster involving Phe-182 (magenta), Leu-340 (orange), Tyr-337 (red), and Phe-339 (magenta) of the human thrombin receptor complexed with the Phe residue (white) of the agonist peptide.

the transmembrane helices, via Kyte–Doolittle hydropathy analysis,[19] placed these four residues near the tops of the respective transmembrane regions. Since the peptide structure–activity relationship for SFLLRN indicates that Phe-43 is the most critical element in the agonist peptide for agonist function,[24,25] it was crucial to have an involvement of the benzene ring of the side chain with a recognition domain in the receptor. The hydrophobic pocket was consistent with our imposed conformation for bound SFLLRN—a bent conformation resembling a 3_{10} helix, since the extended structure had Phe-43 in an arrangement that could not suitably reach the transmembrane cavity (the peptide just lies across the extracellular surface). The extended conformation of SFLLRN is disfavored because it does not allow interactions with any transmembrane amino acids, and might therefore not be able disrupt interactions internal to the receptor that could propagate through the membrane for signal transduction. Interactions solely with loop residues would probably not be strong enough to cause the necessary helix shifts.

To test the hydrophobic cluster hypothesis, Phe-182, Leu-340, Tyr-337, and Phe-339 were individually mutated to Ala, and the receptor functional activities of the mutants were measured. Only one of these mutants, Y337A, showed any effect on function, and that effect was seen only in the presence of agonist peptide, but not with thrombin as agonist. However, *the double mutant F182A/Y337A demonstrated a marked reduction in functional activity with both thrombin and the agonist peptide*. This finding lends support to the existence of a hydrophobic region for interaction with Phe-43. It may be that the Ala side chain for a single replacement can still contribute a sufficient hydrophobic surface to the binding pocket, but that such compensation is no longer possible when two aromatic side chains are replaced.

B. Interactions of the N-Terminal Extension

A moderately important amino acid in the agonist peptide is Leu-45 in position four (SFLLRN). This position seems to require a bulky, hydrophobic side chain. Maintaining the two interactions already set up in the agonist peptide/receptor complex, the terminal ammonium group with the Glu-347 carboxylate ion and the Phe-43 benzene in the hydrophobic pocket, another docking interaction was probed. With SFLLRN in a "turn-like" conformation, and with the electrostatic and hydrophobic interactions already set up, Leu-45 seemed poised to interact with Phe-87 of the N-terminal extension of the receptor. An Ala scan of the N-terminal extracellular surface, however, revealed that the F87A mutant, behaved like wild-type receptor in the presence of thrombin or agonist peptide. Only one Ala mutation in the N-terminus, I88A, caused a dramatic reduction in receptor function with thrombin and peptide agonist, possibly reflecting some contact between Leu-45 and the residue adjacent to Phe-87 in the N-terminus. During the course of our work, Nanevicz et al.[26] reported mutagenesis studies involving human/*Xenopus* thrombin receptor chimeras. The human and *Xenopus* thrombin receptors respond only to their own agonist peptide motifs, SFLLRN and TFRIFD, respectively. These workers found that merely two substitutions in the *Xenopus* receptor, a Phe for Asn-87 in the N-terminus and a Glu for Leu-260 in EC2 (the homologous native human amino acids), caused a switch in specificity to the human agonist. Although our F87A mutational data do not validate our proposed Leu-45/Phe-87 interaction, the I88A mutant results from our laboratories and the N87F/L260E double mutation results from Nanevicz et al.[26] help to support a hydrophobic interaction of Leu-45 with a confined section of the N-terminus. The F87A mutant may still provide a hydrophobic enough environment for the Leu-45 isopropyl group to maintain agonist activity.

IV. CONCLUSION

We have constructed a three-dimensional theoretical model of the human thrombin receptor (PAR-1) based on de novo modeling techniques involving computer–graphical model construction in conjunction with site-directed mutagenesis. Our modeling strategy was predicated on sequence conservation among the GPCR superfamily and certain rules for helix–helix packing. The agonist peptide was set to a turn-like conformation and docked into the receptor at a region which included the extracellular loops, the receptor N-terminus, and the extracellular surfaces of selected transmembrane helices. The disulfide bridge constructed between EC1 and EC2 constrained the extra-cellular space such that binding orientations for the agonist peptide were limited.

Manual docking operations and energy minimization led to a few hypotheses for the binding of SFLLRN to the thrombin receptor. The model that was most closely correlated with the results of site-directed mutagenesis experiments and functional activity data involved two important interactions between the peptide and the receptor. The first was the hydrophobic interaction of the SAR-critical Phe-43 of the agonist peptide with the side chains of Phe-182, Leu-340, Tyr-337, and Phe-339. The second was the electrostatic interaction of the N-terminal ammonium group of the agonist peptide, also very important in the SAR profile, with Glu-347 in EC3. Although the Arg-46 residue is also relevant, it did not appear to interact with any charged carboxylate side chain in our model. This brought us to suppose that the trimethylene segment of Arg-46, rather than its charged guanidinium group, might be responsible for the loss of functional activity observed when the Arg residue is replaced by Ala in the agonist peptide. This outcome is consistent with the observation that replacement of Arg with Phe or citrulline substantially retains agonist potency.[24]

Overall, our results provide a working hypothesis for the molecular recognition that occurs between the human thrombin receptor (PAR-1) and its agonist peptide SFLLRN. Similar features may contribute to the interactions of the *X. laevis* homologue of PAR-1 with TFRIFD and of human PAR-2 with SLIGKV, although future modeling studies are needed for an assessment of these systems. Such structural information can be exploited to assist the design of novel peptide and nonpeptide modulators of protease-activated receptors. In particular, given the fact that thrombin-induced platelet activation and cell proliferation are important to vascular disorders, such as thrombosis and restenosis, thrombin receptor antagonists are a prime drug discovery target.

ACKNOWLEDGMENTS

The authors thank Rik Moereels (Janssen Research Foundation, Beerse, Belgium) for his direction during the de novo model building stage of the project and Harold Almond, Jr., for helpful discussions. We also thank Marco Ceruso for his advice and analysis of the final receptor model; in particular, he examined phi–psi space for angle violations and relaxed the relevant regions through molecular dynamics simulations.

REFERENCES AND NOTES

1. (a) Jackson, T. *Pharmac. Ther.* **1991**, *50*, 425–442. (b) Strader, C. D.; Fong, T. M.; Tota, M. R.; Underwood, D.; Dixon, R. A. F. *Annu. Rev. Biochem.* **1994**, *63*, 101–132. (c) Strader, C. D.; Fong, T. M.; Graziano, M. P.; Tota, M. R. *FASEB J.* **1995**, *9*, 745–754. (d) Gudermann, T.; Nürnberg, B.; Schultz, G. *J. Mol. Med.* **1995**, *73*, 51–63.

2. Henderson, R.; Baldwin, J. M.; Ceska, T. A.; Zemlin, F.; Beckmann, E.; Downing, K. H. *J. Mol. Biol.* **1990**, *213*, 899–929.

3. (a) Schertler, G. F. X.; Villa, C.; Henderson, R. *Nature* **1993**, *362*, 770–772. (b) Havelka, W. A.; Henderson, R.; Oesterhelt, D. *J. Mol. Biol.* **1995**, *247*, 726–738. (c) Baldwin, J. M.; Schertler, G. F. X.; Unger, V. M. *J. Mol. Biol.* **1997**, *272*, 144–164. (d) Du, P.; Salon, J. A.; Tamm, J. A.; Hou, C.; Cui, W.; Walker, M. W.; Adham, N.; Dhanoa, D. S.; Islam, I.; Vaysse, P. J. J.; Dowling, B.; Shifman, Y.; Boyle, N.; Rueger, H.; Schmidlin, T.; Yamaguchi, Y.; Branchek, T. A.; Weinshank, R. L.; Gluchowski, C. *Protein Eng.* **1997**, *10*, 109–117.

4. Grigorieff, N.; Ceska, T. A.; Downing, K. H.; Baldwin, J. M.; Henderson, R. *J. Mol. Biol.* **1996**, *259*, 394–421.

5. Trumpp-Kallmeyer, S.; Hoflack, J.; Bruinvels, A.; Hibert, M. F. *J. Med. Chem.* **1992**, *35*, 3448–3462.

6. (a) Juppner, H.; Schipani, E.; Bringhurst, F. R.; McClure, I.; Keutmann, H. T.; Potts, J. T.; Kronenberg, H. M.; Abou-Samra, A. B.; Segre, G. V.; Gardella, T. J. *Endocrinology* **1994**, *134*, 879–884. (b) Stroop, S. D.; Kuestner, R. E.; Serwold, T. F.; Chen, L.; Moore, E. E. *Biochemistry* **1995**, *34*, 1050–1057. (c) Vilardaga, J.-P.; De Neef, P.; Di Paolo, E.; Bollen, A.; Waelbroeck, M.; Robberecht, P. *Biochem. Biophys. Res. Commun.* **1995**, *211*, 885–891.

7. Vu, T.-K. H.; Wheaton, V. I.; Hung, D. T.; Coughlin, S. R. *Cell* **1991**, *64*, 1057–1068.

8. Reviews on the thrombin receptor: (a) Van Obberghen-Schilling, E.; Chambard, J.-C.; Vouret-Craviari, V.; Chen, Y.-H.; Grall, D.; Pouysségur, J. *Eur. J. Med. Chem.* **1995**, *30S*, 117–130. (b) Coughlin, S. R.; Scarborough, R. M.; Vu, T.-K. H.; Hung, D. T. *Cold Spring Harbor Symp. Quant. Biol.* **1992**, *57*, 149–154. (c) Coughlin, S. R. *Thromb. Haemostasis* **1993**, *66*, 184–187. (d) Coughlin, S. R. *Trends Cardiovasc. Med.* **1994**, *4*, 77–83. (e) Ogletree, M. L.; Natarajan, S.; Seiler, S. M. *Persp. Drug Discovery Des.* **1993**, *1*, 527–536. (f) Dennington, P. M.; Berndt, M. C. *Clin. Exp. Pharmacol. Physiol.* **1994**, *21*, 349–358.

9. (a) Nystedt, S., Emilsson, K., Wahlestedt, C., Sundelin, J. *Proc. Natl. Acad. Sci. USA* **1994**, *91*, 9208–9212. (b) Ishihara, H.; Connolly, A. J.; Zeng, D.; Kahn, M. L.; Zheng, Y. W.; Timmons, C.; Tram, T.; Coughlin, S. R. *Nature* **1997**, *386*, 502–506.

10. Moereels, H.; Janssen, P. A. J. *Med. Chem. Res.* **1993**, *3*, 335–343.

11. Kontoyianni, M.; DeWeese, C.; Penzotti, J. E.; Lybrand, T. P. *J. Med. Chem.* **1996**, *39*, 4406–4420.

12. Elling, C. E.; Schwartz, T. W. *EMBO J.* **1996**, *15*, 6213–6219.

13. Rasmussen, U. B.; Vouret-Craviari, V.; Jallet, S.; Schlesinger, Y.; Pages, G.; Pavirani, A.; Lecocq, J. P.; Pouyssegur, J.; Van Obberghen-Schilling, E. *FEBS Lett.* **1991**, *288*, 123–128.

14. Zhong, C.; Hayzer, D. J.; Corson, M. A.; Runge, M. S. *J. Biol. Chem.* **1992**, *267*, 16975–16979.

15. Gerszten, R. E.; Chen, J.; Ishii, M.; Wang, L.; Nanevicz, T.; Turck, C. W.; Vu, T.-K.; Coughlin, S. R. *Nature* **1994**, *368*, 648–651.

16. The mouse thrombin receptor sequence was obtained by direct submission to GenBank, Accession No. 267129, 1 April 1993.

17. Moereels, H.; Lewi, P. J.; Koymans, L. M. H.; Janssen, P. A. J. *Recept. Channels* **1996**, *4*, 19–30.

18. Zhou, W.; Flanagan, C.; Ballesteros, J. A.; Konvicka, K.; Davidson, J. S.; Weinstein, H.; Millar, R. P.; Sealfon, S. C. *Mol. Pharmacol.* **1994**, *45*, 165–170.

19. Kyte, J.; Doolittle, R. F. *J. Mol. Biol.* **1982**, *157*, 105–132.

20. Surles, M. C.; Richardson, J. S.; Richardson, D. C.; Brooks, F. P. *Protein Sci.* **1994**, *3*, 198–210.

21. (a) Nystedt, S.; Emilsson, K.; Larsson, A.-K.; Stroembeck, B.; Sundelin, J. *Eur. J. Biochem.* **1995**, *232*, 84–89. (b) Nystedt, S.; Larsson, A.-K.; Aberg, H.; Sundelin, J. *J. Biol. Chem.* **1995**, *270*, 5950–5955.

22. Lerner, D. J.; Chen, M.; Tram, T.; Coughlin, S. R. *J. Biol. Chem.* **1996**, *271*, 13943–13947.

23. Bahou, W.; Coller, B.; Potter, C.; Norton, K.; Kutok, J.; Goligorsky, M. *J. Clin. Invest.* **1993**, *91*, 1405–1413.

24. (a) Scarborough, R. M.; Naughton, M.; Teng, W.; Hung, D. T.; Rose, J.; Vu, T.-K. H.; Wheaton, V. I.; Turck, C. W.; Coughlin, S. R. *J. Biol. Chem.* **1992**, *267*, 13146–13149. (b) Hui, K. Y.; Jakubowski, J. A.; Wyss, V. L.; Angleton, E. L. *Biochem. Biophys. Res. Commun.* **1992**, *184*, 790–796. (c) Sabo, T.; Gurwitz, D.; Motola, L.; Brodt, P.; Barak, R.; Elhanaty, E. *ibid.* **1992**, *188*, 604–610. (d) Chao, B. H.; Kalkunte, S.; Maraganore, J. M.; Stone, S. R. *Biochemistry* **1992**, *31*, 6175–6178. (e) Vassallo, R. R., Jr.; Kieber-Emmons, T.; Cichowski, K.; Brass, L. F. *J. Biol. Chem.* **1992**, *267*, 6081–6085. (f) Bernatowicz, M. S.; Klimas, C. E.; Hartl, K. S.; Peluso, M.; Allegretto, N. J.; Seiler, S. M. *J. Med. Chem.* **1996**, *39*, 4879–4887. Note: Although N-acetyl analogues of TRAP were found to have reasonably potent activities in the platelet aggregation assay [1.80 μM EC_{50} for Ac-βAFLLR-NH$_2$ and 0.27 μM EC_{50} for Ac-βAF(f)A(N-2)LR-NH$_2$], when assayed for GTPase activity, they were characterized as partial agonists, possessing only 50–70% and 21% intrinsic potency, respectively. Therefore, the N-terminal ammonium group of SFLLRN-NH$_2$, whose sequence was derived from the native N-terminus of the receptor, and which demonstrated full agonist activity, was considered to be a crucial element for agonist ligand-receptor recognition. (g) Sabo, T.; Gurwitz, D.; Motala, L.; Brodt, P.; Barak, R.; Elhanaty, E. *Biochem. Biophys. Res. Commun.* **1992**, *188*, 604–610.

25. (a) Feng, D.-M.; Veber, D. F.; Connolly, T. M.; Condra, C.; Tang, M.-J.; Nutt, R. F. *J. Med. Chem.* **1995**, *38*, 4125–4130. (b) Natarajan, S.; Riexinger, D.; Peluso, M.; Seiler, S. M. *Int. J. Pept. Protein Res.* **1995**, *45*, 145–151. (c) Nose, T.; Shimohigashi, Y.; Ohno, M.; Costa, T.; Shimizu, N.; Ogino, Y. *Biochem. Biophys. Res. Commun.* **1993**, *193*, 694–699.

26. Nanevicz, T.; Ishii, M.; Wang, L.; Chen, M.; Chen, J.; Turck, C. W.; Cohen, F. E.; Coughlin, S. R. *J. Biol. Chem.* **1995**, *270*, 21619–21625.

27. (a) Cook, J. V. F.; Eidne, K. A. *Endocrinology* **1997**, *138*, 2800–2806. (b) Ohyama, K.; Yamano, Y.; Sano, T.; Nakagomi, Y.; Hamakubo, T.; Morishima, I.; Inagami, T. *Regul. Pept.* **1995**, *57*, 141–147. (c) Savarese, T. M.; Wang, C. D.; Fraser, C. M. *J. Biol. Chem.* **1992**, *267*, 11439–11448. (d) Moereels, H.; Leysen, J. E. *Recept. Channels* **1993**, *1*, 89–97. (e) Farahbakhsh, Z. T.; Hideg, K.; Hubbell, W. L. *Science* **1993**, *262*, 1416–1419.

28. Kunkel, T. A.; Roberts, J. D.; Zakour, R. A. *Meth. Enzymol.* **1987**, *154*, 367–382.
29. Blackhart, B. D.; Emilsson, K.; Nguyen, D.; Teng, W.; Martelli; A. J.; Nystedt, S.; Sundelin, J.; Scarborough, R. M. *J. Biol. Chem.* **1996**, *271*, 16466–16471.
30. Chen, C.; Okayama, H. *Mol. Cell. Biol.* **1987**, *7*, 2745–2752.
31. Shapiro, H. M. *Practical Flow Cytometry*, 2nd ed.; Wiley-Liss: New York, 1988.
32. We tried to dock SFLLRN in a reverse manner to have the Arg side chain interact with Glu-347. However, in this mode there was no acidic residue for interaction with the N-terminal ammonium group and no hydrophobic pocket for Phe-43.

FARNESYL TRANSFERASE INHIBITORS: DESIGN OF A NEW CLASS OF CANCER CHEMOTHERAPEUTIC AGENTS

Theresa M. Williams and Christopher J. Dinsmore

Advances in Medicinal Chemistry
Volume 4, pages 273–314.
Copyright © 1999 by JAI Press Inc.
ISBN: 0-7623-0064-7

ABSTRACT

This review chronicles the strategies and hurdles encountered in farnesyl transferase inhibitor drug discovery programs, with the major focus on substrate-based inhibitor design. Recently reported enzyme structure studies and their potential impact on drug design are also discussed. The anti-proliferative effect that FTIs exhibit in both genetically engineered and human tumor cell lines suggest that this type of drug may be efficacious in treating certain cancers.

I. INTRODUCTION

Following the discovery that the protein product of the *ras* oncogene is a substrate for the enzyme farnesyl transferase (FTase), there has been considerable interest in exploring the utility of an FTase inhibitor (FTI) in suppressing tumor growth. FTase became an attractive target by virtue of biochemical studies which showed that of several posttranslational processing steps, only farnesylation of Ras was absolutely required for its biological activity.[1-6] The approaches taken to discover and optimize FTIs are fairly representative of the strategy and tools available to pharmaceutical research in the 1990s and include: screening sample collections and natural products; utilizing substrate-based inhibitor design; employing molecular modeling to generate pharmacophore models and ideas about enzyme-bound inhibitor conformation in conjunction with NMR and crystallography; and last but not least, synthesizing single compounds and/or libraries of compounds for testing. In this review we will be summarizing the strategies and tactics which resulted in the identification of several classes of highly potent FTIs.

II. BACKGROUND

Ras is part of a cell signal-transduction pathway relaying extracellular growth signals to nuclear target proteins which are involved with transcriptional activation.[7,8] Regulatory signaling is accomplished through a complex series of protein–protein interactions spanning the region from the exterior surface of the cell membrane to the nucleus. In order for Ras to function in the cell, it must be associated with the cell membrane. Farnesylation of Ras allows the protein to associate with the interior of the cell membrane, either through nonspecific lipid–lipid association or via a specific interaction with a membrane-bound farnesyl protein receptor. Blocking Ras farnesylation renders the protein cytosolic, abrogating its biological activity.[9]

Ras is a member of the small-GTP binding protein family (molecular weight ~21,000); the Ras–GTP complex interacts with its effector proteins, e.g. the threonine–serine kinase Raf,[10,11] to propagate a growth signal. Ras–GTP is converted to the inactive Ras-GDP form via interaction with the GTPase activating protein (GAP).[12–16] Conversely, Ras–GDP is reactivated to its GTP-bound form by nucleotide exchange factors (NEFs) in response to stimulated cell-surface growth factor receptors.[17–20] Mutations in the Ras protein cause cellular transformation (that is, acquisition of the properties of a cancer cell) and generally occur in regions associated with GAP and nucleotide binding, impairing hydrolysis of GTP to GDP. In this way Ras becomes constitutively activated; extracellular growth signals no longer affect signaling through the Ras protein, and signaling through Raf is unregulated.[21] Since a fair percentage (30–35%) of human tumors harbor mutations in the *ras* gene,[22,23] the use of an FTI to prevent Ras membrane association and block *ras*-dependent tumor growth could have a significant impact on the disease.

Several major issues loomed at the outset of farnesyl transferase inhibitor programs. In terms of biology, investigators needed to demonstrate growth inhibition by an FTI in *ras*-transformed cells in vivo, and to correlate this with the extent of protein prenylation inhibition. Efficacy in suppressing *ras*-dependent tumors in animal models needed to be demonstrated; a key point was whether tumor growth could be blocked without significant toxicity to normal cells. With respect to chemistry, suitable drug candidates needed to be discovered and developed. Perhaps as a testament to the complexity of the problem, all of these issues continue to be studied in some form. There has been, however, encouraging progress towards obtaining a drug which inhibits tumor growth via farnesyl transferase inhibition.

Prenylated proteins have characteristic C-terminal sequences. For example, the three allelic Ras proteins (H-Ras, K-Ras, and N-Ras) expressed in mammalian tissues contain a C-terminal tetrapeptide which begins with cysteine, and ends with either methionine or serine. This part of the molecule is referred to as the "CaaX box" where C = cysteine, a = an aliphatic amino acid, and X = a prenylation specificity residue. The first step in the posttranslational processing of Ras proteins utilizes FTase and farnesyl diphosphate (FPP) to covalently attach a farnesyl group to the cysteine thiol of the CaaX box. While subsequent processing events involve proteolytic removal of the aaX tripeptide and methylation of the resulting C-terminal carboxylate group, only the farnesyl modification is required for mutant Ras proteins to associate with the cell membrane and transform a cell.[2–6]

Alternative prenylation pathways also exist. The geranylgeranyl transferase enzyme (GGTase) utilizes GGPP to preferentially transfer a geranylgeranyl group to proteins possessing a CaaX box where X is leucine or phenylalanine; therefore one of the prime determinants for farnesylation vs. geranylgeranylation lies with the identity of the X residue. Selective inhibition of FTase vs. the closely related GGTase was judged to be highly desirable, since the majority of prenylated proteins are geranylgeranylated, and fewer side effects might be realized with a selective FTI.

Upon the initiation of screening programs, a number of polycarboxylic acid natural products such as zaragozic acid were found to be quite good inhibitors of FTase. The mechanism of inhibition generally involved competition with FPP, but not Ras, substrate. Since FPP is also utilized by many other enzymes, a decision was made early in our program to pursue Ras-competitive enzyme inhibitors in order to stack the deck in favor of greater specificity. Indeed, zaragozic acid is also an excellent inhibitor of the FPP-utilizing enzyme squalene synthase. As our main focus was to develop Ras-competitive FTIs, we will be reviewing for the most part Ras-competitive inhibitors in this chapter. Synthetic and natural product FPP-competitive inhibitors, as well as FTIs of unknown mechanism, have been discussed in earlier reviews.[24–26]

III. FTI DESIGN BASED ON THE RAS CaaX BOX

A. Tetrapeptide Inhibitors

It turned out that tetrapeptides based on the CaaX sequence of a variety of mammalian and yeast proteins inhibited farnesylation of the full length parent proteins, mostly by acting as alternative substrates (Table 1).[27–29] Tetrapeptides corresponding to the C-terminal sequences of H-Ras (CVLS), K_{4A}-Ras (CIIM) and its alternative splice product K_{4B}-Ras (CVIM), as well as N-Ras (CVVM) were all inhibitors of FTase. Studies on the structure–activity relationships (SAR) of tetrapeptide FTIs provided fertile ground for the cultivation of several generations of inhibitors. In these early substrate-based inhibitors, the N-terminal amino acid of the tetrapeptide had to be cysteine in order for the tetrapeptide to be active. Alkylation of the sulfhydryl group in CVIM with iodoacetamide abolished inhibi-

Table 1. CaaX Tetrapeptide Inhibition of FTase and GGTase

	IC_{50} (μM)	
CaaX	FTase	GGTase I
CAIM	0.13	99
CAIA	10	450
CAIS	1.2	340
CAIT	22	140
CAIL	18	10
CVIL	11	2.3
CVIM	0.09	32

tory activity, suggesting a critical role for the thiol group in binding to the enzyme.[29] Similarly, replacing cysteine with serine, asparagine, or glutamic acid decreased activity by 4 orders of magnitude.[27,30] As substrates, however, peptide inhibitors such as CVIM were converted to S-farnesylated products which were poorer enzyme inhibitors, thus leading to metabolic inactivation of the tetrapeptide inhibitor in vivo.[28] The identification of nonsubstrate peptide FTase inhibitors opened the door for the rational design of FTIs. One key discovery centered around the observation that Ca_1a_2X peptides having an aromatic residue at the a_2 position were inhibitors, but not substrates, of FTase. The peptides CVFM, CVYM, and CVWM inhibited FTase prenylation of H-Ras with IC_{50}s between 25–400 nM but were not turned over, suggesting that a nonproductive binding mode was accessible by these thiol-containing inhibitors.[31] Additional studies demonstrated that substrate vs. nonsubstrate specificity was also determined by the presence or absence of an acylated N-terminal amino group[30,32,33] as well as the degree of steric bulk adjacent to the cysteine thiol.[31] In these compounds, affinity for FTase required the presence of both an N-terminal thiol and C-terminal carboxylic acid, as in the protein substrate.

Because farnesyl transferase is an intracellular enzyme, a major hurdle to surmount involved the demonstration of activity in cell culture. An assay measuring the extent of Ras prenylation in NIH 3T3 cells was used to determine the effect of FTase inhibitors on the prenylation state of newly synthesized, metabolically labeled Ras protein. Since the farnesylated fully processed peptide migrates differently on SDS-PAGE than the non-farnesylated protein, the extent of Ras processing inhibition can be determined. In the presence of 100 μM CIFM, approximately 10–30% of v-H-Ras protein is unprocessed after 20 hours. In vitro, CIFM inhibits FTase with an IC_{50} of 27 nM, so it is over 4000-fold less effective in blocking H-Ras farnesylation in cell culture. The tetrapeptide CIFM was unstable when incubated with cytosol derived from NIH 3T3 cells, most likely due to its degradation by aminopeptidases. The susceptibility of peptides to proteolytic degradation via peptidase activity is a well-known limitation of their use as therapeutic agents in vivo.

B. Isosteric Replacement of Tetrapeptide Amide Bonds

To stabilize the tetrapeptide to aminopeptidase activity, amide bonds were reduced to the corresponding amines. A key observation was that modification of some of these peptide bonds negated the ability of a compound to serve as a substrate; unlike before, this was independent of the identity of the a_1a_2 amino acids. While the peptide CIIhS **1a** was a good substrate for FTase, reduction of the first two amide bonds (cysteine, and a_1-isoleucine) produced compound **1d** which was a potent inhibitor (IC_{50} 20 nM) but was not a substrate (Table 2).[33] Reduction of only one amide bond to give compounds **1b** and **1c** was also well tolerated in terms of inhibition potency, but both compounds were now substrates for FTase. This

Table 2. Inhibition of FTase by Tetrapeptides Incorporating Reduced Amide Bonds

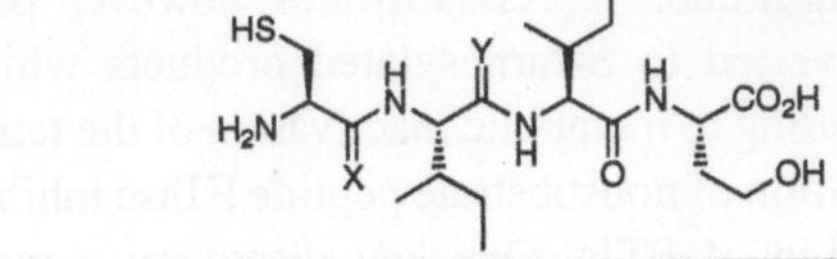

Compound	X	Y	FTase IC_{50} (nM)	GGTase I IC_{50} (nM)	FTase Substrate[a]
1a	O	O	330	ND	ND
1b	H_2	O	42	8,000	S
1c	O	H_2	150	ND	S
1d	H_2	H_2	20	45,000	N

Note: [a]ND = Not determined; S = substrate, N = nonsubstrate.

suggested that the presence of an amide bond was necessary, but perhaps not always sufficient, for substrate activity. Reduced amide bonds were also systematically introduced into CIFM (see Table 3).[30,33] Reduction of the C-I or the C-I and I-F amide bonds gave **2a** and **2c**, respectively, both of which were more potent than the parent tetrapeptide; similar results were also obtained with backbone modifications of the closely related CVFM tetrapeptide.[34] Moreover, by virtue of incorporation of phenylalanine at a_2 and/or its backbone modifications, **2c** was not a substrate for the enzyme. Most significantly, however, it was demonstrated that the doubly reduced peptide isostere **2c** was stable to incubation with NIH 3T3 cell lysates, in contrast to the parent peptide.[30] Despite the resistance of **2c** to intracellular degradation, the ratio of IC_{50}s for Ras processing and FTase inhibition was still 20,000, so it appeared that additional factors were restricting cell activity.

Prodrug CaaX Analogues Show Increased Cell Activity

It seemed reasonable to expect that the ionized carboxylic acid was limiting cell penetration and adversely affecting the biological activity of FTIs in cell culture. A strategy relying on masking the negative charge with a metabolically labile group was used to increase cell membrane penetration. An ester of the carboxylic acid would allow the molecule to move freely through the cell membrane; once inside, the ester would be hydrolyzed by intracellular esterases to the carboxylic acid. Applying the prodrug strategy to the reduced peptide isosteres **2a** and **2c**, the esters **2b** and **2d** increased cell membrane permeability as measured by their enhanced ability to block Ras farnesylation in cell culture.[33] The IC_{50}s for inhibition of Ras processing in cell culture by the esters **2b** and **2d** were now only 200–500 fold

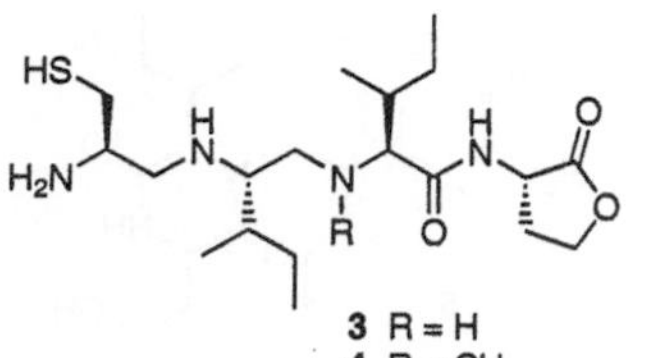

HS
H₂N

3 R = H
4 R = CH₃

higher than the in vitro IC_{50}s of the parent carboxylic acids **2a** and **2c**, even though ester **2b** was 1500-fold *less* active than its parent carboxylic acid **2a** at inhibiting FTase. Homoserine as the C-terminal amino acid can be masked as the homoserine lactone **3**.[33] While the hydroxy acid **1d** was essentially inactive in cell culture at 100 μM, at a similar concentration the lactone **3** inhibited 50–90% of Ras processing (Table 3). The lactone itself is a relatively poor FTase inhibitor (IC_{50} 280 nM), but once inside the cell, hydrolysis to the hydroxy acid **1d** liberates a compound 10 times more potent.

Table 3. Inhibition of FTase In Vitro and In Vivo by Tetrapeptides Incorporating Reduced Amide Bonds

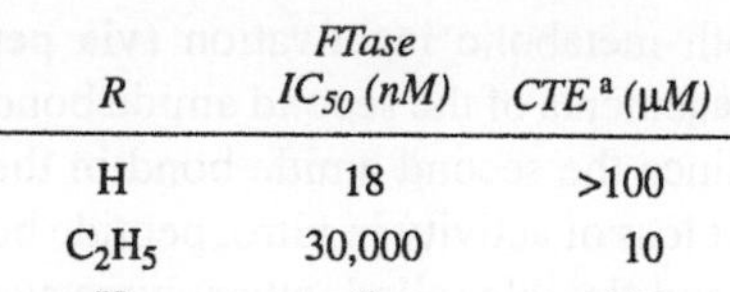

HS
H₂N
OR
SCH₃

Compound	X	R	FTase IC_{50} (nM)	CTE[a] (μM)	Inhibition of Ras Processing in Cells[b] Dose (μM)	Score[c]
2a	O	H	18	>100	100	+/–
2b	O	C₂H₅	30,000	10	10	+/–
2c	H₂	H	5	>100	100	+/–
2d	H₂	CH₃	ND[d]	1	1	+/–
1d	—	—	20	>100	100	—
3	—	—	280	>1,000	100	+++
lovastatin	—	—	—	—	15	+++

Notes. [a]Highest nontoxic concentration of inhibitor in cultured NIH 3T3 cells as determined by viable staining with MTT.

[b]Inhibition of post-translational processing of v-Ras protein in cultured NIH3T3 cells.

[c]Scoring: +++, ≥90% inhibition; +, 30–60% inhibition; +/–, 10–30% inhibition.

[d]ND = not determined.

Having demonstrated that FTIs can indeed block Ras processing in cell culture, it was crucial to demonstrate that this translated into growth inhibition of Ras transformed cells. One of the main phenotype changes occurring in cancerous cells is the ability to sustain anchorage-independent growth. While normal cells are quiescent when suspended in soft agar, transformed cells grow and form colonies under the same conditions. To ascertain its activity in a growth inhibition assay, lactone 3 was incubated with H-Ras transformed NIH 3T3 cells suspended in soft agar. In the first demonstration of this activity, colony formation was inhibited by 3 in a dose-dependent manner.[35] Suppression of Ras-dependent cell growth was achieved with approximately 1 mM 3. The same concentration of 3 had no effect on the growth properties of v-Raf or v-Mos transformed cells, but since neither of these proteins requires farnesylation to effect transformation and both act independently of Ras, the Raf and Mos cell lines should be insensitive to the effects of FTIs.[36–39]

The cellular effects of FTase inhibition with 3 were observed with concentrations 5000–50,000 higher than the in vitro IC_{50} for FTase inhibition by carboxylic acid 1d. Incomplete hydrolysis of the lactone in vivo could be partially responsible for this discrepancy in activity. However, it was also found that the lactone prodrug used in the context of the doubly reduced peptide isostere, i.e. 3, was chemically unstable at physiological pH. Rapid cyclization to the diketopiperazine 5 significantly reduced FTase inhibitory activity.[40] Simple N-alkylation of the reactive secondary amine to give 4 led to loss of activity vs. FTase. To simultaneously protect the compound from both metabolic inactivation (via peptidases) and chemical instability, isosteric replacements of the second amide bond other than methylene–amino were explored. Since the second amide bond in the tetrapeptide inhibitors could be reduced without loss of activity in vitro, peptide bond replacements which were both rigid (olefin) and flexible (alkyl, ether) were synthesized.

Olefin Amide Bond Isosteres

The E-olefin was prepared as a *trans*-amide bond replacement. A number of compounds incorporating substituents to mimic both natural and unnatural amino acid sidechains were prepared by adapting chemistry developed by Ibuka for the synthesis of E-olefin peptide isosteres (see Scheme 1).[40,41] The key step involved anti-S_N2 displacement of vinyl mesylate 8 by boron trifluoride-activated cuprate addition. Compounds containing butyl, propyl, and benzyl substituents at the allylic positions to mimic the a_1 and a_2 sidechains produced potent FTase inhibitors (Table 4).

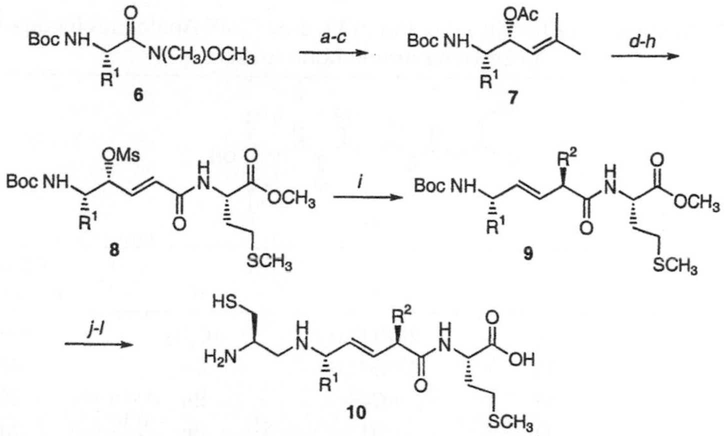

Scheme 1. Synthesis of Olefin Isosteres of CaaX Analogues. *a)* $(CH_3)_2C= CHMgBr$; *b)* $NaBH_4$; *c)* Ac_2O; *d)* O_3, $(CH_3)_2S$; *e)* $Ph_3P=CHCO_2 CH_3$; *f)* LiOH; *g)* EDC, HOBt, L-methionine methyl ester; *h)* CH_3SO_2Cl; *i)* $R^2MgCuCNCl \cdot BF_3$; *j)* HCl; L-N-Boc-S-tritylcysteine aldehyde, $NaCNBH_3$; *k)* TFA, $(C_2H_5)_3SiH$; *l)* NaOH.

Table 4. Inhibition of FTase In Vitro and In Vivo by CaaX Analogues Incorporating an Olefin Amide Bond Isostere

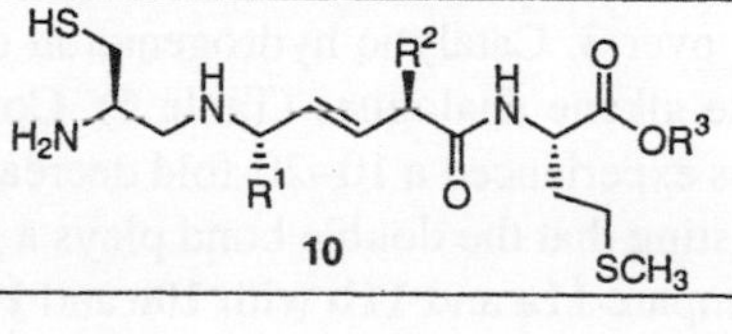

Compound	R^1	R^2	R^3	FTase IC_{50} (nM)	CTE [a] (μM)	Inhibition of Ras Processing in Cells [b] IC_{50} (μM)
10a	2-(R)-C_4H_9	Bn	H	1.9	ND [c]	ND
10b	2-(R)-C_4H_9	Bn	CH_3	80	25	1
10c	i-C_3H_7	i-C_3H_7	H	4.5	ND	ND
10d	i-C_3H_7	i-C_3H_7	i-C_3H_7	ND	50	0.10
10e	2-(R)-C_4H_9	i-C_3H_7	H	20	ND	ND
10f	2-(R)-C_4H_9	i-C_3H_7	i-C_3H_7	ND	25	0.10

Notes: [a]Highest nontoxic concentration of inhibitor in cultured NIH 3T3 cells as determined by viable staining with MTT.

[b]Inhibition of post-translational processing of v-Ras protein in cultured NIH 3T3 cells.

[c]ND = not determined.

Table 5. Inhibition of FTase In Vitro and In Vivo by CaaX Analogues Incorporating an Ethylene Amide Bond Isostere

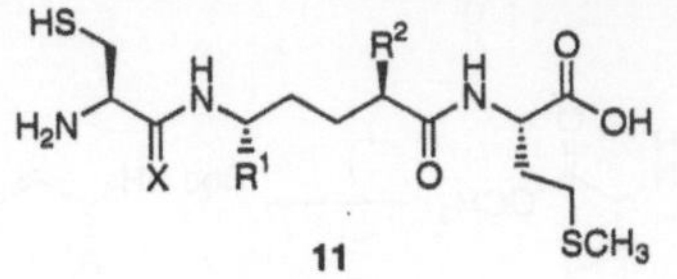

Compound	X	R^1	R^2	FTase IC_{50} (nM)
11a	H_2	2-(R)-C_4H_9	i-C_3H_7	470
11b	H_2	2-(R)-C_4H_9	Bn	40
11c	O	i-C_3H_7	Bn	26
11d	O	H	Bn	20
11e	H_2	H	H	20,000

Many of the olefin isosteres displayed good activity in cell culture. Inhibition of Ras processing in v-H-Ras transformed NIH 3T3 cells was dramatically improved over the earlier peptide and methylene–amino pseudopeptide inhibitors, reflecting both the enzymatic and chemical stability of the linking element relative to the earlier compounds. Compounds **10d** and **10f** blocked v-H-Ras processing in NIH 3T3 cells with IC_{50} values of 100 nM, only 5 to 20-fold higher than the IC_{50}s for FTase inhibition of their respective parent acids in vitro. This level of inhibition of cellular Ras processing was unprecedented, and represented approximately a 1000-fold improvement over **3**. Catalytic hydrogenation of olefin intermediate **9** was used to generate the alkane analogues (Table 5). Compounds incorporating both a_1 and a_2 side chains experienced a 10–20-fold decrease in potency relative to the parent alkene, suggesting that the double bond plays a positive role in terms of conformational bias (compare **11a** and **11b** with **10e** and **10a**). In the olefin series, replacing the a_1 substituent with hydrogen decreased affinity approximately 10-fold; in contrast, removing the a_1 substituent had no effect on potency in the alkane series (compounds **11c** and **11d**, Table 5). By incorporating either methionine or homoserine at the C-terminus, the carbon isosteres selectively inhibited FTase vs. GGTase. As in the methylamino pseudotetrapeptides, these modifications to the peptide backbone yielded inhibitors which were not substrates for the enzyme.

Methylene–Oxy Amide Bond Isosteres

The methyleneoxy peptide isostere was also used to replace the a_1–a_2 amide bond. The requisite ether isostere dipeptides were synthesized according to Scheme 2.[42] High levels of 1,4-asymmetric induction were observed in the alkylation of morpholinone **14**; pseudo-allylic $A_{1,3}$ strain in acylmorpholinone **14** caused the

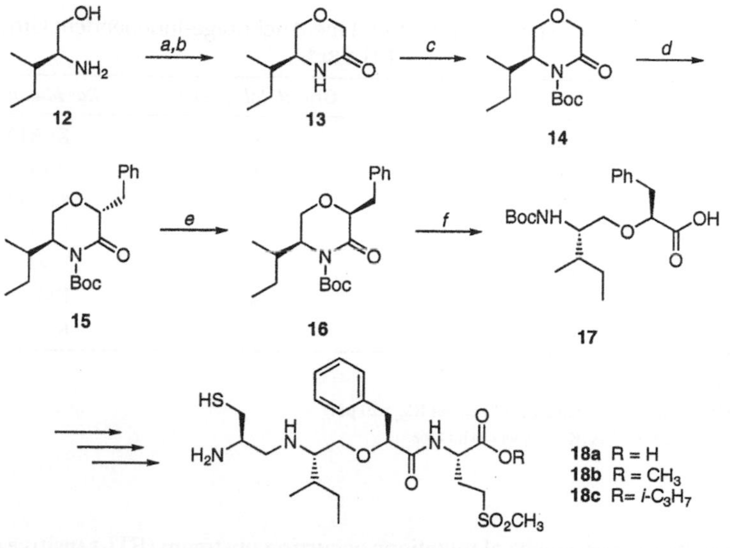

Scheme 2. Synthesis of Methylene-Oxy Isosteres of CaaX Analogues. *a)* ClCH$_2$COCl, Et$_3$N; *b)* NaH, THF; *c)* Boc$_2$O, DMAP; *d)* NaHMDS, THF/DME −78°C; *e)* NaHMDS, THF/DME −78°C; NH$_4$Cl(aq) −78°C; *f)* LiOOH, THF/H$_2$O

isobutyl group to assume an axial disposition, thus directing the incoming electrophile to the opposite face. Kinetic quench of the enolate derived from the initial alkylation product **15** served to invert the stereochemistry of the substituent, giving in the final product the stereochemistry associated with the natural or L-amino acid. The resulting methylene-oxy peptide isostere produced the potent FTase inhibitor **18a**, IC$_{50}$ 2 nM (Scheme 2).[43] In cell culture, the isopropyl ester prodrug **18c** blocked the posttranslational processing of H-Ras with an IC$_{50}$ between 0.100–1 µM, and suppressed colony formation of H-Ras transformed cells in soft agar at a minimum inhibitory concentration (MIC) of 1 µM.[44]

The FTase inhibitor **18a** and its prodrugs were used to explore more of the biological consequences of FTase inhibition. At a concentration of 1 µM, the prodrug **18c** inhibited the anchorage-independent growth of H-*ras*-, but not *raf*-, transformed cells.[44] Aside from showing efficacy in genetically engineered cell lines, it was important to demonstrate efficacy in the presence of multiple genetic defects such as are present in human tumors. At a concentration of 20 µM, **18c** inhibited the colony formation in soft agar of approximately 70% of 60 human tumor cell lines tested (Table 6).[45] An encouraging result from this experiment was that cells with multiple genetic defects responded to treatment. Unexpectedly, however, efficacy did not correlate with *ras* mutational status. Cell lines with either wild-type or mutant *ras* were sensitive as well as refractory to treatment with **18c**.

Table 6. Inhibition of Human Tumor Cell Line Anchorage-Independent Growth by the FTI **18c**[a]

Cell Line	Type	Growth Inhibition[b]	Ras Mutation[c]
PSN-1	Pancreatic	+	Ki-R12
Colo 205	Colon	+	wt
DLD-1	Colon	−	Ki-D13
MDA MB 468	Breast	+	wt
HeLa	Cervical	−	wt
LoVo	Colon	+	Ki-D12
LS180	Colon	−	Ki-D12

Notes: [a]From ref. 45.

[b]"+" denotes $IC_{50} <20$ µM; "−" denotes $IC_{50} >20$ µM.

[c]Ki = Kirsten Ras (K-Ras), wt = wild-type.

This may be a consequence of mutations occurring upstream (FTI-sensitive phenotype), downstream (FTI-insensitive phenotype), or independent of Ras signaling.

In addition to Ras, a number of other proteins are known to be substrates for FTase (many of unknown identity or function, determined by 2-D gel shift experiments in the presence and absence of FTI); these may also play a part in determining relative susceptibility to FTase inhibition.[46] In particular, RhoB, a farnesylated protein involved with the formation of actin stress fibers and cytoskeletal organization, has been implicated in the growth-inhibitory consequences of FTase inhibition.[47,48] Another study using the FTI **19** found that growth inhibition in soft agar of a panel of human tumor cell lines correlated with the mutational status of the H- and N-Ras proteins, but not with K-Ras mutations.[49] In this study also, cells with genetic defects in addition to *ras* were sensitive to FTase inhibition.

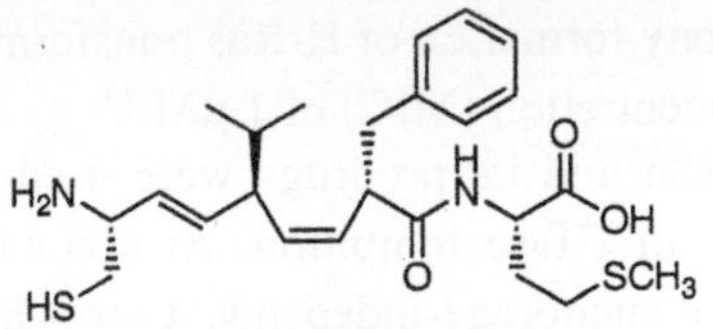

19

C. Conformationally Constrained Tetrapeptide FTIs

Introduction of the conformationally constrained amino acid tetrahydroisoquinolinecarboxylic acid (TIC) into the Ca_1a_2X motif produced potent FTIs. Incorporating TIC at a_2 gave FTIs **20a–c**, with IC_{50} values ranging from 0.6–20 nM (Table 7).[50–52] The carboxylic acids **20c** and **21** inhibited the anchorage-independent growth of H-Ras transformed NIH 3T3 cells with IC_{50} values of 5 µM and 190 nM,

Table 7. Inhibition of FTase By Tetrahydroisoquinoline (TIC) CaaX Analogues

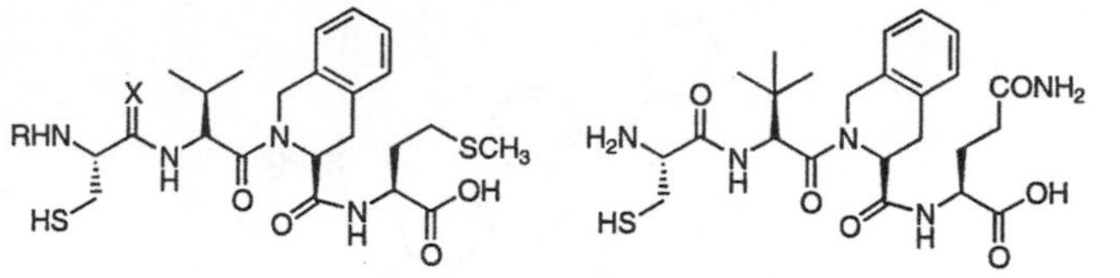

Compound	R	X	FTase IC$_{50}$ (nM)
20a	Ac	O	1
20b	Lysyl	O	20
20c	H	H$_2$	0.6
21	—	—	2.8

respectively. The potency of **21** in soft agar was particularly striking, since a prodrug ester was not required for excellent activity in cell culture.

D. Nonpeptide Replacements for the a₁a₂ Dipeptide

An important area of research evolved to elucidate the gross structural requirements for high enzyme affinity. Two different and complementary designs were reported where the central a_1a_2 dipeptide was completely excised and replaced with a rigid spacer template. Both models sought to impose conformational restrictions on the relatively flexible tetrapeptides and pseudotetrapeptide inhibitors, in order to specifically orient the critical thiol- and carboxylate-binding elements. However, while one model sought to constrain the inhibitor in an extended conformation, the other was designed to mimic a turn conformation bringing the carboxylate and thiol groups in close proximity to each other.

FTase was known to be a zinc metalloenzyme, and it was known that the zinc was essential for catalytic activity.[53] It was attractive to speculate that the FTI cysteinyl thiol group functioned as a zinc ligand in the enzyme–inhibitor complex. Since the carboxylate group was also required for high binding affinity, it was possible that it was functioning as a second zinc ligand. This model was structurally analogous to the CXXC motif found in X-ray crystal structures of zinc-binding proteins and zinc–peptides, wherein the two cysteine thiols coordinate a single zinc cation. The Ca₁a₂X tetrapeptides would need to assume a turn conformation in order to bring both putative ligands in close proximity to each other. Somewhat later, transfer NOE NMR studies supported the hypothesis that the peptides were binding in some type of turn conformation (*vide infra*).[54,55]

A suitably functionalized benzodiazepine was designed to provide a scaffold for the correct positioning of thiol and carboxylate groups, according to the turn

Table 8. Inhibition of FTase By Benzodiazepinone CaaX Analogues

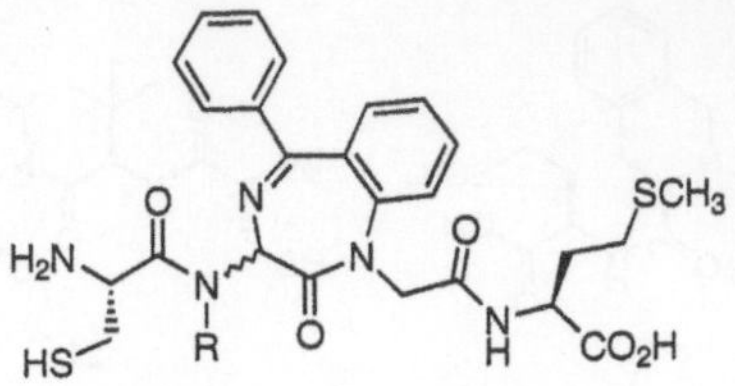

Compound	Stereochemistry	R	FTase IC_{50} (nM)
22a	S	H	415
22b	R	H	440
23a	S	CH_3	370
23b	R	CH_3	0.8

model.[56] The a_1a_2 dipeptide was replaced with 3-amino-1-carboxymethyl-5-phenylbenzodiazepin-2-one. The diastereomeric benzodiazepines **22a** and **22b** were separated chromatographically, and inhibited FTase with IC_{50}s of 415 and 440 nM (Table 8). N-Methylation of the pendant amino group selectively increased the activity of the R isomer **23b** to 0.8 nM, suggesting that a *cis*-amide bond coupled with the correct stereochemistry was necessary for high activity. This compound was not a substrate for farnesylation, and was selective for inhibition of FTase over GGTase. The methyl ester prodrug of **23b** suppressed protein farnesylation in cell culture at concentrations 10-fold lower than the parent carboxylic acid, and inhibited 90% of the growth of H-*ras* transformed cells in monolayer culture at a concentration of 25 μM. Thus, a highly active benzodiazepine inhibitor incorporated a rigid template which could mimic a peptide backbone turn conformation.

A second approach was designed to test whether an extended conformation was compatible with enzyme binding. Unplugging the central dipeptide from Ca_1a_2X and inserting a *meta*-substituted aminobenzoic acid or aminomethylbenzoic acid gave **24a–c**, which although lacking the amino acid sidechains of the central dipeptide, still inhibited FTase with IC_{50} values of 100–6450 nM (Table 9).[57] The *para*-substituted analogues **25a–c** also exhibited potencies in the 50–2550 nM range. In **24** and **25**, the IC_{50} depended on the overall distance between the thiol and carboxylic acid groups, as well as the presence of a cysteine amide bond. In the frame-shifted **26**, the reduced amide bond isostere **26b** was more potent than the amide bond-containing **26a**.[41] These observations indicated that small changes in the backbone structure were capable of dramatic effects on activity. Substituting 5-aminopentanoic acid for a_1a_2 gave an inactive compound, suggesting either that the alkane analogue was too flexible or that the aromatic template is itself a binding element. As observed in other series, these modifications to the peptide backbone

Table 9. Inhibition of FTase By Benzamide and Phenylacetamide CaaX Analogues

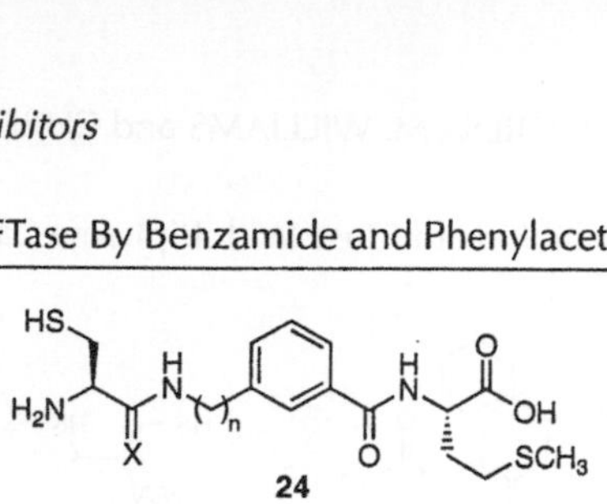

Compound	X	n	FTase IC_{50} (nM)
24a	O	0	6,450
24b	O	1	100
24c	H_2	1	1,200
25a	O	0	50
25b	O	1	2,550
25c	H_2	0	200
26a	O	—	3,300
26b	H_2	—	240

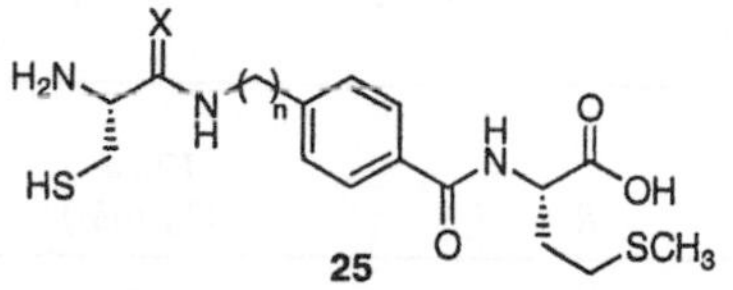

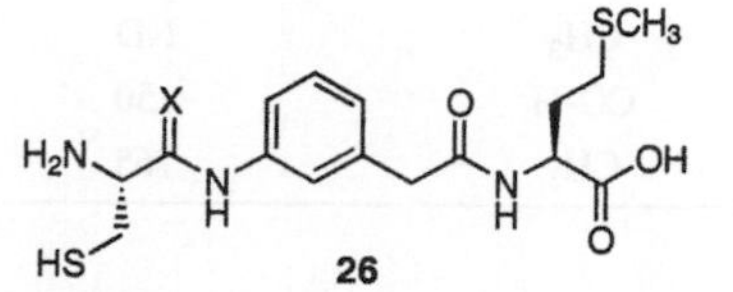

produced compounds which were not substrates for the enzyme, and the C-terminal methionine provided specificity for inhibition of FTase over GGTase. Inhibition of Ras processing in cell culture was demonstrated with high concentrations (500 μM) of the methyl ester prodrug of **24b**.[58]

To maximize hydrophobic interactions with the enzyme, an additional aromatic residue was introduced, yielding biphenyl **27a** (Table 10). This compound was an extremely potent inhibitor of FTase (IC_{50} 0.5 nM) with modest selectivity vs. GGTase (IC_{50} 50 nM).[59] The methyl ester **27b** also exhibited good potency in cell culture, inhibiting Ras processing in H-*ras* transformed NIH 3T3 cells with an IC_{50} of 100 nM. The activity of the rigid aromatic template unequivocally demonstrated that an extended conformation was at least one of the preferred binding modes of thiol–carboxylate inhibitors.

A napthalene ring was also used to replace the $a_1 a_2$ of $Ca_1 a_2 X$ with a rigid spacer element. Both 1,5 (**28**, IC_{50} 5.6 nM) and 1,6 (**29**, IC_{50} 1.8 nM) isomers were quite potent FTIs, further confirming that at least one binding mode of CaaX analogs utilized an extended conformation.[60]

Table 10. Inhibition of FTase With Biphenyl CaaX Analogs

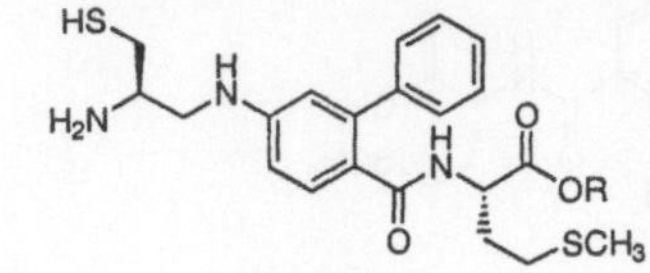

Compound	R	FTase IC$_{50}$ (nM)	Inhibition of Ras Processing in Cells IC$_{50}$ (μM)
27a	H	0.5	ND[a]
27b	CH$_3$	ND	0.100
30a	CO$_2$H	150	ND
30b	CH$_3$	765	100

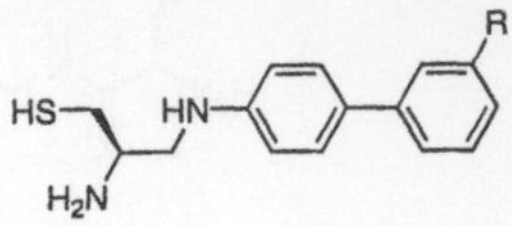

The entire C-terminal a_1a_2X tripeptide could also be replaced with biphenylcarboxylic acid, giving **30a** (IC$_{50}$ 150 nM), a compound which no longer incorporated any amide bonds.[61] Since the specificity-determining C-terminal residue is no longer present in **30a**, it is interesting to note that selectivity is observed with respect to GGTase inhibition (IC$_{50}$ 100 μM). In contrast, substituting leucine for the methionine of **27a** gave a compound which preferentially inhibited GGTase I (IC$_{50}$ 5 nM) over FTase (IC$_{50}$ 25 nM).[62] It may be that specific residues such as leucine act as a signal sequence for binding to GGTase, but that in their absence the default is selective binding to FTase.

Inserting a spacer element between the biphenyl rings of **30a** produced a series of diaryl ethers **31** and **32**. In this series, both the *meta,meta* (**31a–b**) and *para,meta* (**32a–b**) analogues preferred a reduced cysteinyl amide bond (Table 11).[63] The *meta,para* and *para,para* analogues behaved similarly, although these isomers were on average 10-fold less active than **31** and **32**.

Two interesting observations were that replacing the carboxyl group of **30a** with a methyl group only resulted in a fivefold loss of potency (**30c** IC$_{50}$ 765 nM), and

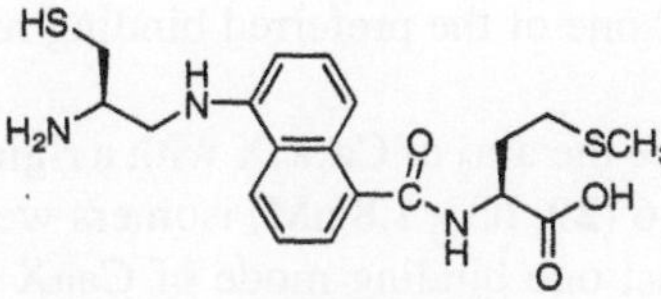

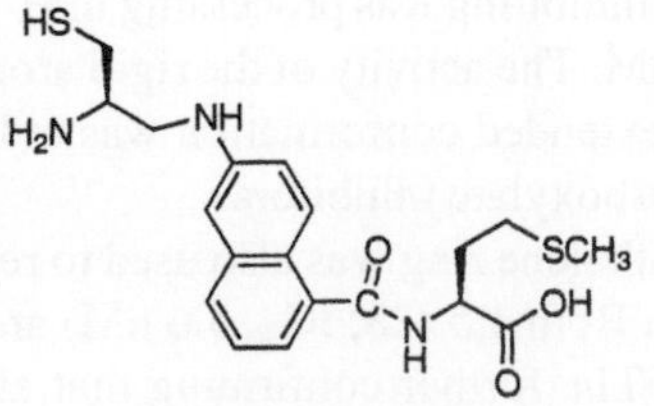

Table 11. Inhibition of FTase With Diaryl Ether CaaX Analogues

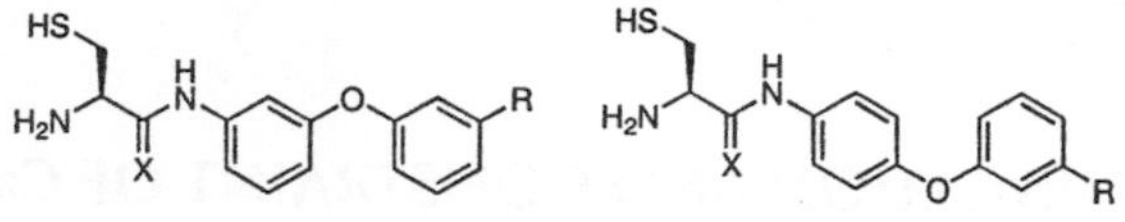

Compound	X	R	FTase IC_{50} (nM)
31a	O	CO_2H	6,900
31b	H_2	CO_2H	250
31c	H_2	CH_3	330
32a	O	CO_2H	16,000
32b	H_2	CO_2H	403
32c	H_2	CH_3	615

replacing the carboxylic acid group of **31b** or **32b** with a methyl group had no effect on potency (**31c** and **32c**). This is in stark contrast to earlier compounds, which typically lost 100-fold in potency when the carboxylate was modified or deleted.[61,63] Since it was speculated (later proven, *vide infra*) that the carboxylate participated in a specific ionic binding interaction with the enzyme, **30a**, **31b**, and **32b** most probably were not optimally positioning their carboxylate groups to take advantage of this interaction. As will be discussed in more detail below, it was found that a carboxylic acid was not an absolute requirement for significant inhibition of FTase after all, and the overall activity of the des-carboxylates **30b**, **31c**, and **32c** (IC_{50}s between 330 and 765 nM), can be viewed in this light as well.

E. Functional Group Limitations with CaaX Analogue Drugs

Designing inhibitors of FTase based upon the structure of its substrates enabled much of the early biochemistry and cell biology of FTase inhibition to be elucidated. The first demonstrations of inhibition of Ras processing in vivo,[64] the selective inhibition of anchorage independent growth in *ras*-transformed cells,[35,56] and efficacy in a mouse model[43] were all achieved with inhibitors designed to mimic and exploit the essential binding elements of the enzymatic substrates. These milestones were tremendously informative in ascertaining and understanding the potential of FTIs as antitumor agents. As important as these early compounds were to the development of the field, several serious drawbacks would ultimately limit their potential as drugs. Among these, poor pharmacokinetics in several animal species proved difficult to overcome. Moreover, there were concerns about potential thiol-related adverse effects, especially given that large doses of drug were generally required to see efficacy in vivo (*vide infra*). While improved pharmacokinetics would eventually lead to lower doses, it was clear that the optimal situation would be to have an FTI which was neither peptidic in nature, nor dependent upon a thiol or carboxylate group for activity. Progress towards this goal is summarized in the following sections.

IV. TRUNCATION AND CONSTRAINT OF Ca₁a₂X ANALOGUES

While extremely potent Ca_1a_2X analogue inhibitors were being designed as pro-drugs, the inherent limitations of this approach spurred research into FTIs lacking a carboxylic acid. Instead of trying to fine-tune complex pharmacokinetic parameters with the appropriate choice of prodrug, it was of considerable interest to obtain FTIs which did not require a prodrug at all for cell permeability.

C-Terminal truncation of pseudotetrapeptide **2a** (C(ΨCH₂)IFM) by deletion of the methionine residue proceeded through pseudotripeptide amide **33a** (C(ΨCH₂)IF–NH₂) to pseudodipeptide phenethylamide **33b** (C(ΨCH₂)IF–NHCH₂CH₂Ph) (Table 12).[65] While the potency of these truncated C-terminal amides was significantly less than that of the parent carboxylic acid **2a**, the structure had been considerably simplified. Reduction of the remaining amide bond in **33b** resulted in a 10-fold decrease in potency. Deletion of either the sulfhydryl or primary amino group in **33b** resulted in a complete loss of activity. SAR studies on phenethyl amide **33b** included examining variations on the easily modified amide

Table 12. Inhibition of FTase By Pseudodipeptide Amides

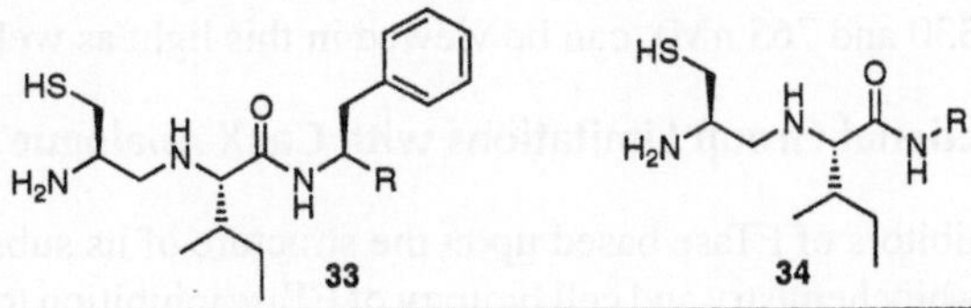

Compound	R	FTase IC_{50} (nM)	GGTase I IC_{50} (nM)	CTE^a (μM)	Inhibition of Ras Processing in Cells^b IC_{50} (μM)
2a	—	18	310	>100	>100
33a	CO₂NH₂	2,100	ND^c	ND	ND
33b	H	707	100,000	10	>10
34a	2,3-Cl₂Bn	23	6,800	5	>5
34b	2,3-(CH₃)₂Ph	40	8,200	10	10

Notes: ^{a}Highest nontoxic concentration of inhibitor in cultured NIH 3T3 cells as determined by viable staining with MTT.

^{b}Inhibition of post-translational processing of v-Ras protein in cultured NIH3T3 cells.

^{c}ND = not determined.

N-substituent. A 30-fold improvement in potency was achieved when the phenethyl amide was replaced by 2,3-dichlorobenzylamide giving **34a**, or by 2,3-dimethylphenylamide to give **34b**. The dipeptide amide **34a** was the first example of a compound lacking the specificity-determining X residue which nonetheless selectively inhibited FTase vs. GGTase (IC_{50} 23 nM vs. 6.8 µM). Because of nonspecific cytotoxicity, the most potent compound reported, **34a**, could not be tested at a high enough concentrations in cell culture to observe inhibition of Ras processing. The somewhat less potent **34b** clearly inhibited Ras processing in cell culture at its maximally tolerated concentration of 10 µM.

The pseudodipeptide amides represented a major advance in the substrate-based design of FTIs. They removed the metabolic liability of the prodrug, were cell active, and were considerably simpler chemically. However, the majority of pseudodipeptide amides were not potent enough and too cytotoxic to demonstrate activity in cell culture. Although the potency issue was perceived as solvable, the dark cloud of nonspecific cytotoxicity hovered over this particular class of compounds like an unwelcome visitor. Changes to the molecule which virtually destroyed FTase activity had no effect on the level of cytotoxicity, indicating that the observed toxicity was unlikely to be mechanism related.

The introduction of a conformational constraint was considered as a way to improve the intrinsic potency of the conformationally flexible pseudodipeptide inhibitors. Unfortunately, no structural information existed on binding conformation and no preferred low-energy conformation was evident, save for the *trans*-amide bond. Since the sulfhydryl and primary amino groups were absolutely required for activity, it seemed prudent to allow more conformational degrees of freedom in this region of the molecule and to direct initial effort towards constraining the peptide amide portion of the molecule. One of the first modifications considered was forcing an approximately extended conformation in this part of the molecule, as shown in Eq. 1. To accomplish this, an ethylene bridge was introduced into **35**

$$(1)$$

between the amide and secondary amine nitrogens, forming keto piperazine **36**. This conformational constraint promptly led to a 10-fold decrease in potency (from 900 to 9000 nM).[66] Comparing piperazine **36** with **35**, however, it was apparent that in **36** the amide bond was constrained in the geometry corresponding to a *cis*-amide bond in **35**. Simply translocating the carbonyl group to the other side of the ring, i.e. **37**, increased potency by two orders of magnitude (IC_{50} 88 nM). In **37** the relative relationship between the butyl substituent and the carbonyl group more approximated their relationship in the *trans*-amide acyclic inhibitors. The increased activity may be the result of a specific interaction between the carbonyl group of the inhibitor and the enzyme, or alternatively it may reflect conformational influences on the pendant benzyl group. In any case, the constrained analogue was able to adopt a conformation compatible with good enzyme affinity. Moving the carbonyl group to an exocyclic position gave N-acylpiperazine 38 (IC_{50} 24 nM) and an approximate fourfold further improvement in potency. SAR studies were pursued by truncation of the butyl substituent, which led to progressive loss of activity for the propyl, ethyl, and methyl analogues, consistent with this group occupying the a_1 binding pocket. Improvements in potency were realized by varying the N-acyl substituent (Table 13). The *ortho*-toluamide 39 was essentially equipotent to the unsubstituted benzamide 38, but *meta*-substitution as in 40 increased potency

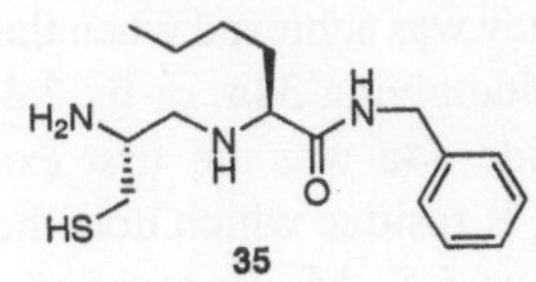

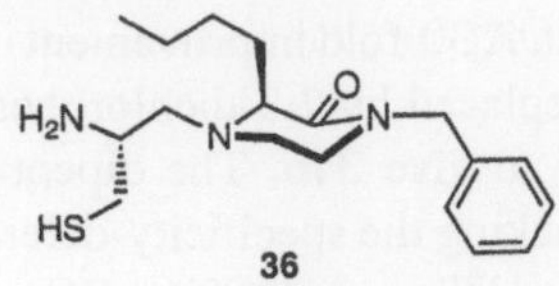

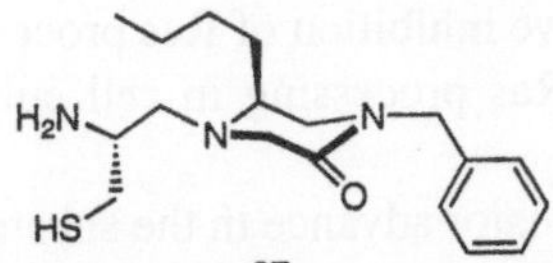

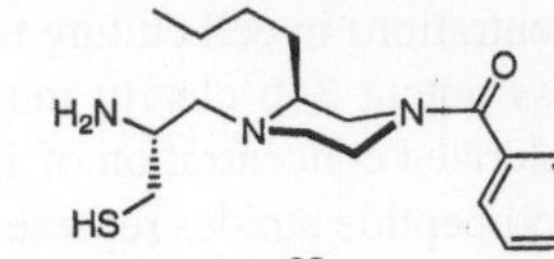

Table 13. Inhibition of FTase By N-Acylpiperazines

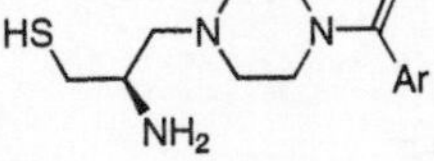

Compound	Ar	FTase IC_{50} (nM)	Inhibition of Ras Processing in Cells[a] IC_{50} (μM)
38	Ph	24	ND[b]
39	o-CH$_3$Ph	27	ND
40	m-CH$_3$Ph	4	0.5
41	p-CH$_3$Ph	5,000	ND
42	2,3-(CH$_3$)$_2$Ph	3	0.5
43	1-naphthyl	1	0.5
44	2-naphthyl	450	ND

Notes: [a]Inhibition of posttranslational processing of v-Ras protein in cultured RAT-1 cells.
           [b]ND = not determined

Table 14. Effect of *N*-Acylpiperazine FTase Inhibitors on Anchorage Independent Growth of *ras*- and *raf*-Transformed Cells

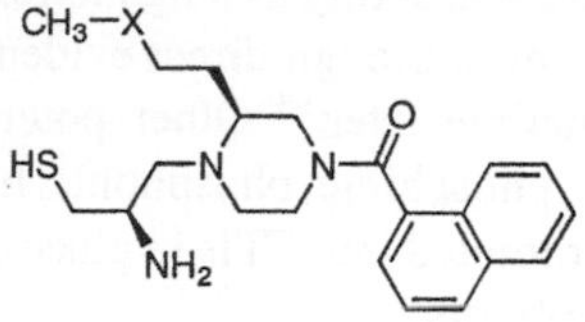

Compound	X	FTase IC_{50} (nM)	CTE [a] (μM)	Soft Agar MIC [b] (μM) ras	raf
43	CH_2	1	10	10	>10
45	O	3	≥100	10	>50

Notes: [a]Highest nontoxic concentration of inhibitor in cultured NIH 3T3 cells as determined by viable staining with MTT.

[b]Minimum inhibitory concentration (MIC) required to achieve a reduction in size and number of colonies of RAT-1 v-*ras* or RAT-1 v-*raf*-transformed cells in soft agar (relative to vehicle).

sixfold. In contrast, *para*-substitution was poorly tolerated (**41**, IC_{50} 5000 nM). The most potent analogue was the 1-naphthoyl derivative **43**, which had an IC_{50} of 1 nM, an overall 25-fold improvement in intrinsic activity over the psuedodipeptide inhibitors. These compounds were not substrates, and were selective for FTase inhibition vs. GGTase inhibition.

Analogous improvement in inhibition of Ras processing and Ras-dependent cell growth compared to the pseudodipeptides was also observed. Piperazine **43** blocked Ras processing with an IC_{50} of 0.5 μM (Table 13), and inhibited H-*ras* transformed cell growth in soft agar with an MIC of 10 μM (Table 14). Inhibition of the control Raf-transformed cell line exhibited an MIC >10 μM, the highest concentration tested since nonspecific cytotoxicity was apparent in a nontransformed cell line at concentrations >10 μM. A related compound, methoxyethyl ether **45**, was as potent as **43** in cell culture vs. *ras*-transformed cells, but did not exhibit nonspecific cytotoxicity up to 100 μM, and did not inhibit Raf-transformed cell growth at >50 μM. Although the cell potency of *N*-acylpiperazine FTIs were not yet as good as the most potent prodrugs, it at least demonstrated that compounds lacking a carboxylic acid could be as potent as the best CaaX analogues in vitro.

V. CYSTEINE SURROGATES IN Ca₁a₂X FTIs

FTIs requiring the presence of a thiol group for high inhibition potency have been described so far. There was also a significant effort to remove or replace the thiol group in our inhibitors. Although therapeutic agents containing thiol groups exist, as a class they have a tendency to exhibit undesirable side effects.[67] It seemed reasonable that the thiol group was acting as a ligand for a zinc ion embedded near or at the active site of FTase, even though direct evidence for such an association was not obtained until sometime later.[68] Other potential zinc ligands such as imidazole and alkoxy groups; phosphinic, phosphonic, hydroxamic, and carboxylic acid groups were either incorporated into FTIs in place of the thiol, or examined in the context of bisubstrate inhibitors.

The histidine-containing tetrapeptide HVFM (IC_{50} 6.8 μM) was substantially less active than the corresponding cysteine-containing peptide CVFM (IC_{50} 37 nM) in FTase inhibition.[30] As discussed earlier, replacing phenylalanine in CVFM with tetrahydroisoquinolinecarboxylic acid (TIC) improved potency 37-fold,[52] and the same substitution was examined in the histidine series.[69] While the potency of **46** increased approximately 10-fold relative to HVFM, deleting the N-terminal primary amino group of **46** led to a substantial increase in potency; nearly 4 orders of magnitude relative to HVFM (Table 15). Thus **47** was prepared by appending imidazoleacetamide to the N-terminus of Val-Tic-Met, giving a very effective inhibitor with an IC_{50} of 0.8 nM. The mechanism of inhibition was not reported. As the free carboxylic acid, **47** inhibited Ras processing with an IC_{50} of 5 μM, and suppressed colony formation of H-Ras transformed NIH 3T3 cells in soft agar with an EC_{50} of 3.8 μM.

In an experiment where CVIM inhibited Ras farnesylation by FTase with an IC_{50} of 150 nM, the serine analogue SVIM was inactive (IC_{50} >100 μM).[31] Assuming the thiol was binding as the thiolate anion, the inactivity of the serine analogue could be accounted for by the large difference in pK_as between a mercapto group

Table 15. Tripeptide FTIs Incorporating Imidazole as a Cysteinyl Replacement

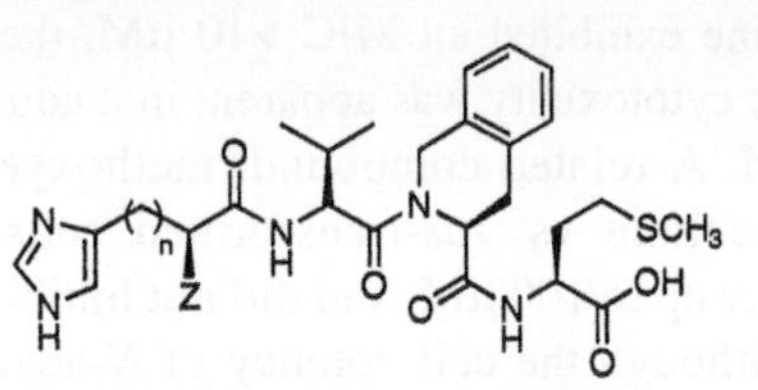

Compound	Z	n	FTase IC_{50} (nM)	GGTase IC_{50} (nM)	MTD^a (μM)	Soft Agar EC_{50} (μM)^b H-ras
46	NH_2	1	520	>360,000	≥150	100
47	H	0	0.8	234	≥150	3.8

Notes: ^{a}Maximum tolerated dose (MTD) of inhibitor in cultured NIH 3T3 cells as determined by viable staining with MTT.

^{b}Effective concentration (EC) required to achieve a 50% reduction in size and number of colonies of RAT-1 v-H-*ras* transformed cells in soft agar (relative to vehicle).

Table 16. Tripeptide FTIs Incorporating Phenol as a Cysteinyl Replacement

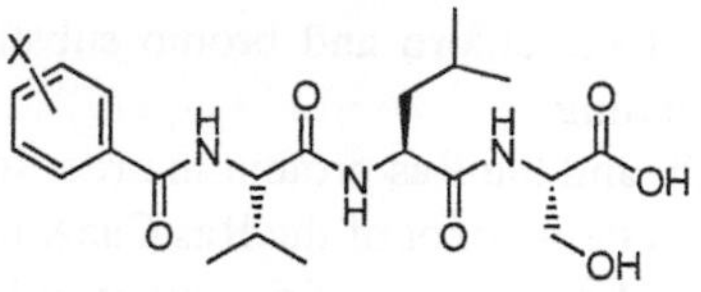

Compound	X	FTase IC$_{50}$ (nM)
48	2-OH	>360,000
49	3-OH	29,000
50	4-OH	>360,000

(pK_a 10.3) and a hydroxyl group (pK_a 15.5). Decreasing the pK_a of the alkoxy group was adopted as a strategy to obtain nonthiol inhibitors, leading to phenol (pK_a ~10) and substituted phenol FTIs. The cysteine residue of CVLS (IC$_{50}$ 3 μM) was deleted and replaced with *ortho*-, *meta*- and *para*-hydroxybenzamide to give compounds 48–50 in Table 16; in this series, only the *meta*-isomer 49 had measurable activity.[70] A similar strategy replaced the cysteine residue in CVFM (IC$_{50}$ 37 nM) with hydroxybenzyl substituents (Table 17).[71] In this series, the *ortho*-hydroxybenzyl isomer was the preferred regioisomer (52, IC$_{50}$ 2.4 μM); activity could be improved 10-fold by placing either a 5-chloro (56 IC$_{50}$ 0.5 μM) or 5-bromo substituent (57 IC$_{50}$ 0.45 μM) onto the phenol ring. It is possible that the further

Table 17. Tripeptide FTIs Incorporating Hydroxybenzyl as a Cysteinyl Replacement

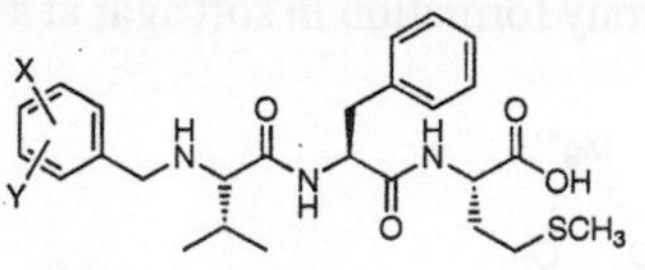

Compound	X	Y	FTase IC$_{50}$ (nM)
51	H	H	21,000
52	2-OH	H	2,400
53	3-OH	H	13,000
54	4-OH	H	17,000
55	2-OH	5-F	1,600
56	2-OH	5-Cl	500
57	2-OH	5-Br	450

decrease in IC_{50} was due to additional lowering of the phenol pK_a by halogen substitution, but if this were the case, the fluoro analogue **55** (IC_{50} 1.6 μM) should have been the most potent analogue. Since **55** was the least potent of the halo analogues, it is likely that the chloro and bromo substituents were themselves contributing to enzyme binding.

FTase binds to both FPP and the Ras protein in order to catalyze the transfer of the farnesyl group to the cysteine thiol of the Ras CaaX box. Consideration of the reaction mechanism led to the synthesis of a number of bisubstrate and transition state mimetic FTIs with the intent of eliminating the requirement for a free thiol, with all its attendant chemical and biological liabilities. A hypothetical model for thiol alkylation catalyzed by FTase involves attack of the cysteine thiol on FPP in a reaction with considerable S_N1 character, but proceeding with complete inversion of configuration at the prenyl reaction site (Figure 1).[72,73] Since both magnesium and zinc are required for enzymatic activity,[53] it is possible that the magnesium coordinates with the FPP diphosphate oxygens to activate the diphosphate as a leaving group. The zinc may be important in binding the thiol group in the proper orientation to capture the allylic cation generated by dissociation of the diphosphate group. Several types of bisubstrate analogues were prepared incorporating a farnesyl group to occupy the isoprenoid binding site and a peptide to occupy the Ras protein a_1a_2X binding site. The two pieces were connected via a tether or spacer element, forming a bisubstrate inhibitor. If the tether contained a phosphate or phosphate mimic the compound could potentially function as either a bisubstrate inhibitor or a transition state mimic, depending how the inhibitor bound relative to the true transition state.

Farnesylphosphonyl and farnesylphosphinyl groups were appended to the N-acylated tripeptide VVM to give compounds **58a–b** (Table 18).[74,75] Both compounds were 6 nM inhibitors of FTase in vitro, with greater than 1000-fold selectivity for inhibition of FTase vs. GGTase. Although the parent carboxylic acids did not affect the growth of H-*ras* transformed NIH-3T3 cells, a methyl ester prodrug **58c** completely suppressed colony formation in soft agar at a concentration of 100 μM.

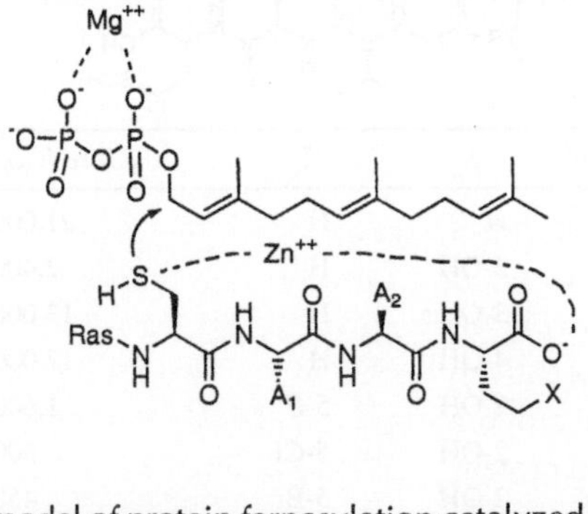

Figure 1. Active site model of protein farnesylation catalyzed by FTase (from ref. 76).

Table 18. Bisubstrate FTIs Incorporating Both Farnesyl
and Peptide Binding Motifs

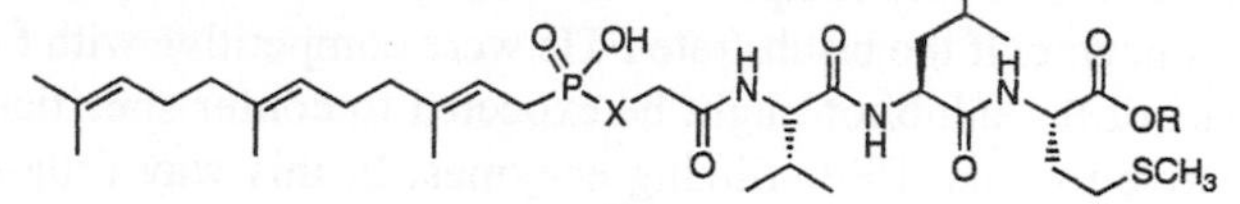

58

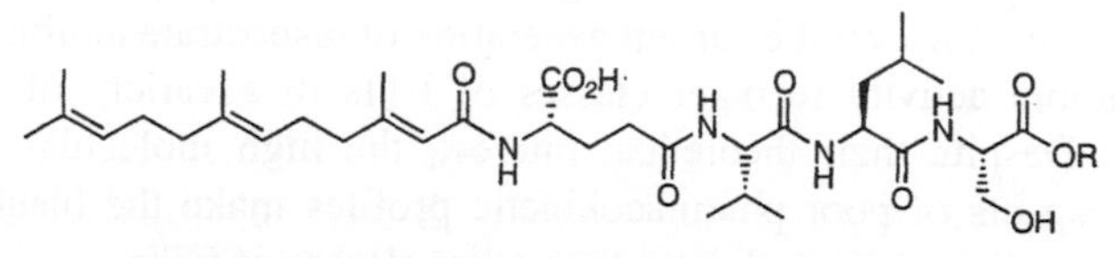

59

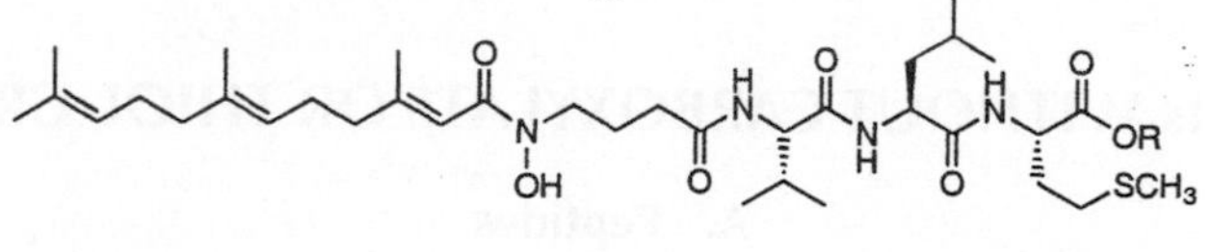

60

Compound	X	R	FTase IC$_{50}$ (nM)
58a	O	H	6
58b	CH$_2$	H	6
58c	CH$_2$	CH$_3$	ND
59	—	H	33
60	—	H	180

In contrast, growth inhibition of H-Ras-CVLL (a genetically engineered substrate for GGTase; the geranylgeranylated form is biologically active) or myristoylated H-Ras (transforms cells independent of farnesylation) transfected NIH 3T3 cells was not observed with **58c** in soft agar. Analysis of total cellular extracts showed inhibition of Ras protein processing only in the cell lines which were sensitive to the effects of **58c** in soft agar.

Similarly designed bisubstrate inhibitors utilized a carboxylic acid as a sulfhydryl mimic in analogy to the design of inhibitors of the zinc metalloenzyme angiotensin converting enzyme.[76] The farnesylated γ-Glu-Val-Leu-Ser **59** had an IC$_{50}$ of 33 nM, which is 10-fold more active than the sulfhydryl-containing tetrapeptide CVLS. Although the carboxylic acid was originally included as a sulfhydryl mimic, it could also be mimicking the negatively charged phosphate, especially in concert with the adjacent amide carbonyl group. Another bisubstrate analogue utilized a hydroxamic acid to link the lipid and protein segments, in an effort to suitably position another zinc ligand.[77] The most potent analogue, **60**, had an IC$_{50}$ of 180 nM. Activity in cell culture was not reported for the carboxylic acid or hydroxamic acid bisubstrate inhibitors.

Kinetic analyses of these bisubstrate inhibitors were not reported, and so it is not clear if the inhibitors in fact competed with both protein and lipid substrates for binding to the enzyme. If the bisubstrate FTIs were competitive with FPP binding, the protein part of the inhibitor might be expected to confer specificity to FTase inhibition relative to other FPP-utilizing enzymes. In this way drug-related side effects which might result from blocking FPP utilization by a variety of enzymes could be avoided. However the current generation of bisubstrate inhibitors does not have comparable activity to other classes of FTIs in a variety of cell culture experiments. Despite their theoretical interest, the high molecular weight and attendant prospects of poor pharmacokinetic profiles make the bisubstrate analogues less attractive drug candidates than other classes of FTIs.

VI. FTIs WITHOUT CARBOXYLATE OR THIOL GROUPS

A. Peptides

The combined advantages of a non-carboxylate FTI and a non-thiol FTI represents the Holy Grail of farnesyl transferase inhibitors. The power of random or directed screening of collections and libraries of compounds is evident in the first reports of non-carboxylate, non-thiol FTIs. Pentapeptide 61 was one of the first reasonably potent FTIs lacking both a carboxylic acid and thiol group; it was identified through screening of a peptide library.[78] Pentapeptide 61 specifically inhibited FTase (IC$_{50}$ 10 nM, 30 mM phosphate buffer) vs. GGTase (IC$_{50}$ 1.25 μM), but was inactive in cell culture. The working hypothesis that cell activity was precluded by high molecular weight (MW 976) led to the synthesis of smaller, lower molecular weight analogues. The compound 62 (MW 676) was essentially equipotent with 61 in vitro, but was shown to inhibit Ras processing in H-*ras* transformed

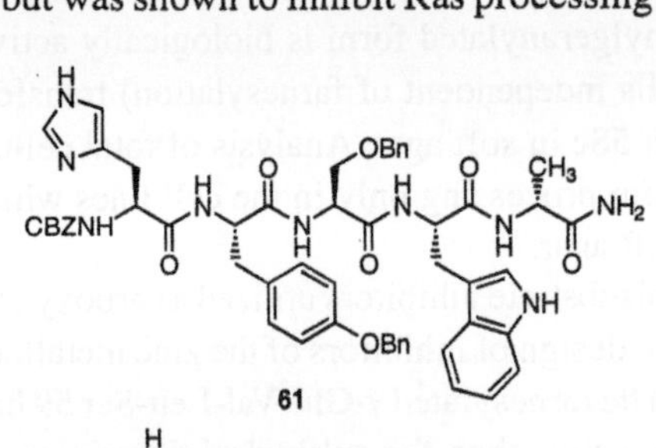

61

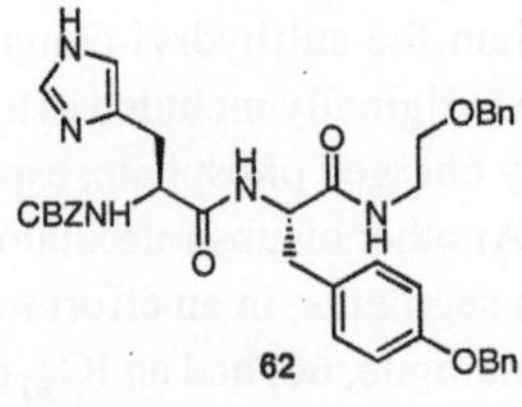

62

Rat1 cells at a concentration of 25 µM. Interestingly, the mechanism of inhibition of these compounds has been reported to be competitive with FPP substrate, and non-competitive with respect to the Ras substrate. A surprising observation was that the IC_{50} values for many of these compounds depended on the concentration of phosphate anion in the assay. In the absence of phosphate, **61** and **62** had IC_{50} values of 1.6 and 6.1 µM, respectively, vs. FTase. In the presence of added phosphate (5 mM), the IC_{50} values decreased to 160 and 165 nM, respectively. Kinetic analyses indicated that the inhibitors and phosphate anion were binding synergistically to FTase. Despite this interesting kinetic profile, the substantial peptide character of **62** and related compounds, in addition to the relatively weak Ras processing activity, indicate that further modifications are warranted to obtain a suitable drug candidate.

B. Nonpeptides

A screening program also led to the discovery that tricyclic compounds related to Loratadine were FTIs.[79] Unlike **61** and **62**, kinetic studies demonstrated that **63a** competed with Ras and not FPP for FTase binding, showing that it was possible to eliminate the thiol and carboxylate moieties from a Ras-competitive inhibitor. In vitro, **63a** inhibited FTase with an IC_{50} of 280 nM while in vivo, Ras processing in H-*ras* transfected Cos-7 monkey kidney cells was inhibited with an IC_{50} of 3 µM (Table 19). Compound **63a** was not an inhibitor of GGTase ($IC_{50} > 114$ µM) and as such, inhibition of Ras processing in a cell line transfected with the artificial GGTase substrate H-Ras-CVLL had an $IC_{50} > 11.8$ µM. Monolayer growth of H-*ras* transformed cells was suppressed with an IC_{50} of 1–4 µM, which was very similar to the IC_{50} for inhibition of Ras processing and about 10-fold higher than the in vitro IC_{50}. This is in contrast to earlier thiol-containing and ester prodrug FTIs which typically exhibited 200–2000 fold differences between in vivo and in vitro potency. The improved in vivo/in vitro potency ratio for **63a** could be attributed to increased stability and/or increased membrane permeability. Related compounds **64** and **65** showed similar levels of inhibition, with the resolution of **65** revealing that greater activity resided in the *S*-isomer (**65b**, IC_{50} 140 nM) than in the *R*-isomer (**65c**, IC_{50} 490 nM; Table 19).[80] Addition of a methyl group to the pyridine ring of **63a** gave **63b**, which was sevenfold more active than **63a** in vitro (IC_{50} 40 nM), and threefold more active than **63a** in blocking Ras processing in Cos-7 monkey kidney cells (IC_{50} 1 µM).[81] Removing the pyridylacetamide group of **63a** and replacing it with either aminoacetyl or sulfonamide functionality generally led to decreases in intrinsic potency.[82]

Pharmacokinetic studies of **63a** and **65a** in mice demonstrated that the compounds had modest bioavailability (21% and 13%, respectively) but short plasma half-lives ($t_{1/2} < 10$ min).[80] While pharmacokinetic parameters in other species have not been reported, it seems likely that issues relating to the absorption, distribution, metabolism, and excretion of this first generation of tricyclic inhibitors needs to be

Table 19. Tricyclic Nonpeptide FTIs: 8-Chlorobenzocycloheptapyridine
Substituted Piperazine and Piperidine Analogues

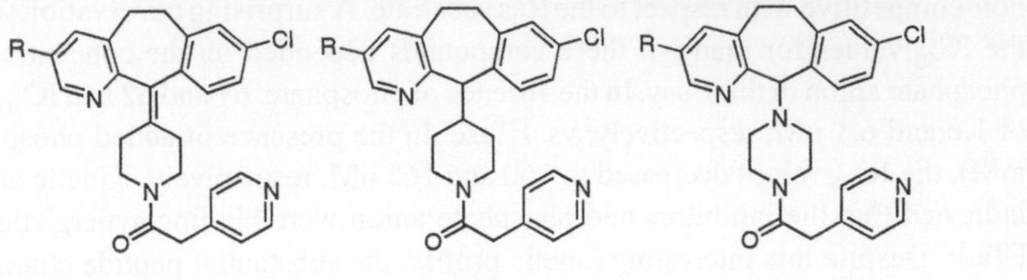

Compound	R	Stereochemistry	FTase IC$_{50}$(nM)	GGTase IC$_{50}$ (nM)	Inhibition of Ras Processing in Cells[a] IC$_{50}$ (μM)
63a	H	NA[b]	280	>114,000	1–3
63b	CH$_3$	NA	40	40,000	1
64	H	R,S	160	>46,000	1.2
65a	H	R,S	180	>46,000	3.7
65b	H	S	140	ND[b]	ND
65c	H	R	490	ND	ND

Notes: [a]Inhibition of H-Ras processing in Cos-7 monkey kidney cells.
[b]NA = not applicable; ND = not determined.

addressed in order to obtain a suitable oral drug candidate. The tricyclic FTIs **64** and **65a** showed efficacy in a mouse tumor model (nude mice inoculated with the human colorectal tumor cell line containing the K-Ras mutation) when administered orally at a dose of 100 mg/kg twice daily.

VII. STRUCTURE AND FUNCTION OF FTase: POTENTIAL GUIDES FOR THE DESIGN OF NOVEL FTIs

An accurate representation of enzyme structure, although not essential for the design of small molecule enzyme inhibitors as drugs, can be a remarkably powerful guide. While some drug discovery programs have employed an understanding of active site structure to aid in the development of potent ligands, essentially the opposite has been the case for farnesyl transferase inhibitor design. Our knowledge to date of the binding interactions between the CaaX substructure of Ras and the catalytic site of FTase has been largely inferred from the structure–activity relationships of CaaX mimetic ligands. Clarification of FTase structure and function continues to mature, but the advance of FTase inhibitors toward the later stages of development might precede our ability to employ detailed structure-aided design.

Farnesyl protein transferase is a zinc-containing enzyme containing nonidentical α- and β-subunits which are 48 kDa and 46 kDa in size, respectively.[83,84] While both subunits are required for catalytic activity of FTase, the β-subunit fulfills the role of binding to both the protein and FPP substrates. Kinetic studies have revealed that bovine FTase appears to bind the Ras and FPP substrates in a random sequential mechanism, while the human enzyme employs the sequential binding of FPP prior to Ras.[85–87] The chemical mechanism for the farnesyl transfer to the thiol moiety of the cysteine residue has been characterized as an electrophilic alkylation, rather than nucleophilic displacement.[72] The presence of one enzyme-bound zinc ion is required for FTase activity as well as protein substrate binding, but not for iso-prenoid binding.[53] A recent study of metal coordination within FTase was performed using the Co^{2+} rather than Zn^{2+}-substituted form, a catalytically active species.[68] Optical absorption spectroscopy of the ternary complex, Co^{2+}–FTase·FPP·TKCVIM, indicated metal binding to both the thiol group of the CaaX substrate and to the prenylsulfide product, suggesting its presence at the enzyme active site and by inference implicating a catalytic rather than structural role for zinc. The function of Mg^{2+} ion however, required in millimolar concentration for enzyme activity, is less clear.[53,83]

A recent X-ray crystal structure of rat FTase, which shares 97% sequence identity with the human enzyme, has significantly enhanced our three-dimensional view of the mammalian enzyme and its interactions with the FPP and protein substrates.[88] The secondary structure is comprised of two unusual domains. The crescent-shaped α-domain, which envelops part of the β-subunit, is an arrangement of seven successive pairs of α-helices grouped in right-handed antiparallel coiled coils which are folded into a double-layered right-handed superhelix. The β-subunit is made up of 12 α-helices which are folded into an α–α-barrel. Near the interface between the two subunits is the distorted pentacoordinate zinc ion, ligated by the conserved residues Asp-279β (in bidentate fashion), Cys-299β, His-362β, and a well-ordered water molecule. Cross-linking studies of photoactivatable FPP and CaaX peptide substrates have revealed that binding occurs in the β-subunit.[84,89,90] It is interesting to note, therefore, that the crystal structure reveals two clefts in the β-subunit which intersect at the zinc-bound active site, and have inner surface properties which complement the structures of the enzyme substrates. One cleft, parallel to the rim of the α–α-barrel, is lined with hydrophilic positively charged residues, and is thought to bind the CaaX motif of the Ras substrate. A second cleft formed by the hydrophobic cavity down the center of the α–α-barrel is just deep enough to accommodate the farnesyl chain of FPP, but appears too shallow to enable binding of GGPP.

The cocrystallization of an enzyme and inhibitor can give invaluable information with regard to specific intermolecular contacts which may be exploited in new compound design. Although a CaaX substrate or mimetic was not bound in the active site of the crystallized rat FTase, an interesting crystallization artifact enabled some degree of speculation about the mode of binding which FTIs may exploit.

The nine-residue C-terminus of a neighboring enzyme's β-subunit was found to insert itself into the hydrophilic cleft at the active site. Thus, although the last four residues of the bound sequence Ala-9β-Val-Thr-Ser-Asp-Pro-4β-Ala-Thr-Asp-1β-CO_2H bear little resemblance to CaaX, some mimicry of normal CaaX peptide binding was suggested. Notably, a hydrogen bond between the CO_2H-terminus and Lys-164α was consistent with the previous finding that this residue's mutation disabled FTase activity.[91] In addition, Pro-4β of the nonapeptide, which would correspond to the reactive cysteine of CaaX, is directly adjacent to both the zinc ion and the presumed cleft for the FPP binding. Finally, an additional observation which may have implications for FTase inhibitor design was that the Pro-4β-Ala-Thr-Asp-1β-CO_2H sequence had adopted a type I β turn, reminiscent of earlier NMR-based structure determinations for FTase-bound CaaX peptides (*vide infra*).

The insertion of the C-terminal FTase nonapeptide sequence into the active site crystal structure, while provocative and compelling as a potential CaaX structural mimic, may in principle present a significant caveat to an interpretation of the critical interresidue interactions which may govern the specific binding of an enzyme inhibitor to FTase. Indeed, recent site-directed mutagenesis studies of several conserved residues in the β-subunit of human FTase[92] suggests that artifactual perturbation of the rat crystal structure active site should be considered. While independent point mutations of the three residues found to ligate the zinc ion in the crystal structure were deleterious to enzyme activity as one might expect,[88,93] an additional FTase mutant αβD359A was defective in Zn^{2+} binding, even though Asp-359β did not appear to be a zinc ligand in the rat crystal structure. Instead, Asp-359β had formed a direct interaction with Thr-7β of the bound nonapeptide C-terminus. The seemingly conflicting data may suggest that Asp-359β has a direct or indirect effect on zinc binding.

Mutational studies to identify residues which interact with the C-terminal CO_2H of CaaX mimetic FTIs also suggest that the crystal structure interpretation of the molecular basis for CaaX recognition by FTase may warrant modification. Whereas the crystal structure implicated Lys-164α, Arg-291β, and Lys-294β in binding to the terminal Asp-1β-CO_2H residue of the bound nonapeptide, the X-residue specificity of a CaaX substrate was found to be affected by mutations of the more distal Tyr-362β residue in yeast FTase (homologous to Tyr-361β in rat and human FTase).[94] In fact, mutations at this site were found to switch the protein substrate specificity to that of GGTase I. Furthermore, another amino acid (Arg-202β) located adjacent to Tyr-361β in the crystal structure has been found to interact with the CaaX carboxylate.[92] From the X-ray crystal data, Arg-202β had been postulated to be the residue which interacts with the phosphate moieties of FPP. However, a structure–activity relationship study of a series of carboxyl- and non-carboxyl containing FTIs using the human FTase αβR202A mutant revealed the residue's role in binding to the C-terminus of CaaX. While carboxylic acid containing FTIs exhibited significant reduction in inhibitory activity toward R202A-FTase relative to the wild-type enzyme, the non-carboxylates were insensitive to the mutational

Table 20. Inhibition of Wild Type and αβR202A FTase by Carboxyl-Containing and Non-Carboxyl-Containing FTase Inhibitors

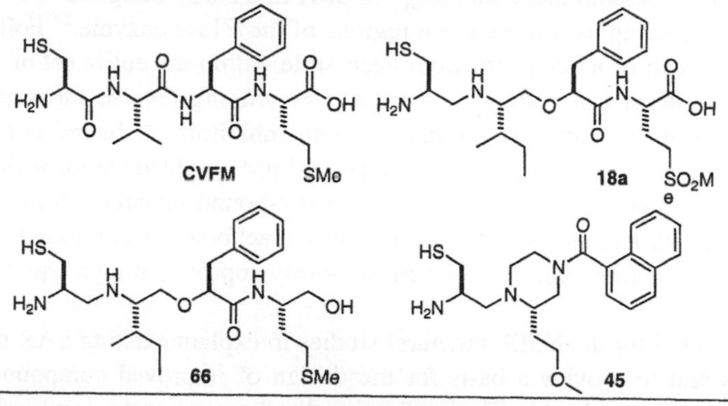

Compound	FTase IC$_{50}$ (nM)		
	Wild Type	*R202A Mutant*	*Ratio R202A/wt*
CVFM	45	11,000	244
18a	3	290	94
66	1,600	6,200	4
45	3	3	1

change (Table 20). Furthermore, the lower IC$_{50}$ value for a C-terminal carboxylate derivative (**18a**) in comparison to alcohol derivative (**66**) is consistent with the X-residue carboxylate binding to Arg-202β through an ionic interaction.

Prior to the publication of the FTase crystal structure, only indirect methods for probing the enzyme active-site structure had been accomplished. Among these, two reports described NMR studies to elucidate the structural conformations of enzyme-bound inhibitors. The two-dimensional nuclear Overhauser effect spectroscopic technique (TRNOESY) is well suited for elucidating the bound structures of relatively weakly bound inhibitors, since fast exchange of the ligand with the macromolecule enables transfer of the negative NOEs from the bound state to the resonances of the unbound species in solution, where they can be observed. The NOE data is converted to hydrogen–hydrogen internuclear distances which serve as constraints in subsequent computer modeling experiments.

In an initial study,[54] the heptapeptide KTKCVFM (K$_i$ 4.5 μM) was employed as a mimic of the K-Ras protein. Several intraresidue TRNOEs were found in the terminal tetrapeptide sequence, including the notable NH$_i$/NH$_{i+1}$ interactions at Val-Phe and Phe-Met, as well as between the Val-Phe sidechains. After incorporation of the distance constraints, semiquantitative and qualitative manipulations provided a structure which adopts a type I β-turn about the CVFM fragment, aligning the Cys and Met residues in proximity to each other. A subsequent report described the TRNOE-derived structures the peptide inhibitor CVWM (IC$_{50}$ 525 nM) and the peptidomimetic **66** (IC$_{50}$ 1.6 μM) in a study designed to correct for nonspecific binding to nonactive site regions of the FTase enzyme.[55] Following a computer search of all conformations accessible within the entire set of TRNOE distance constraints, a variety of structures containing reverse turns, extended structures, and half turns was found for both inhibitors. A ligand competition experiment using the tightly binding **18a** (IC$_{50}$ 2 nM) enabled the identification of those TRNOEs derived only from the active site-bound ligand conformations. In the adjusted structures, both of the inhibitor backbones were found to adopt nonideal reverse turn conformations most closely approximating a type III β-turn (Figure 2).

The potential for the NMR structural studies to explain existing SAR of earlier inhibitors and to provide a basis for the design of improved compounds is, in principle, quite considerable. The implication that the cysteine thiol and carboxylate groups of CaaX bind to the same or adjacent features in the active site may bring to mind a host of β-turn mimetics as templates for preparing new FTIs. Indeed, SAR within the series of benzodiazepine a$_1$a$_2$ dipeptide mimetics (*vide supra*), which demonstrated the value of a *cis*-amide conformation to inhibitory activity, are consistent with this design approach.[56] In a related study,[95] replacement of phenylalanine in acetyl-CVFM with L-1,2,3,4-tetrahydro-3-isoquinolinecarboxylic acid (TIC) was argued to aid in constraining the backbone in a turned conformation, providing a 20-fold boost in potency.

The notion that FTase-bound CaaX inhibitors adopt turned conformations is not without controversy, since several studies have implicated extended conformations, and thus a more distant relationship between the thiol and carboxylate moieties. In fact, a different interpretation with regard to the influence of the TIC constraint on conformational control was offered based on molecular modeling studies of

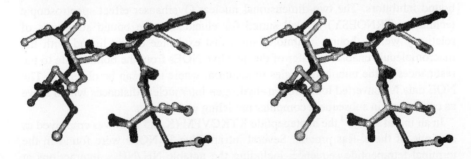

Figure 2. Best-fit superposition (cross-eyed stereoview) of the FTase-bound conformation of **CVWM** (dark) and **66** (light) as determined by 1H NMR transferred NOE spectroscopy.

KCa_1a_2M pentapeptides.[51] Molecular dynamics simulations (with the C- and N-termini uncharged to avoid artifactual turn conformations) gave a population of KCV(TIC)M conformers consisting mostly of extended (80%), and to a lesser extent bent (10%) and turned (10%) structures. Subsequent analysis of a series of similar pentapeptides, spanning a wide range of conformational distributions (% extended structures), revealed an interesting direct relationship to enzyme activity. Thus, the progression of a_1a_2 replacements from Val-Phe to Val-TIC to (N-Me)Val-TIC gave increasingly potent FTIs with increasingly higher proportion of extended conformations. This same conformational issue had been probed in a comparison of FTIs where a_1a_2 is replaced by 3-aminomethylbenzamide (**24b**, IC_{50} 100 nM) and 4-aminobenzamide (**25a**, IC_{50} 50 nM; *vide supra*).[57] Both analogues are able to maintain a similar degree of separation between the sulfhydryl and carboxyl groups, analogous to the 10.8 Å distance found for the extended conformation of CVIM. Although their inhibitory activities are similar, the flexibility of the spacers is very different; it is not possible for the more extended **25a** to adopt a β-turn conformation.

Regardless of any similarity or dissimilarity of the TRNOE-derived FTase bound structures to existing structural series of FTIs, their use as guiding tools in the search for novel structures with improved biological properties has ample merit. It is worthy of note, for example, that although C-terminal polar groups presumably played an important role in the binding and orientation of the turned CaaX NMR substrates (perhaps by interacting with Arg-202β), the resulting structures could nonetheless be applied to a search for the highly desirable non-carboxyl-containing FTIs which might share only a portion of the same binding interactions. An overlay of highly potent non-carboxylate **45**[66] with the CVWM TRNOE structure[55] illustrates the overlap of cysteine residues, a_1-hydrophobic residues, and the aromatic groups (Figure 3).

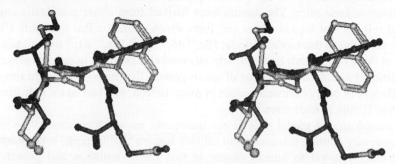

Figure 3. Best-fit superposition (cross-eyed steroview) of the FTase-bound conformation of **CVWM** (dark) and energy-minimized **45** (light) as determined by ^{1}H NMR transferred NOE spectroscopy (see ref. 113).

The same CVWM NMR structure was found to fit within a pharmacophore model generated from a series of non-carboxylate non-thiol tricyclic FTIs using the Catalyst computer program.[80] Although the goal of elucidating a potential similarity between the binding mode of the tricyclic compounds (e.g. 65a) and that of peptide inhibitors (e.g. CVWM) requires further study, a nice overlay within four hydrophobic regions and one hydrogen bond acceptor region was achieved. At the very least, the notion that the bound conformation of a peptide inhibitor can mimic a three-dimensional database of nonpeptides was validated. This suggests that the converse should hold; use of the active site structures may play an important role in the design of structurally unique compounds with improved biological properties.

VIII. BIOLOGICAL CONSEQUENCES OF FARNESYL TRANSFERASE INHIBITION

A. Anti-Proliferative Effect of FTIs in Animal Tumor Models

Two animal models of Ras-dependent tumor growth were used to evaluate the efficacy of prodrugs of 18a in vivo. Nude mice, which are deficient in T-cells and therefore lack a cell-mediated immune response, were subcutaneously implanted with ras-transformed cells which multiply to form a tumor. Treatment with FTase inhibitor 18b (20 mg/kg, administered once daily s.c.) was initiated 2 days following tumor cell implantation, and continued for 5 days. After cessation of treatment, the tumors were allowed to grow for an additional 6 days before harvesting. For mice with tumors derived from either the H-, K-, or N-ras oncogenes, 18b was found to reduce tumor size by 58–66% compared to vehicle-treated control animals.[43] The average tumor weight in 18b-treated mice injected with raf- or mos-transformed cells, however, was not statistically different from the control tumors.

A number of other FTIs have been reported to inhibit tumors containing ras mutations in nude mice. The tumors were derived from either genetically engineered cell lines or human tumor cell lines which contain a Ras mutation. FTIs which are active in this model include: 18c,[44] 45,[66] 19,[49] 27a,[96] 21,[52] 64, and 65a.[80] Most of these compounds had to be administered at high doses in order to observe a statistically significant inhibition of tumor growth. The high doses might simply be reflective of poor pharmacokinetics in mice. Indeed, 65a had an i.v. half-life of less than 10 min in nude mice.[80]

A second animal model uses H-ras transgenic mice (Oncomouse[TM]) which spontaneously develop mammary and salivary tumors.[97] This model is considered to be more relevant to human cancer in that tumor initiation and growth is endogenous, in contrast to nude mouse tumor explants. Mice with palpable tumors between 50 and 350 mm^3 were treated with either the ester prodrug 18c (40 mg/kg, once daily, s.c.) or vehicle, and the size of the tumors monitored over time.[44] While tumors steadily increased in size in the control group, treated animals experienced a decrease in tumor size. Averaged over a 30 day treatment period, the control group of mice experienced a 16 mm^3/day mean growth rate for their primary tumors, while **18c**-treated mice showed −5.4 mm^3/day mean growth rate. Upon cessation of treatment the tumor would reappear; this "new" tumor was responsive to renewed treatment with drug. The relatively high dose more than likely reflects poor pharmacokinetics in mice; even at these high doses, however, adverse effects on the mice were not noted.

B. The Relationship Between GGTase and K-Ras Tumor Growth

As mentioned above, selective inhibition of FTase vs. GGTase was pursued to avoid affecting the major pool of prenylated proteins. Under normal physiological conditions, H-Ras, K$_{4A}$-Ras, K$_{4B}$-Ras, and N-Ras proteins are exclusive substrates of FTase. While H-Ras, K-Ras, and N-Ras share sequence homology and are similarly processed, mutations found in human cancer most frequently occur in the K-*ras* gene. For technical reasons, many early studies utilized H-*ras* transfected cells under the assumption that the different Ras proteins would behave similarly biochemically. As research progressed, it was observed that cells transformed by K$_{4B}$-*ras* were less sensitive to growth inhibition by FTIs **23c**, **27b**, or **18c**, than cells transformed by H-Ras.[62,98] One explanation for this difference is that the K$_{4B}$-Ras protein has 20-fold higher affinity for FTase than H-Ras, and therefore should be less sensitive to the effects of FTIs in vivo.[99,100] A second scenario is also plausible, however. Genetic studies in yeast found that some of the preferred substrates for FTase can be cross-prenylated by GGTase under certain conditions.[101] The K$_{4A}$-Ras, K$_{4B}$-Ras, and N-Ras isoforms can in fact be geranylgeranylated by GGTase, albeit 5- to 400-fold less efficiently than they are farnesylated by FTase.[98,100] When farnesylation by FTase is blocked in cells, the alternative geranylgeranylation reaction occurs. Since geranylgeranylated forms of Ras were also capable of inducing cell transformation,[36,99,102] it was suggested that geranylgeranylation of K$_{4B}$-Ras by GGTase provided a mechanism for cells harboring a K$_{4B}$-Ras mutation to escape growth inhibition by an FTI. In fact, direct chemical evidence for the occurrence of the cross-prenylation reaction in vivo has been obtained with the FTIs **63a** and **19**.[103,104] The impact of cross prenylation on tumor biology remains to be fully elucidated, however. Coadministration of FTI **27a** and a selective GGTI to nude mice inoculated with K$_{4B}$-Ras-containing human tumor cell lines did not result in an enhanced antitumor activity relative to the FTI alone.[105] As mentioned above, the selective FTI **18b** inhibited the anchorage-independent growth of a variety of human tumor cell lines independent of the mutational status of the Ras protein. Since 2-D gel studies showed that the FTI **23c** blocked the prenylation of at least 18 different proteins,[46] it is quite possible that other substrates for FTase are involved in the anti-proliferative effect of FTIs, and that the relative sensitivity of these substrates towards FTIs plays a crucial role in determining efficacy in vivo.

C. Effect of FTIs on Nontumor Tissue

Early studies in the field indicated that Ras-transformed cell growth could be inhibited by an FTI at concentrations well below that which caused cell death in normal cells. This experiment compared the MIC for growth inhibition of H-*ras* transformed cells in soft agar with the concentration of FTI compatible with ≥90% cell survival of a nontransformed cell line [the concentration which represents the cytotoxic endpoint (CTE)]. One explanation for this phenomenon may lie in the observation that the pool of unfarnesylated and cytosolic H-Ras competes with the membrane-bound farnesylated form for binding to its effector protein Raf. Sequestering Raf away from the cell membrane effectively negates its biological activity, and disrupts the signaling pathway. This dominant-negative effect of unfarnesylated Ras with activating mutations might be more pronounced than with normal unfarnesylated Ras, especially since the mutant cytosolic Ras would exist predominately in the GTP-bound form (the form which complexes Raf) due to its impaired GTPase activity.

In addition to the CTE assay, a second measure of specificity was provided by the selective growth inhibition of *ras*-transformed cells vs. control cell lines which should be insensitive to the effects of an FTI. For example, cell lines transformed with proteins that are not part of the Ras signaling pathway, or which are part of the Ras signaling pathway but which operate downstream from Ras, should not be affected by an FTI. This type of selectivity was demonstrated for a number of different prodrugs using H-*ras* transformed cells and either *mos*- or *raf*-transformed cells.

Based upon the above observations, it appeared that FTIs could selectively inhibit the growth of only *ras*-transformed cells while not affecting normal cells. Since most cancer chemotherapeutic agents kill both normal and tumor cells, it appeared that an FTI might be efficacious at impeding tumor growth with fewer mechanism-related side effects.

In mouse efficacy studies, treating animals with high doses of FTIs did not engender signs of toxicity or ill health. At the dose levels required to achieve efficacy, **18c** had no visible effect on animal health, in contrast to doxorubicin, one of the mainstays of antitumor therapy and a classic cytotoxic drug.[43] As normal Ras proteins (and other farnesylated proteins) participate in cell processes, the greater sensitivity of tumor tissue (vs. normal tissue) to the antiproliferative effects of FTIs has yet to be unequivocally accounted for, but several hypotheses have been advanced.[106] Susceptibility to FTIs would depend on the exact nature of the protein substrate. While H-Ras is one of the most sensitive substrates, lamins A and B (structural proteins in the cell nuclear membrane which are farnesylated) required higher concentrations of FTIs **23c** and **B581** (the Ile→Val analogue of **2c**) to achieve similar levels of processing inhibition.[34,107] It is also not known what degree of farnesylation is required for maintenance of normal function, and indeed this probably varies among protein substrates. Additionally, cross-prenylation by

GGTase I may rescue the functions of some normally farnesylated proteins or redundant signaling pathways might be able to circumvent the Ras/MAP kinase pathway.

The effect of FTIs on retinal function also needs to be carefully examined. Several proteins involved in retinal signal transduction are farnesylated in vivo, presumably by FTase. These include rod cell cGMP phosphodiesterase α-subunit,[108,109] rod cell transducin γ-subunit,[110,111] and rhodopsin kinase.[112] Since the retina consists of terminally differentiated, nondividing cells, the anti-proliferative properties of FTIs should be inconsequential. Visual function could possibly be affected by alterations in the prenylation of proteins involved in retinal signal transduction, although any changes of this sort should be reversible.

IX. CONCLUSIONS

Both pharmaceutical and academic research have led to a substantial amount of progress both in the design and discovery of FTIs, and in the elucidation of the biological consequences of FTase inhibition in vitro and in vivo. In this complex area of research, surely there await additional major discoveries relating to protein prenylation and tumor growth. Despite the success of FTIs in treating tumors in mouse models, however, the final gauge of success lies with controlling human cancer. With the evolution of better and more powerful FTIs, hopefully the clinical outcome will be the ultimate validation of farnesyl transferase inhibitors as antitumor agents.

ACKNOWLEDGMENTS

We would like to acknowledge the many contributions of our colleagues and co-workers, whose work comprises this review. We would especially like to express our gratitude to our co-workers in Cancer Biology, and to Dr. Sam Graham, Dr. George Hartman, and Dr. Bob Smith of Medicinal Chemistry for their guidance, insight, and contributions to the program at Merck. We would like to acknowledge Dr. Chris Culberson in Molecular Systems for modeling studies and for providing computer generated structures for this review. Finally, we are indebted to Mrs. Joy Hartzell for her help in manuscript preparation.

REFERENCES AND NOTES

1. Casey, P. J.; Solski, P. A.; Der, C. J.; Buss, J. E. *Proc. Natl. Acad. Sci. USA* **1989**, *86*, 8323–8327.
2. Schafer, W. R.; Kim, R.; Sterne, R.; Thorner, J.; Kim, S.-K.; Rine, J. *Science* **1989**, *245*, 379–385.
3. Hancock, J. F.; Magee, A. I.; Childs, J. E.; Marshall, C. J. *Cell* **1989**, *57*, 1167–1177.
4. Jackson, J. H.; Cochrane, C. G.; Bourne, J. R.; Solski, P. A.; Buss, J. E.; Der, C. J. *Proc. Natl. Acad. Sci. USA* **1990**, *87*, 3042–3046.
5. Kim, R.; Rine, J.; Kim, S.-H. *Mol. Cell. Biol.* **1990**, *10*, 5945–5949.
6. Kato, K.; Cox, A. D.; Hisaka, M. M.; Graham, S. M.; Buss, J. E.; Der, C. J. *Proc. Natl. Acad. Sci. USA* **1992**, *89*, 6403–6407.

7. Lowry, D. R.; Willumsen, B. M. *Annu. Rev. Biochem.* **1993**, *62*, 851–891.
8. McCormick, F. *Nature* **1993**, *363*, 15–16.
9. Der, C. J.; Cox, A. D. *Cancer Cells* **1991**, *3*, 331–340.
10. Zhang, X.-F.; Settleman, J.; Kyriakis, J. M.; Takeuchi-Suzuki, E.; Elledge, S. J.; Marshall, M. S.; Bruder, J. T.; Rapp, U. R.; Avruch, J. *Nature* **1993**, *364*, 308–313.
11. Warne, P. H.; Viciana, P. R.; Downward, J. *Nature* **1993**, *364*, 352–355.
12. McCormick, F. *Cell* **1989**, *56*, 5–8.
13. Hall, A. *Cell* **1990**, *61*, 921–923.
14. Xu, G.; Lin, B.; Tanaka, K.; Dunn, D.; Wood, D.; Gesteland, R.; White, R.; Weiss, R.; Tamanoi, F. *Cell* **1990**, *63*, 835–841.
15. Martin, G. A.; Viskochil, D.; Bollag, G.; McCabe, P. C.; Crosier, W. J.; Haubruck, H.; Conroy, L.; Clark, R.; O'Connell, P.; Cawthon, R. M.; Innis, M. A.; McCormick, F. *Cell* **1990**, *63*, 843–849.
16. Ballester, R.; Marchuk, D.; Boguski, M.; Saulino, A.; Letcher, R.; Wigler, M.; Collins, F. *Cell* **1990**, *63*, 851–859.
17. West, M.; Kung, H.-F.; Kamata, T. A. *FEBS Lett.* **1990**, *259*, 245–248.
18. Wolfman, A.; Macara, I. G. *Science* **1990**, *248*, 67–69.
19. Downward, J.; Riehl, R.; Wu, L.; Weinberg, R. A. *Proc. Natl. Acad. Sci. USA* **1990**, *87*, 5998–6002.
20. Mizuno, T.; Kaibuchi, K.; Yamamoto, T.; Kawamura, M.; Sakoda, T.; Fujioka, H.; Matsuura, Y.; Takai, Y. *Proc. Natl. Acad. Sci. USA* **1991**, *88*, 6442–6446.
21. Grand, R. J. A.; Owen, D. *Biochem. J.* **1991**, *279*, 609–631.
22. Barbacid, M. *Annu. Rev. Biochem.* **1987**, *56*, 779–827.
23. Bos, J. L. *Cancer Res.* **1989**, *49*, 4682–4689.
24. Graham, S. L. *Exp. Opin. Ther. Patents* **1995**, *5*, 1269–1285.
25. Graham, S. L.; Williams, T. M. *Exp. Opin. Ther. Patents* **1996**, *6*, 1295–1304.
26. Manne, V.; Ricca, C. S.; Brown, J. C.; Tuomari, A. V.; Yan, N.; Patel, D.; Schmidt, R.; Lynch, M. J.; Ciosek, C. P.; Carboni, J. M.; Robinson, S.; Gordon, E. M.; Barbacid, M.; Seizinger, B. R.; Biller, S. A. *Drug Devel. Res.* **1995**, *34*, 121–137.
27. Reiss, Y.; Goldstein, J. L.; Seabra, M. C.; Casey, P. J.; Brown, M. S. *Cell* **1990**, *62*, 81–88.
28. Moores, S. L.; Schaber, M. D.; Mosser, S. D.; Rands, E.; O'Hara, M. B.; Garsky, V. M.; Marshall, M. S.; Pompliano, D. L.; Gibbs, J. B. *J. Biol. Chem.* **1991**, *166*, 14603–14610.
29. Reiss, Y.; Stradley, S. J.; Gierasch, L. M.; Brown, M. S. *Proc. Natl. Acad. Sci. USA* **1991**, *88*, 732–736.
30. Leftheris, K.; Kline, T.; Natarajan, S.; DeVirgilio, M. K.; Cho, Y. H.; Pluscec, J.; Ricca, C.; Robinson, S.; Seizinger, B. R.; Manne, V.; Meyers, C. A. *Bioorg. Med. Chem. Lett.* **1994**, *4*, 887–892.
31. Goldstein, J. L.; Brown, M. S.; Stradley, S. J.; Reiss, Y.; Gierasch, L. M. *J. Biol. Chem.* **1991**, *266*, 15575–15578.
32. Brown, M. S.; Goldstein, J. L.; Paris, K. J.; Burnier, J. P.; Marsters, J. C. *Proc. Natl. Acad. Sci. USA* **1992**, *89*, 8313–8316.
33. Graham, S. L.; deSolms, S. J.; Giuliani, E. A.; Kohl, N. E.; Mosser, S. D.; Oliff, A. I.; Pompliano, D. L.; Rands, E.; Breslin, M. J.; Deana, A. A.; Garsky, V. M.; Scholz, T. H.; Gibbs, J. B.; Smith, R. L. *J. Med. Chem.* **1994**, *37*, 725–732.
34. Garcia, A. M.; Rowell, C.; Ackermann, K.; Kowalczyk, J. J.; Lewis, M. D. *J. Biol. Chem.* **1993**, *268*, 18415–18418.
35. Kohl, N. E.; Mosser, S. D.; deSolms, S. J.; Giuliani, E. A.; Pompliano, D. L.; Graham, S. L.; Smith, R. L.; Scolnick, E. M.; Oliff, A. I.; Gibbs, J. B. *Science* **1993**, *260*, 1934–1937.
36. Cox, A. D.; Hisaka, M. M.; Buss, J. E.; Der, C. J. *Mol. Cell. Biol.* **1992**, *12*, 2606–2615.
37. Feig, L. A.; Cooper, G. M. *Mol. Cell. Biol.* **1988**, *8*, 3235–3243.

38. Stacy, D. W.; Roudebush, M.; Day, R.; Mosser, S. D.; Gibbs, J. B.; Feig, L. A. *Oncogene* **1991**, *6*, 2297–2304.

39. Smith, M. R.; DeGudicibus, S. J.; Stacey, D. W. *Nature* **1986**, *320*, 540–543.

40. Wai, J. S.; Bamberger, D. L.; Fisher, T. E.; Graham, S. L.; Smith, R. L.; Gibbs, J. B.; Mosser, S. D.; Oliff, A. I.; Pompliano, D. L.; Rands, E.; Kohl, N. E. *Bioorg. Med. Chem.* **1994**, *2*, 939–947.

41. Harrington, E. M.; Kowalczyk, J. J.; Pinnow, S. L.; Ackermann, K.; Garcia, A. M.; Lewis, M. D. *Bioorg. Med. Chem. Lett.* **1994**, *4*, 2775–2780.

42. Anthony, N. J.; Gomez, R. P.; Holtz, W. J.; Murphy, J. S.; Ball, R. G.; Lee, T.-J. *Tetrahedron Lett.* **1995**, *36*, 3821–3824.

43. Kohl, N. E.; Wilson, F. R.; Mosser, S. D.; Giuliani, E. A.; deSolms, S. J.; Conner, M. W.; Anthony, N. J.; Holtz, W. J.; Gomez, R. P.; Lee, T.-J.; Smith, R. L.; Graham, S. L.; Hartman, G. D.; Gibbs, J. B.; Oliff, A. *Proc. Natl. Acad. Sci. USA* **1994**, *91*, 9141–9145.

44. Kohl, N. E.; Omer, C. A.; Conner, M. W.; Anthony, N. J.; Davide, J. P.; deSolms, S. J.; Giuliani, E. A.; Gomez, R. P.; Graham, S. L.; Hamilton, K.; Handt, L. K.; Hartman, G. D.; Koblan, K. S.; Kral, A. M.; Miller, P. J.; Mosser, S. D.; O'Neill, T. J.; Shaber, M. D.; Gibbs, J. B.; Oliff, A. *Nature Med.* **1995**, *1*, 792–797.

45. Sepp-Lorenzino, L.; Ma, Z.; Rands, E.; Kohl, N.; Gibbs, J. B.; Oliff, A.; Rosen, N. *Cancer Res.* **1995**, *55*, 5302–5309.

46. James, G. L.; Goldstein, J. L.; Pathak, R. K.; Anderson, R. G. W.; Brown, M. S. *J. Biol. Chem.* **1994**, *269*, 14182–14190.

47. Lebowitz, P. F.; Davide, J. P.; Prendergast, G. C. *Mol. Cell. Biol.* **1995**, *15*, 6613–6622.

48. Lebowitz, P. F.; Casey, P. J.; Prendergast, G. C.; Thissen, J. A. *J. Biol. Chem.* **1997**, *272*, 15591–15594.

49. Nagasu, T.; Yoshimatsu, K.; Rowell, C.; Lewis, M. D.; Garcia, A. M. *Cancer Res.* **1995**, *55*, 5310–5314.

50. Marsters, J. C.; McDowell, R. S.; Reynolds, M. E.; Oare, D. A.; Somers, T. C.; Stanley, M. S.; Rawson, T. E.; Struble, M. E.; Burdick, D. J.; Chan, K. S.; Duarte, C. M.; Paris, K. J.; Tom, J. Y. K.; Wann, D. T.; Xue, Y.; Burnier, J. P. *Bioorg. Med. Chem.* **1994**, *2*, 949–957.

51. Clerc, F.-F.; Guitton, J.-D.; Fromage, N.; Lelievre, Y.; Duchesne, M.; Tocque, B.; James-Surcouf, E.; Commercon, A.; Becquart, J. *Bioorg. Med. Chem. Lett.* **1995**, *5*, 1779–1784.

52. Leftheris, K.; Kline, T.; Vite, G. D.; Cho, Y. H.; Bhide, R. S.; Patel, D. V.; Patel, M. M.; Schmidt, R. J.; Weller, H. N.; Andahazy, M. L.; Carboni, J. M.; Gullo-Brown, J. L.; Lee, F. Y. F.; Ricca, C.; Rose, W. C.; Yan, N.; Barbacid, M.; Hunt, J. T.; Meyers, C. A.; Seizinger, B. R.; Zahler, R.; Manne, V. *J. Med. Chem.* **1996**, *39*, 224–236.

53. Reiss, Y.; Brown, M. S.; Goldstein, J. L. *J. Biol. Chem.* **1992**, *267*, 6403–6408.

54. Stradley, S. J.; Rizo, J.; Gierasch, L. M. *Biochemistry* **1993**, *32*, 12586–12590.

55. Koblan, K. S.; Culberson, J. C.; deSolms, S. J.; Giuliani, E. A.; Mosser, S. D.; Omer, C. A.; Pitzenberger, S. M.; Bogusky, M. J. *Protein Sci.* **1995**, *4*, 681–688.

56. James, G. L.; Goldstein, J. L.; Brown, M. S.; Rawson, T. E.; Somers, T. C.; McDowell, R. S.; Crowley, C. W.; Lucas, B. K.; Levinson, A. D.; Marsters, J. C., Jr. *Science* **1993**, *260*, 1937–1942.

57. Qian, Y.; Blaskovich, M. A.; Saleem, M.; Seong, C.-M.; Wathen, S. P.; Hamilton, A. D.; Sebti, S. M. *J. Biol. Chem.* **1994**, *269*, 12410–12413.

58. Qian, Y.; Blaskovich, M. A.; Seong, C. M.; Vogt, A.; Hamilton, A. D.; Sebti, S. M. *Bioorg. Med. Chem. Lett.* **1994**, *4*, 2579–2584.

59. Lerner, E. C.; Qian, Y.; Blaskovisch, M. A.; Fossum, R. D.; Vogt, A.; Sun, J.; Cox, A. C.; Der, C. J.; Hamilton, A. D.; Sebti, S. M. *J. Biol. Chem.* **1995**, *270*, 26802–26806.

60. Burns, C. J.; Guitton, J.-D.; Baudoin, D.; Lellevre, Y.; Duchesne, M.; Parker, F.; Fromage, N.; Commercon, A. *J. Med. Chem.* **1997**, *40*, 1763–1767.

61. Vogt, A.; Yimin, Q.; Blaskovich, M. A.; Fossum, R. D.; Hamilton, A. D.; Sebti, S. M. *J. Biol. Chem.* **1995**, *270*, 660–664.

62. Lerner, E. C.; Qian, Y.; Hamilton, A. D.; Sebti, S. M. *J. Biol. Chem.* **1995**, *270*, 26770–26773.

63. Dinsmore, C. J.; Williams, T. M.; Hamilton, K.; O'Neill, T. J.; Rands, E.; Koblan, K. S.; Kohl, N. E.; Gibbs, J. B.; Graham, S. L.; Hartman, G. D.; Oliff, A. I. *Bioorg. Med. Chem. Lett.* **1997**, *7*, 1345–1348.

64. Gibbs, J. B.; Pompliano, D. L.; Mosser, S. D.; Rands, E.; Lingham, R. B.; Singh, S. B.; Scolnik, E. M.; Kohl, N. E.; Oliff, A. I. *J. Biol. Chem.* **1993**, *268*, 7617–7620.

65. deSolms, S. J.; Deana, A. A.; Giuliani, E. A.; Graham, S. L.; Kohl, N. E.; Mosser, S. D.; Oliff, A. I.; Pompliano, D. L.; Rands, E.; Scholz, T. H.; Wiggins, J. M.; Gibbs, J. B.; Smith, R. L. *J. Med. Chem.* **1995**, *38*, 3967–3971.

66. Williams, T. M.; Ciccarone, T. M.; MacTough, S. C.; Bock, R. L.; Conner, M. W.; Davide, J. P.; Hamilton, K.; Koblan, K. S.; Kohl, N. E.; Kral, A. M.; Mosser, S. D.; Omer, C. A.; Pompliano, D. L.; Rands, E.; Schaber, M. D.; Shah, D.; Wilson, F. R.; Gibbs, J. B.; Graham, S. L.; Hartman, G. D.; Oliff, A. I.; Smith, R. L. *J. Med. Chem.* **1996**, *39*, 1345–1348.

67. Jaffe, I. A. *Am. J. Med.* **1986**, *80*, 471–476.

68. Huang, C.-C.; Casey, P. J.; Fierke, C. A. *J. Biol. Chem.* **1997**, *272*, 20–23.

69. Hunt, J. T.; Lee, V. G.; Leftheris, K.; Seizinger, B.; Carboni, J.; Mabus, J.; Ricca, C.; Yan, N.; Manne, V. *J. Med. Chem.* **1996**, *39*, 353–358.

70. Patel, D. V.; Patel, M. M.; Robinson, S. S.; Gordon, E. M. *Bioorg. Med. Chem. Lett.* **1994**, *4*, 1883–1888.

71. Kowalczyk, J. J.; Ackermann, K.; Garcia, A. M.; Lewis, M. D. *Bioorg. Med. Chem. Lett.* **1995**, *5*, 3073–3078.

72. Dolence, J. M.; Poulter, C. D. *Proc. Natl. Acad. Sci. USA* **1995**, *92*, 5008–5011.

73. Mu, Y.; Omer, C. A.; Gibbs, R. A. *J. Am. Chem. Soc.* **1996**, *118*, 1817–1823.

74. Patel, D. V.; Gordon, E. M.; Schmidt, R. J.; Weller, H. N.; Young, M. G.; Zahler, R.; Barbacid, M.; Carboni, J. M.; Gullo-Brown, J. L.; Hunihan, L.; Ricca, C.; Robinson, S.; Seizinger, B. R.; Tuomari, A. V.; Manne, V. *J. Med. Chem.* **1995**, *38*, 435–442.

75. Manne, V.; Yan, N.; Carboni, J. M.; Tuomari, A. V.; Ricca, C. S.; Brown, J. G.; Andahazy, M. L.; Schmidt, R. J.; Patel, D.; Zahler, R.; Weinmann, R.; Der, C. J.; Cox, A. D.; Hunt, J. T.; Gordon, E. M.; Barbacid, M.; Seizinger, B. R. *Oncogene* **1995**, *10*, 1763–1779.

76. Bhide, R. S.; Patel, D. V.; Patel, M. M.; Robinson, S. P.; Hunihan, L. W.; Gordon, E. M. *Bioorg. Med. Chem. Lett.* **1994**, *4*, 2107–2112.

77. Young, M. G.; Patel, D. V.; Simon, R.; Gordon, E. M. *National Medicinal Chemistry Symposium*; Salt Lake City, UT, 1994.

78. Leonard, D. M.; Shuler, K. R.; Poulter, C. J.; Eaton, S. R.; Sawyer, T. K.; Hodges, J. C.; Su, T.-Z.; Scholten, J. D.; Gowan, R. C.; Sebolt-Leonard, J. S.; Doherty, A. M. *J. Med. Chem.* **1997**, *40*, 192–200.

79. Bishop, W. R.; Bond, R.; Petrin, J.; Wang, L.; Patton, R.; Doll, R.; Njoroge, G.; Catino, J.; Schwartz, J.; Windson, W.; Syto, R.; Schwarz, J.; Carr, D.; James, L.; Kirschmeier, *J. Biol. Chem.* **1995**, *270*, 30611–30618.

80. Mallams, A. K.; Njoroge, F. G.; Doll, R. J.; Snow, M. E.; Kaminski, J. J.; Rossman, R. R.; Vibulbhan, B.; Bishop, W. R.; Kirschmeier, P.; Liu, M.; Bryant, M. S.; Alvarez, C.; Carr, D.; James, L.; King, I.; Li, Z.; Lin, C.-C.; Nardo, C.; Petrin, J.; Remiszewski, S. W.; Taveras, A. G.; Wang, S.; Wong, J.; Catino, J.; Girijavallabhan, V.; Ganguly, A. K. *Bioorg. Med. Chem.* **1997**, *5*, 93–99.

81. Njoroge, F. G.; Doll, R. J.; Vibulbhan, V.; Alvarez, C. S.; Bishop, W. R.; Petrin, J.; Kirschmeier, P.; Carruthers, N. I.; Wong, J. K.; Albanese, M. M.; Piwinski, J. J.; Catino, J.; Girijavallabhan, V.; Ganguly, A. K. *Bioorg. Med. Chem. Lett.* **1997**, *5*, 101–113.

82. Njoroge, F. G.; Vibulbhan, B.; Alvarez, C. S.; Bishop, W. R.; Petrin, J.; Doll, R. J.; Girijavallabhan, V.; Ganguly, A. K. *Bioorg. Med. Chem. Lett.* **1996**, *6*, 2977–2982.

83. Chen, W.-J.; Andres, D. A.; Goldstein, J. L.; Brown, M. S. *Proc. Natl. Acad. Sci. USA* **1991**, *88*, 11368–11372.

84. Omer, C. A.; Kral, A. M.; Diehl, R. E.; Prendergast, G. C.; Powers, S.; Allen, C. M.; Gibbs, J. B.; Kohl, N. E. *Biochemistry* **1993**, *32*, 5167–5176.

85. Furfine, E. S.; Leban, J. J.; Landavazo, A.; Moomaw, J. F.; Casey, P. J. *Biochemistry* **1995**, *34*, 6857–6862.

86. Pompliano, D. L.; Schaber, M. D.; Mosser, S. D.; Omer, C. A.; Shafer, J. A.; Gibbs, J. B. *Biochemistry* **1993**, *32*, 8341–8347.

87. Pompliano, D. L.; Rands, E.; Schaber, M. D.; Mosser, S. D.; Anthony, N. J.; Gibbs, J. B. *Biochemistry* **1992**, *31*, 3800–3807.

88. Park, H.-W.; Boduluri, S. R.; Moomaw, J. F.; Casey, P. J.; Beese, L. S. *Science* **1997**, *275*, 1800–1804.

89. Reiss, Y.; Seabra, M. C.; Armstrong, S. A.; Slaughter, C. A.; Goldstein, J. L.; Brown, M. S. *J. Biol. Chem.* **1991**, *266*, 10672–10677.

90. Ying, W.; Sepp-Lorenzino, L.; Cai, K.; Aloise, P.; Coleman, P. S. *J. Biol. Chem.* **1994**, *269*, 470–477.

91. Andres, D. A.; Goldstein, J. L.; Ho, Y. K.; Brown, M. S. *J. Biol. Chem.* **1993**, *268*, 1383.

92. Kral, A. M.; Diehl, R. E.; deSolms, S. J.; Williams, T. M.; Kohl, N. E.; Omer, C. A. *J. Biol. Chem.* **1997**, *272*, 27319–27323.

93. Fu, H. W.; Moomaw, J. F.; Moomaw, C. R.; Casey, P. J. *J. Biol. Chem.* **1996**, *271*, 28541–28547.

94. Del Villar, K.; Mitsuzawa, H.; Yang, W.; Sattler, W.; Tamanoi, F. *J. Biol. Chem.* **1997**, *272*, 680–687.

95. Marsters, J. C., Jr.; McDowell, R. S.; Reynolds, M. E.; Oare, D. A.; Somers, T. C.; Stanley, M. S.; Rawson, T. E.; Struble, M. E.; Burdick, D. J.; Chan, K. S., et al. *Bioorg. Med. Chem.* **1994**, *2*, 949–957.

96. Sun, J.; Qian, Y.; Hamilton, A. D.; Sebti, S. M. *Cancer Res.* **1995**, *55*, 4243–4247.

97. Sinn, E.; Muller, W.; Pattengale, P.; Tepler, I.; Wallace, R.; Leder, P. *Cell* **1987**, *49*, 465–475.

98. James, G.; Goldstein, J. L.; Brown, M. S. *Proc. Natl. Acad. Sci. USA* **1996**, *93*, 4454–4458.

99. James, G. L.; Goldstein, J. L.; Brown, M. S. *J. Biol. Chem.* **1995**, *270*, 6221–6226.

100. Zhang, F. L.; Kirschmeier, P.; Carr, D.; James, L.; Bond, R. W.; Wang, L.; Patton, R.; Windsor, W. T.; Syto, R.; Zhang, R.; Bishop, R. W. *J. Biol. Chem.* **1997**, *272*, 10232–10239.

101. Trueblood, C. E.; Ohya, Y.; Rine, J. *Mol. Cell. Biol.* **1993**, *13*(7), 4260–4275.

102. Hancock, J. F.; Cadwallader, K.; Paterson, H.; Marshall, C. J. *EMBO J.* **1991**, *10*, 4033–4039.

103. Whyte, D. B.; Kirschmeier, P.; Hockenberry, T. N.; Nunez-Oliva, I.; James, L.; Catino, J. J.; Bishop, R. B.; Pai, J.-K. *J. Biol. Chem.* **1997**, *272*, 14459–14464.

104. Rowell, C. A.; Kowalczyk, J. J.; Lewis, M. D.; Garcia, A. M. *J. Biol. Chem.* **1997**, *272*, 14093–14097.

105. Lerner, E. C.; Hamilton, A. D.; Sebti, S. M. *Anti-Cancer Drug Design* **1997**, *12*, 229–238.

106. Gibbs, J. B.; Oliff, A. I. *Annu. Rev. Pharmacol. Toxicol.* **1997**, *37*, 143–166.

107. Dalton, M. B.; Fantle, K. S.; Bechtold, H. A.; DeMaio, L.; Evans, R. M.; Krystosek, A.; Sinensky, M. *Cancer Res.* **1995**, *55*, 3295–3304.

108. Anant, J. S.; Ong, O. C.; Xie, H.; Clarke, S.; O'Brien, P. J.; Fung, B. K.-K. *J. Biol. Chem.* **1992**, *267*, 687–690.

109. Qin, N.; Pittler, S. J.; Baehr, W. *J. Biol. Chem.* **1992**, *267*, 8458–8463.

110. Fukada, Y. T.; Takao, T.; Ohguro, H.; Yoshizawa, T.; Akino, T.; Shimonishi, Y. *Nature* **1990**, *346*, 658–660.

111. Perez-Sala, D.; Tan, E. W.; Canada, F. J.; Rando, R. J. *Proc. Natl. Acad. Sci. USA* **1991**, *88*, 3043–3046.

112. Inglese, J.; Glickman, J. F.; Lorenz, W.; Caron, M. G.; Lefkowitz, R. J. *J. Biol. Chem.* **1992**, *267*, 1422–1425.

113. Conformations were generated using metric matrix distance geometry algorithm JG (S. Kearsley, Merck & Co., unpublished). The conformations were subjected to energy-minimization within

Macromodel (ref. 114) using the MM2* force field. Overlay of generated energy minimized conformations was done using SQ (ref. 115).

114. Mohamadi, F.; Richards, N. G. J.; Guida, W. C.; Liskamp, R.; Caufield, C.; Chang, G.; Hendrickson, T.; Still, W. C. *J. Comput. Chem.* **1990**, *11*, 440–467.
115. Miller, M.; Kearsley, S.; Culberson, J. C.; Prendergast, K. American Chemical Society 210th National Meeting and Exposition, Chicago, IL, August 20–25, 1995.

INDEX

315

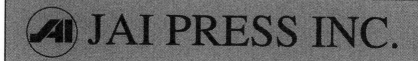

Advances in Medicinal Chemistry

Edited by **Bruce E. Maryanoff** and **Allen B. Reitz,**
*Drug Discovery, R.W. Johnson Pharmaceutical Research
Institute, Spring House, PA*

This series presents first hand accounts of industrial and academic research
projects in medicinal chemistry. The overriding purpose is representation of the
many organic chemical facets of drug discovery and development. This would
include: de novo drug design, organic chemical synthesis, spectroscopic studies, process development and engineering, structure-activity relationships,
chemically based drug mechanisms of action, the chemistry of drug metabolism, and drug physical-organic chemistry.

100 Prospect Street, P. O. Box 811, Stamford, CT 06904-0811
Tel: (203) 323-9606   Fax: (203) 357-8446

JAI PRESS INC.

Volume 3, 1995, 187 pp. $109.50/£70.00
ISBN 1-55938-798-X

REVIEW: "In summary, the third volume of this series continues to offer well-written, interesting accounts of topics important to the discipline of medicinal chemistry and with up-to-date references. Some of the chapters seem well suited as special toipcs in graduate courses. This series should appeal to a broad audience of researchers, teachers, and students."

— *Journal of American Chemical Society*

CONTENTS: Preface, *Bruce E. Maryanoff and Cynthia A. Maryanoff.* Novel Antipsychotics with Unique $D_{\bar{A}}$/5-HT$_{1A}$ Affinity and Minimal Extrapyramidal Side Effect Liability, *Allen B. Reitz and Malcolm K. Scott.* Antiplatelet and Antithrombotic Agents: From Viper Venom Proteins, to Peptides and Peptidomimetics, to Small Organic Molecules, *Peter L. Barker and Robert R. Webb, II.* Discovery and Preclinical Development of the Serotonin Reuptake Inhibitor Sertraline, *Willard M. Welch.* Boronic Acid Inhibitors of Dipeptidylpeptidase IV: A New Class of Immunosuppressive Agents, *Roger J. Snow.* Index.

Volumes 1-3 were published under the editorship of Bruce E. Maryanoff and Cynthia A. Maryanoff, R.W. Johnson Pharmaceutical Research Institute.

100 Prospect Street, P. O. Box 811, Stamford, Connecticut 06904-0811
Tel: (203) 323-9606   Fax: (203) 357-8446

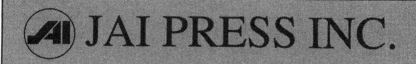
JAI PRESS INC.

Advances in Biophysical Chemistry

Edited by **C. Allen Bush,** *Department of Chemistry and Biochemistry, The University of Maryland, Baltimore County*

100 Prospect Street, P. O. Box 811, Stamford, CT 06904-0811
Tel: (203) 323-9606　Fax: (203) 357-8446

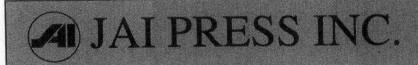

JAI PRESS INC.

CONTENTS: Introduction to the Series: An Editor's Foreword, *Albert Padwa*. Preface, *C. Allen Bush*. Probing the Unusually Similar Metal Coordination Sites of Retroviral Zinc Fingers and Iron-Sulfur Proteins by Nuclear Magnetic Resonance, *Paul R. Blake and Michael F. Summers*. Mass Spectrometry Studies of Primary Structures and Other Biophysical Properties of Proteins and Peptides, *Catherine Fenselau*. Multidimensional NMR Experiments and Analysis Techniques for Determining Homo-and Heteronuclear Scalar Coupling Constants in Proteins and Nucleic Acids, *Clelia Biamonti, Carlos B. Rios, Barbara A. Lyons and Gaetano T. Montelione*. Mechanistic Studies of Induced Electrostatic Potentials on Antigen-Antibody Complexes for Bioanalytical Applications, *Chen S. Lee and Ping Yu Huang*. Conformation and Dynamics of Surface Carbohydrates in Lipid Membranes, *Harold C. Jarrell and Beatrice G. Winsborrow*. Structural Analysis of Lipid A and Re-Lipopolysaccharides by NMR Spectroscopic Methods, *Pawan K. Agrawal, C. Allen Bush, Nilofer Qureshi and Kuni Takayama*. Index.

Volume 5, 1995, 263 pp. $112.50/£72.50
ISBN 1-55938-978-8

CONTENTS: Preface, *C. Allen Bush*. Sequence Context and DNA Reactivity: Application to Sequence-Specific Cleavage of DNA, *Albert S. Benight, Frank J. Gallo, Teodoro M. Paner, Karl D. Bishop, Brian D. Faldasz, and Michael J. Lane*. Deciphering Oligosaccharide Flexibility Using Fluorescence Energy Transfer, *Kevin G. Rice*. NMR Studies of Cation-Binding Environments on Nucleic Acids, *William H. Braunlin*. The Cytochrome *c* Peroxidase Oxidation of Ferrocytochrome *c:* New Insights into Electron Transfer Complex Formation and the Catalytic Mechanism from Dynamic NMR Studies, *James E. Erman and James D. Satterlee*. Statistical Thermodynamic Modeling of Hemoglobin Cooperativity, *Michael L. Johnson*. Measurement of Protein-Protein Association Equilibria by Large Zone Analytical Gel Filtration Chromatography and Equilibrium Analytical Ultracentrifugation, *Dorothy Beckett and Elizabeth Nenortas*. Index.

Volume 6, 1997, 253 pp. $112.50/£72.50
ISBN 0-7623-0060-4

CONTENTS: Preface, *C. Allen Bush*. Thermodynamic Solvent Isotope Effects and Molecular Hydrophobicity, *Terrence G. Oas and Eric J. Toone*. Membrane Interactions of Hemolytic and Antibacterial Peptides, *Karl Lohner and Richard M. Epand*. Spin-Labeled Metabolite Analogs as Probes of Enzyme Structure, *Chakravarthy Narasimhan and Henry M. Miziorko*. Current Perspectives on the Mechanism of Catalysis by the Enzyme Enolase, *John M. Brewer and Lukasz Lebioda*. Protein-DNA Interactions: The Papillomavirus E2 Proteins as a Model System, *Rashmi S. Hedge*. NMR-Based Structure Determination for Unlabeled RNA and DNA, *Philip N. Borer, Lucia Pappalardo, Deborah J. Kerwood, and István Pelczer*. Evolution of Mononuclear to Binuclear Cu_A: An EPR Study, *William E. Antholine*. Index.

100 Prospect Street, P. O. Box 811, Stamford, Connecticut 06904-0811
Tel: (203) 323-9606   Fax: (203) 357-8446